AF517019

Irrigation Resources

Editors

Prof. S: walia Bihari Verma
Faculty of Commerce
SNS College, Muzaffarpur
BRA Bihar University, Muzaffarpur
Former Reader
Nagaland (Central) University, Kohima

Dr. Arvind Kumar Shrivastawa
Head, Deptt. of Psychology
Nandangarh College, Lauria
West Champaran, Bihar

Dr. Jeebu Kumar Jha
Government Inter College (Zila School), Muzaffarpur
Santiniketan Senior Secondary School
Muzaffarpur (Bihar)

Published by:

Scientific Publishers (India)
5 A, New Pali Road, P.O. Box 91
Jodhpur 342 001 (India)

Branch Office
Scientific Publishers (India)
4806/24, Ansari Road, Daryaganj
New Delhi - 110 002 (India)

E-mail: info@scientificpub.com
Website: www.scientificpub.com

ISBN: 978-81-7233-914-2

Typeset by Rajesh Ojha
Printed in India

PREFACE

India's agriculture sector consumers over 90 per cent of the total water resources yet the full potential of irrigation has not been achieved with increasing demand from other sectors and need to augment food production, efficient ways to the water resources are being work out.

India draws 80 percent of its irrigation water from ground water which often Leads to drying up of wells and crates groundwater shortages. Experts have cautioned that the entire replenishable ground water of the world could be utilized by 2025 AD.

Due to lack of adequate storage and related infrastructure, around 48 per cent of the rainfall ends up in the rivers due to lack of storage and related infrastructure.

While efforts are being made to store the runaway water to rivers, there are studies which say, that the rivers themselves are drying up due to climate changes. A recent study of 900 rivers in the world has found that the Ganga is one of the world's rapidly shrinking rivers.

Despite the constraints and problems associated with the scarce water resource the irrigation potential of India has been increasing over the years. The irrigation potential created up to the beginning of the first year plan was only 23 million hectares which has increased to 103 hectares by end of the tenth plan.

The 11th plan envisages creation of an additional potential of 16 million hectares at an estimated required out lay of about 2,10,000 crore rupees. Since irrigation is a state subject, most of this amount has been earmarked for financing by states and an analysis of states own preliminary 11th plan allocations shows that this might actually be exceed.

Irrigation, is one of the important inputs of agriculture, has experienced a steedy decline in public investment in irrigation over the years. A large number of irrigation related projects are facing financial constraints and the investments already made in these projects are new treated as ``sunk investments''. It is estimated that there were 171 major, 259 medium and 72 Extension, renovation and modernisation (ERM) ongoing irrigation projects in the country at the end of 8th five year plan. These projects need timely completion through prioritization of execution of projects and effective and adequate resource allocation. The effort under "Bharat Nirman" is, therefore, directed towards identifying all such projects and targeting their completion. The government has already identified major and medium irrigation projects covering four million hectares as well as minor irrigation projects covering 2.8 million hectares.

In the conventional irrigation methods like flooding, border, furrow methods as irrigation, the opportune time for infiltration of water is more at one end where irrigation started and less at another end, leading to more water intake at another end. In drip irrigation system, distribution water in all directions of the field is uniform as the opportune time is same for the entire field.

The micro irrigation technology is yet to become popular in India. It was introduced in the 1980s. It aims at water conservation. It is the most important instrument to bring down the severity of drought.

Water co-operative societies have promising future in the command area of irrigation projects of all kinds. It offers ideal solution of the rather complex problems as distribution of irrigation water on the basis of equity. It also makes easy introduction as the discipline of rotational water distribution and sale of water in bulk on volumetric basis. Water co-operatives should be encouraged to grow very fast and in great number with a view to optimize the benefits of irrigation projects and there by increase the rate of overall status of social life and economic prospects from good to better and best in days to come.

The key strategy to the irrigation component of the Bharat Nirman also seeks to reduce the gap in the creation and utilisation of irrigation potential. For this, it proposes to restore and utilize irrigation potential of one million hectares in rough extension, renovation and modernisation of the existing schemes and by introducing more efficient water management practices in the irrigation command areas.

I, Dr. S.B. Verma, articulate my deep sense of gratitude to my father, late Baban Pd. Srivastava, Mother Late Mrs. Lalmuni; as well as my wife Dr. Hema Verma. who although are not alive today but their invisible soul always inspired me to reach this venture.

The Editor cum Anthologist, Dr. S.B. Verma, is thankful to 'Yojana' Kurukshetra, Employment News and so on for giving their permission to reprint the articles published by them.

Sawalia Bihari Verma
Arvind Kumar Shrivastawa
Jeebu Kumar Jha

Place - Shivpuri, Muzaffarpur (Bihar)

CONTRIBUTORS

A. Gayathri
Assistant Professor in Economics, Distance education, Annamalai University Annamalai Nagar, Tamil Nadu

Altaf Hussain
Researcher, University of Agriculture Science and Technology, Soil Sciences Division, Srinagar, Jammu and Kashmir.

Amod K. Thakur
Water Technology, Centre for Eastern Region Chandrasekharpur, Buhbneshwar.

Arpita Sharma
UGC-JRF in Agricultural Communication, G.B. Pant University of Agriculture & Technology, Pantnagar

B. Chandra Sekhar
Divisional Forest Officer, DWWM, Anantapur

B. Hemalatha
Central Research Institute for Dryland Agriculture Santoshnagar, Hyderabad

B. Venktateshwarlu
Central Research Institute for Dryland Agriculture, Hyderabad

Dhurjati Mukherjee
Freelance Writer, Kolkata

Deepa Kylasam Iyer
Editor, Web Portal based in Peris for the people of Indian Origin

Dinesh Kumar
Senior Scientist in Agronomy, Indian Agriculture Institute, New Delhi

Dr. Achelladurai
Reader in Economics, Adityanagar College of Arts and Science Tiruchendry, Tamil Nadu.

Dr. Amrit Patel
Ex. Deputy General Manager, Bank of Baroda.

Dr. Anita Modi
Head Deptt. of Economics, GSS Girls PG College, Chirawa.

Dr. Anshu Rahal
Assistant Professor in Animal Nutrition, College of Veterinary and Animal Science, G.B. Pant University of Agriculture and Technology, Pantnagar

Dr. B. Manihar Sharma
Professor of Life Science, Manipur University, Imphal.

Dr. Gopal Kalkoti
Professor of Business Economics Vice Principal and Member Board of Studies, Mumbai

Dr. Harender Raj Gautam
Scientists, Deptt. of Mycology and Plant pathology. Dr Y.S. Parmar University of Horticulture and Forestry, Nauni, Solan, Himachal Pradesh.

Dr. K. Srinivasa Rao
Deptt. of Commerce Vivek vardhini College of P.G. Studies, Jambagh Koti, Hyderabad

Dr. L. Rathakrishnan
Professor of RIM
Gandhigram Rural Institute, Gandhigram

Dr. Manasmani Dev Goswami
Engineer, Water Resource, Govt. of Assam.

Dr. N.T. Somashekaraiah
Head, Deptt. of Economics, Govt. First Grade College, Channarayapatna.

Dr. P. Jayashree
Associate Professor, Deptt. of DOS in Geography, University of Masore, Masore.

Dr. R.S. Sapehia
Scientist, Deptt. of Soil Science, Dr. Y.S. Parmar University of Horticulture and Forestry, Solan, H.P.

Dr. Ramu
Project coordinator,
Deptt. of Geography,
Maharaja's College, Mysore

Dr. S. Manicham
Professor of Economics, Manonmaniam Sundaranar University, Tirunelveli

Dr. Yashbir Singh Shivay
Scientist in Agronomy, Indian Agricultural Research Institute, New Delhi.

Dr. Ajay Kumar
Lecturer in Commerce
Dr. J.M. College, Muzaffarpur
BRA Bihar University, Muzaffarpur (Bihar)

Dr. Jeebu Kumar Jha
Government Inter College, (Zila School), Muzaffarpur & Principal, Santiniketan Senior Secondary School, Muzaffarpur

Dr. Sanjeet Kumar Gupta
Associate Professor
University Deptt. of Commerce and Coordinator, NSS programme
DDU Gorakhpur University
Gorakhpur (U.P.)

Dr. Satish Kumar Arya
Lecturer in Commerce
RLGY College, Muzaffarpur (Bih.)

Dr. A.M. Michael
Professor and Project Director
Water Technology Centre, IARI
New Delhi &
Director, IARI, New Delhi
Vice Chancellor, Kerala Agriculture University, Trichur

D.K. Singh
Indian Institute of Vegetable Research, Shahan Shapura, Jakkeni, Varanasi

R.M. Singh
Deptt. of Farm Engineering
Institute of Agriculture Sciences
Banaras Hindu University, Varanasi

Mihir Shah
Member, Planning Commission, National Advisory Council
Portfolio of Water Resources

Francis Kuriakose
Assistant Professor in Commerce and Management,
M.I. College, University of Kerala
Thiruvananthapuram

G. Sheshagiri S. Subhash
Researcher, IIT. Mumbai

I. Satya Sundaram
Research Director in Economics
Machilipatnam, Andhra Pradesh.

Jaya Chaterji
Senior Project Officer, India Sanada Environment Faculty, Vasant Bihar, New Delhi.

K. Govindappa
Professor, S.K. University, Anantapur.

K. Perumalammal
Professor of Economics, Alagappa Got. Arts College, Kalikudi.

K. Ravi Kumar
IAS Officer in Jharkhand

K. Srinivasulu
Scientist, Regional Agriculture Research Station, Guntur, Andhra Pradesh

K. Srivani
Assistant Professor in Economics, Karimnagar, Warangal

K.G. Suresh
Special Correspondent PTI, Freelance, Journalist.

K.K. Tripathy
Indian Economic Service, Project Manager, planning commission, New Delhi.

K.N. Tiwary
Ex. Director, International plant Nutrition Institute, India

K.P. Manzoor
Assistant Professor in Economics, Irshadiya College, Calicut, Kerala

M. Adinarayana
Scientist, Regional Agriculture Research Station, Guntur, Andhra Pradesh.

M. Lathika
Department of Economics, NSS College for Women, Thiruvananthapuram.

Nivedita Thapliyal
Social Worker and Researcher in the Deptt. of Social Work University of Delhi Delhi

P. Veerachamy
Assistant Professor in Economics, Distance Education, Annamalai University Annamalai Nagar, Tamil Nadu

Prof. Sawalia Bihari Verma
Professor of Commerce, SNS College, Muzaffarpur BRA Bihar University, Muzaffarpur.

Rajeeb K. Mohanty
Water Technology, Centre for Eastern Region Chandrasekharpur, Bubhneshwar.

Rakesh Tiwary
Soil Science Specialist SVBP University of Agriculture and Technology Krishi Vigyan Kendra, Hastinapur, Meerut.

Sanjay Kumar
Media and Communication officer, PIB, New Delhi.

Surinder Sud
Agriculture Journalist

Y.S. Ramakrishna
Central Research Institute for Dryland Agriculture Hyderabad

Y.S. Shivay
Senior Scientists in Agronomy, Indian Agriculture Institute, New Delhi

Y.V.R. Reddy
Central Research Institute for Dryland Agriculture Santoshnagar, Hyderabad

CONTENTS

Part - A
Indian Irrigation Potential Strategies

Part - A
Indian Irrigation Potential Strategies

Chapter - 1

Irrigation Development in India

Dr. A.M. Michael

First Century A.D. to Middle of 20th Century

Irrigation development in India in the first to third century A.D. has been mainly in Southern India which was not influenced by the cultural ingression of tribes from central Asia. The Sangam Literature gives detailed description of the Tamil dynasties of pandya, chera, chola, pallaya, chalukya and Rastitrakuta and their contribution to development projects. Kings of these dynasties took great interest in promting agriculture. Situated in the detaic or valley regions, most of these kingdoms had fertile soil. With the expansion of irrigated agriculture, South India became the rice bowl of India. Sangam literature gives descriptions of the sites chosen for storing run-off water. Both tank and well irrigation were in practice in South India. There are distinct references to 'lift irrigation', where bullocks were used for lifting water from deep wells. An important irrigation work has been the construction of Large embankments (both stone and earth) on kaveri river in the delta region and the vaigai river to store flood water. Karikala, a chola king (A.D. 180) constructed a 160km long-embankment along the Kaveri River primarily to protect his people from recurrent floods. The Grand Anicut in the Kaveri delta built in the second century A.D. is the first major irrigation system in the Indian Sub-continent. He also patronized the construction of a large number of irrigation tanks.

Gupta era (A.D. 300 to 500): The Gupta dynasty made substantial contribution by promoting irrigated agriculture in their empire. The Gupta empire extended over almost two-thirds of the Indian sub-continent. Dykes and embankments were constructed to divert river water for irrigation. Separate canals for irrigation and drainage have been reported.

Pallava Kingdom: The Pallava dynasty, founded in A.D. 550 promoted extensive tank irrigation throughout Tamil Nadu. The Parameswara tank and the. Tiraiyaneri tank were contributions of this dynasty on irrigation. Vayiramega, Gudimallam and Ukkal tanks were constructed during the later part of the 8th Century A.D. The large tanks at Kanakvallieri Solapuram and Kaveripak were constructed during the 9th century. The Kaveripak tank had a 6.4 km long earth dam and is still functional. The Chitramegha tank and the Tandalam tank with sluice gate are other major tanks of this era.

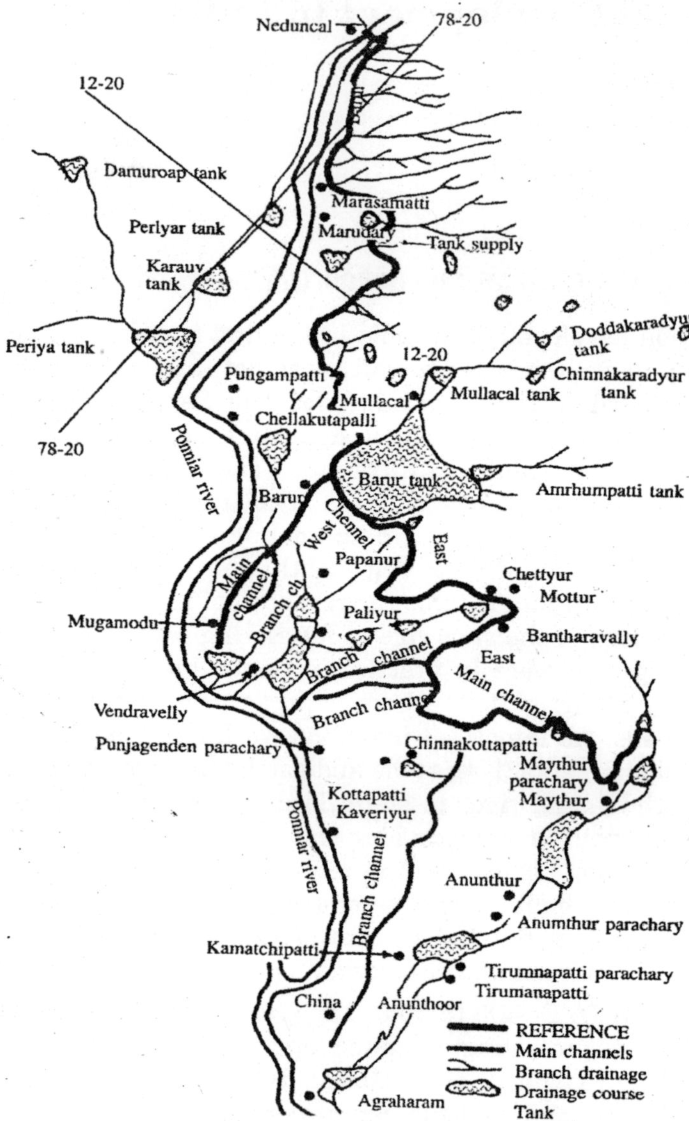

Fig. 1. Layout of system tanks (chain tanks). Barur Tank Project (1883), Salem district, Tamil Nadu.

***Choldempire* (AJD. 985 to 1205):** Chola king's are known for their enormous contribution to irrigated agriculture. It led to unprecedented prosperity and wealth in the region. Among the famous ancient *anicuts* are those in the Thanjavur district. Huge stones were used to build this

329 km long, 12-18 m broad *anicut* built near the Srirangam islands. The Talkad *anicut* built downstream of the Kaveri river is another important structure. A number of channels were dug to cover large areas of land on both sides of the river bank to ensure production of crops throughout the year.

Chain Tanks of Karnataka and *Andhra* Pradesh

Tanks for water storage and diversion were constructed in continuous series like links of an imaginary chain in many of the river valleys. The Telengana region of Andhra Pradesh was referred to as the land of thousand tanks. System tanks (chain tanks) are low dams which are constructed at the upper reaches of a stream to store water; its overflow is allowed to pass downstream to be stored in the next dam in the series. Such tanks have become common in Tamil Nadu, Karnataka and Andhra Pradesh (Fig. 1). Often, the chain tanks are adequate to cover the whole watershed area. The disadvantage was that during heavy rains if a breach occurred in an upstream tank, its water would come with tremendous force to the tanks on the lower reaches which consequently develop breaches unless the downstream tanks are provided with the required factor of safety.

Tanks of Tamil Nadu: Though initiated by the Pallava kings, the chain tanks of Tamil Nadu were extensively promoted by the Chola kings. Some of the important ones are the Cholavaridhi tank, Sodiy ambakkam tank and a large number of tanks in Madurai, Salem and other districts (Fig. 1).

Tanks of Karnataka: A large number of tanks were built in Karnataka by the Chola kings during the 10th Century A.D. They constructed tanks at every feasible point at regular intervals along a stream. Many of these are still in use. The ingenuity and precision with which the sites of tanks were selected are remarkable. The Pallar valley tank system had a chain of 1,000 tanks.

The neighbourhood of Delhi had at one time about 350 tanks (large and small) in use. Some of them like Surajkund are still in good condition. The greatness of a town or an area was based on the tanks it had for its use. The states of Andhra Pradesh, Karnataka, Madhya Pradesh, Maharashtra, Rajasthan, Tamil Nadu and West Bengal are still having a large number of old tanks. Separate management committees were formed at the village level to manage the tanks.

Irrigation Development in Northern and Eastern India during the 9th to 12th century A.D.

Rajput rulers were reigning different parts of the Northern and North-Western India. King Krishnaraja made significant contribution in irrigation development. The Bhopal lake built during his period has 647 sq km of surface area and an extensive system of irrigation channels.

Vijayanagara Kingdom (1336 to 1500 A.D.). The Vrjayanagara Kingdom was established by King Harihara I in A.D. 1336. Large-scale land reclamation and irrigation works were given boost by the kings. The most famous of these is the Anantaraja Sagar tank in Cuddapah district of Andhra Pradesh.

Mughal Period *(14th and 18th centuries AD.).* Major efforts in irrigation development during the period was construction of numerous tanks and dug wells in Peninsular India and Rajasthan by local rulers. Amongst the Mughal emperors Ferozshah Tughlaq built the Western Yamuna Canal in A.D. 1355 to carry Yamuna water to his hunting ground at Safidem in Hisar district (Haryana). The canal was abandoned after his time. Later, Emperor Akbar renovated the canal in 1568 for irrigating lands in Hisar district. Emperor Shahjahan restored the silted and unusable Western Yamuna Canal and took a branch canal to 59 irrigate the gardens in the Red Fort, in Delhi. Emperor Shahjahan built a 180 km canal from Ravi river through the Shalimar Gardens at Lahore. A branch of this canal carried water to the Golden Temple in Amritsar. A large part of Punjab and Kashmir were benefited by it. During the British period, this canal was replaced by the Upper Bari Doab Canal system. The Eastern Yamuna Canal system was constructed by Emperor Mohammad Shah in A.D. 1735. The canal was useful during the initial period, but got silted within a few decades. Development works did not make any major strides during the second-half of the eighteenth century, mainly due to the unstable governments in Delhi.

Irrigation Development in the 19th Century A.D.

There was substantial increase in irrigation in India in the nineteenth century. At the beginning of the 19th century, the net irrigated area was estimated at 0.8 million ha. This rose to 13.3 million ha at the close of the century when 16% of the gross sown area of 82.2 million ha was brought under irrigation. Of this, 45% was irrigated by canals; 35% by wells, 15% by tanks, and 5% by other sources.

Irrigation development in India in the nineteenth century, under the British rule was essentially a part of the famine relief works. The projects were designed to provide protective irrigation to as many areas as possible. The major irrigation projects constructed during the first half of the nineteenth century included the Kaveri delta system in Tamil Nadu in 1834 to irrigate 0.6 million ha, the Godavari *anicut* and canal system in Andhra Pradesh in 1846 to irrigate 0.4 million ha, and the Krishna delta system in Andhra Pradesh in 1852 to irrigate 0.5 million ha. The major credit for the massive delta projects goes to Sir Arthur Cotton, a great engineer. Cotton is held in high esteem in India for his spectacular contributions in irrigation development.

The other major irrigation projects of the 19th century are Upper Ganga Canal system in Uttar Pradesh commissioned in 1854 to irrigate 0.37 m ha, the Sone Canal in Bihar (1879) to irrigate 0.28 million ha and the Orissa Canal system on river Mahanadi (1895) - to irrigate 0.1 million ha. The Upper Ganga Canal, with its headworks at Haridwar was the largest irrigation canal in the world at the. time of its construction. It had a carrying capacity of 191 m^3/sec. It was later remodelled to carry a discharge of 288 m^3/sec. The Western Yamuna Canal in Haryana was remodelled in 1892 to provide an irrigation potential of 0.6 million ha. The lower Ganga Canal was constructed during 1872-78.

A definite policy for the construction of irrigation works was introduced in 1854 with the setting up of the Public Works Department by the then Government of India. The Government of India appointed the First Irrigation Commission in 1901 to report on irrigation as a 'means of protection against famine in India'. A few Famine Commissions had been set up by the government and these Commissions had also recommended the development of irrigation works to contain the adverse impact of frequent famine in India. In its recommendations to the Government in 1903, the Irrigation Commission suggested definite policies regarding the selection, financing and maintenance of productive irrigation works. Apart from the remunerative works, the Commission recommended the extension of irrigation as a means of protection against famines in areas of insecure and precarious cultivation. The Commission observed that such irrigation works may not be directly remunerative to the Government, but are important to mitigate the miseries caused by famines.

In accordance with the recommendations of the First Irrigation Commission, several productive and protective irrigation projects were undertaken in the early part of the twentieth century. Prominent amongst them are the Sarda Canal in Uttar Pradesh (1926) to irrigate 0.6 million ha, Ganga Canal with dam and weir on Satluj river (1928) to irrigate 0.3 million ha in Rajasthan, Damodar Canal (1935) to irrigate 0.1 million ha in West Bengal, Ken Canal (1915) to irrigate 0.1 m ha in Uttar Pradesh; Nira Left Bank Canal (1906), and Right Bank Canal (1938) and Pawara Canal (1938) in Maharashtra with a total irrigation potential of 0.1 m ha; Mahanadi Canal (1931) and Kharung Canal Project (1931) in Madhya Pradesh with a total irrigation potential of 0.2 m ha.

The Triple Canal Project (1905-15), now in Pakistan was the first large-scale trans-basin diversion of water from one river basin to another. It comprised of two large weirs and one barrage across three different rivers and three canals to transfer water from the Jhelum across the Chenab and Ravi rivers to irrigate desert areas. The Satluj Valley Project constructed during 1921-31 was the largest irrigation project of the time. Major irrigation projects of South India constructed during 1921 and 1935 include the Krishnarajasagar Project with dam on Kaveri, the Nizamsagar Project with dam on Majira river, a tributary of Godavari in Andhra Pradesh and the Mettur Dam project with dam across Kaveri in Tamil Nadu constructed in A.D. 1927. A number of small works which could be taken up for famine relief in various provinces were recommended by the Irrigation Commission which gave considerable importance to the development of private irrigation works including wells in the tracts affected by famine.

In 1854, two categories of irrigation works were introduced, namely, minor works and major works. Major irrigation projects constructed during the second-half of the nineteenth century include the Sirhind Canal in Punjab (1873-82) consisting of 6000 km of canals to irrigate 1 million ha, the Lower Sohang and Para Canals (1872-77), now in Pakistan, the Lower Chenab Canal in Punjab, the Lower Ganga Canal, the Agra Canal in Uttar Pradesh (1873) with its headworks over Yamuna at Okhla in Delhi to irrigate 0.16 million Jia, the enlargement of the Eastern Yamuna Canal in Uttar Pradesh to irrigate 0.19 m ha, the Periyar Dam (1887) in Kerala along with the Periyar Vaigai Canal system in Tamil Nadu to irrigate 60,000 ha and the Khadakvasla Dam (1869-79) and the Mutha Canals in Maharashtra. The Periyar Dam project is a remarkable feature involving the diversion of the west-flowing Periyar river with its catchment in the high rainfall Western Ghats region of Kerala to the rain shadow region of Madurai district in Tamil Nadu in the east.

Irrigation Works in India in the Pre-independence Era of Twentieth Century

The indispensability of well planned and rapid development of irrigation was pronounced by the two great famines of 1897-98 and 1899-1900. The economic considerations were out-weighed for irrigation development by the need for increasing food production for sustaining the population.

The result of the large-scale expansion of irrigation works in the Indian sub-continent during the 19th and the first-half of the twentieth century has been largely due to the great efforts of irrigation engineers with the support of the governments, including the rulers of the princely states. The role of three great personalities are immemorable in the history of irrigation development during the period. They are Sir Arthur Cotton, Major P.T. Cautley, both in the 19th century and Sir M. Visveshwariah in the first-half of the 20th century. Major Cautley, a British military engineer assumed the major responsibility for the construction of the Upper Ganga Canal Project in Uttar Pradesh.

Table 1. Net Area Irrigated in the Indian Union from 1910 to 1950 (Quinquennial average) *On ha = million hectares)*

Period	Gout		Total	Irrigated
	Canals	Wells	(all sources)	area as % q,
	m ha	m ha	m ha	Sown area
1910-1 1 to 1914-15	4.4	4.0	14.5	17.9
1915-16 to 1919-20	4.1	4.7	15.7	19.0
1920-21 to 1924-25	4.4	4.7	16.0	17.4
1925-26 to 1929-30	4.6	4.8	16.2	17.1
1930-31 to 1934-35	5.0	4.8	17.1	17.6
1935-36 to 1939-40	5.6	5.2	18.0	18.6
1940-41 to 1944-45	6.0	5.4	19.0	19.2
1945-46 to 1949-50	6.4	5.3	19.4	19.1

(Source: Irrigation Commission, 1972).

Bharata Rathna M. Visveshwariah was largely responsible for the construction of Krishnarajasagar Scheme in 1911 A.D. on River Kaveri near the city of Mysore. The dam was the largest reservoir built in India till then. The scheme created an irrigation potential of 40,000 ha and provided electricity to the people of Mysore and Bangalore. The project converted the Mandya district of Karnataka, which was once covered by shrub forests, into a prosperous region with high-yielding paddy and sugarcane (Bhavanishankar, 2000). Table 1 presents the irrigation development in India during A.D. 1910 to 1950.

Classification of Irrigation Works in India

Post-independence development of irrigation works in India were on the basis of classification of works into major, medium and minor projects. The classification has changed over the years from the basis of the extent of culturable command area of the project to that of expenditure involved in the project and back again to the basis of culturable command area. During the preplan period (prior to A.D. 1951), irrigation projects were classified on the basis of culturable command area of the project. During the period ranging from 1951 to 1978, the classification was based on the cost involved in the schemes. Since 1978, the basis of classification of irrigation projects has again been on the basis of the culturable command area of the project.

Accordingly, irrigation projects in India are presently classified into the following three categories:

(i) **Major Irrigation Projects.** All irrigation schemes with a culturable command area of 10,000 ha or more are classified as major

irrigation projects. They are essentially surface water projects involving large-scale storage/diversion works.

(ii) **Medium Irrigation Projects.** Projects having a culturable command area of 2000 ha to 10,000 ha are classified as medium irrigation projects. Medium irrigation projects are also usually surface water projects, excepting a few large lift irrigation schemes.

(iii) **Minor Irrigation Projects.** All irrigation schemes having a culturable command area upto 2000 ha individually are classified as minor irrigation projects. Minor irrigation schemes are further grouped as (a) surface water minor irrigation schemes, and *(b)* groundwater minor irrigation schemes.

Jurisdiction of water resources. Under the Indian Constitution, water is the responsibility of the states. Thus, the federal states are primarily responsible for the planning, implementation, funding and management of water resources development projects. This responsibility in each state is borne by the Irrigation/Water Resources Departments. The Inter-State Water Disputes Act of 1956 provides a framework for the resolution of possible conflicts between states on the utilization of water resources. At central level, there are three main institutions involved in water resources development and management:

(i) **The Ministry of Water Resources:** This is responsible for laying down policy guidelines and programmes for the development and regulation of the country's water resources. The ministry's technical arm, the Central Water Commission, provides general infrastructure, and technical and research support for water resources development at state level. The Central Water Commission is also responsible for the assessment of water resources.

(ii) The Planning Commission: This is responsible for the allocation of financial resources required for various programmes and schemes of water resources development to the states as well as to the Ministry of Water Resources. It is also actively involved **in** policy formulation related to water resources development at the national level; and the Ministry of Agriculture, which promotes irrigated agriculture through its Department of Agriculture and Cooperation. The research support at the national level is by the Indian Council of Agricultural Research. The Central Pollution Control Board is in charge of water quality monitoring, and the preparation and implementation of action plans to solve pollution problems.

International water treaties/agreements. Under the Indus Water Treaty (1960) between India and Pakistan, all the waters of the eastern rivers, i.e. the Satluj, Beas and Ravi rivers taken together, shall be available for the unrestricted use of India. All the waters, while flowing in Pakistan, or of any tributary which in its natural course joins the Satluj main or the Ravi main after these rivers have crossed into Pakistan, shall be available for the unrestricted use of Pakistan.

India controls the flow of the Ganges River through a barrage completed in 1974 at Farraka, 18 km from the border with Bangladesh. A treaty was signed in December 1996, under which Bangladesh is ensured a fair share of the flow reaching the Farraka barrage, in addition to the releases during the monsoon seasons. Appropriate agreements exist between Nepal and India for the exploitation of the Kosi River (1954, 1966) and the Gandak River (1959) *(Source:* UN Food and Agriculture Organization (FAO), 2001).

Irrigation Development in India During the Five Year Plans

The partition of India in 1947 brought about a drastic reduction in the irrigated area in the post-independent India. A substantial portion of the irrigated area of undivided India went to Pakistan. In 1950-51, the net irrigated area was 20.85 million ha and the gross irrigated area 22:60 million ha. The Government of India gave priority to develop new irrigation projects, realising the urgent need for developing the available water resources of the country to meet the deficits in food production.

The increase in irrigation potential during the five year plans, which commenced in 1951 and The Annual Plans was phenomenal and unprecedented. The gross irrigation potential rose from 22.6 ha in 1951 and was estimated to reach 196.61 million ha in 1997-2012.

National Water Policy

India adopted a National Water Policy in 1987 for the planning and development of water resources to be governed by national perspectives. It emphasizes the need for river basin planning. Amongst the priorities in the allocation of water for various needs, drinking water is given the highest priority, followed by irrigation, hydro-power, navigation, and industrial and other uses. As water resources development is a State responsibility, all the states are required to develop their state water policies within the framework of the National Water Policy. Since the adoption of the policy in 1987, a number of new issues and Challenges emerged in the development and management of water resources. Therefore, in 2002 the National Water Policy (1987) was reviewed and updated. The revised National Water Policy (2002) has the following priorities in water allocation: (1) Drinking water, (2) Irrigation, (3) Hydro-power, (4) Ecology, (5) Industries, (6) Navigation and other uses. However, it has been stipulated that the priorities could be modified or added if required by the area/region specific considerations (Ministry of Water Resources, 2002). The revised policy has stressed that in the planning, implementation and operation of a water resource project, the preservation of the quality of environment and ecological balance should be primary considerations. An integrated and multi-disciplinary approach to the planning, formulation, clearance and implementation of projects, including catchment area treatment, command area development and the rehabilitation of the people adversely affected by the project are envisaged. The drainage system should form the integral part of an irrigation project right from the planning stage. The involvement and participation of beneficiaries and other stakeholders should be encouraged from the project planning stage. Integrated coordinated development of surface and groundwater resources and their conjunctive use should be envisaged right from the project planning stage. There should be periodical assessment of the groundwater potential and its quality. Exploitation of groundwater should be regulated to prevent its use to exceed the recharging capabilities of aquifers and ensure social equity. There should be a close integration of water use and land use policies. Municipal and industrial effluents should be treated to acceptable levels and standards before discharging them into natural streams. Minimum flow should be ensured in the perennial streams for maintaining ecology and social considerations. Drought-prone areas should be made less vulnerable to drought-affected problems through soil moisture conservation measures and water harvesting procedures. The water sharing and distribution amongst states should be guided by a national perspective with due regard to water resources availability and needs within the river basin/sub-basin. Water resources development will have to be planned for a hydrological unit, multi-sectorally. Watershed management through extensive soil and water" conservation in the catchment area, preservation of forests and increasing the area under the forests are to be given priority. Construction of check dams should be prompted wherever feasible.

Energy Requirement in Irrigation

Careful planning of energy sources and their efficient utilization are vital in meeting the water needs of agriculture. Agricultural production in developing countries has been seriously hampered by the non-availability of adequate power to lift water for irrigation from groundwater and surface water sources. In many countries there is lack of co-ordination between the agencies responsible for the development of water resources and those vested with energy development and allocation. Energy requirements in irrigation pumping may be broadly grouped into the following two categories:

(1) ***Pumping from wells:*** Irrigation wells are classified into *(i)* Open wells, (ii) Shallow tubewells, and *(iii)* Deep tubewells. Water is lifted from open wells by manually operated and animal operated water lifts and pumps driven by diesel engines and electric motors. Water lifting from shallow tubewells and deep tubewells is almost exclusively done by pumps operated by diesel engines and electric motors.

(2) River and canal pumping: Pumping from canals and river-pumping for lift irrigation are almost exclusively done by diesel engine or electric motor driven pump-sets. The development of energy sources, especially electrical energy should match with the development of lift irrigation using groundwater and/or surface water sources. Whenever electric power is available, it is the most convenient and efficient source of energy in irrigation pumping. The reliability and efficiency of electrically operated pumping-sets are high, provided they are properly selected, installed and maintained.

PARTICIPATORY IRRIGATION MANAGEMENT

The traditional systems of irrigation the world over have had a built-in system of distribution of water among the beneficiaries, thereby linking utilisation with the creation of irrigation potential. In India, the management of tanks and small scale river diversion schemes have involved the irrigators in the maintenance, distribution as well as in modalilites of conflict resolution. There was an unwritten convention among the farmers under a tank to use water on a weekly rotation. One among the beneficiaries acted as the coordinator and leader for water distribution within the unit. Similarly, supply of water among fields on a particular day was decided by mutual agreement. The system on the whole, worked with minimal government intervention.

Against this background, major canal irrigation project planners of the 20th century did not concern themselves with developing a proper system of water sharing and infrastructure maintenance at the farm level. The responsibility of the irrigation department is deemed to end at the point where water leaves the outlet. The planners and executors of irrigation projects have tended to focus largely on engineering aspects with little attention to socio-economic dimensions and modern methods of water management. There has been a general lack of constructive and functional relationship between irrigation officials and the farmers. Thus, the farmer who had the most important stake in the proper management of irrigation water as the ultimate beneficiary, had little role to play in the management of irrigation water under the prevailing system. This realisation lead the government to emphasise the need for the participation of farmers in the management of irrigation water during the Sixth Plan and its reiteration during the Seventh Plan. In 1987, the Ministry of Water Resources, Govt. of India, issued guidelines to state governments to involve farmers' organizations to take over maintenance, management of water, allocation, and collection of water charges and to develop participatory mode of management on distinct segments of the irrigation system (Joshi, 1997).

Participatory irrigation management refers to the involvement of irrigation users in part or all the aspects of irrigation management and at part or all levels. This implies initial planning and design, construction, supervision, financing, decisions, rules, operation, maintenance, monitoring and evaluation of the system. 'All Levels' refers to the full physical, limits of the irrigation project, comprising of harnessing water at the headworks, conveyance, distribution, and application systems (Singh, 1998).

Even though, the intent to create farmers' organizations has been expressed during the Sixth and Seventh Plans, the visible efforts to involve farmers in the form of Water. Users' Associations (WUAs) in different states have been made in late 1980s and early 1990s to manage irrigation water from outlets, minors, and distributories of the canal ' system. Farmers have also been involved in the management of irrigation water from tanks, and deep tubewells. These organizations in different states are registered under State Cooperative Acts or Irrigation Utilisation and Command Area Development Acts and their variants. The water users' associations under the participatory irrigation management system are charged with the responsibilities

(i) to improve the water delivery service through better operation and maintenance;

(ii) to achieve optimum utilisation of available water from canals, tanks, and tubewells;

(iii) to achieve equity in the distribution of water amongst water users;

(iv) to promote economy in water use and to increase water and land productivity;

(v) to optimise the conjunctive use of precipitation and groundwater with the water, delivered from the canal system;

(vi) to assist farmers in crop planning, and frequency and duration of water supply;

(vii) to inculcate a sense of ownership and responsibility amongst the farmers and to ensure collection of water changes;

(viii) to develop professional relationship between the water users and state irrigation agencies;

(ix) to generate, local resources in cash, kind or labour for operation, and maintenance; and

(x) to assist farmers' association and its members in other spheres of activities (e.g., agro-industries and marketing) for their upliftment.

Thus, the water users associations are charged with the responsibilities of integrated management of water, including conveyance, distribution and application to optimally utilise the water resources for enhancing crop production (Singh, 1998).

International Status

Even though the irrigation systems are meant for the farmers, the countries around the world have promoted the development of irrigation, projects primarily through the involvement of their governments. Only in 1960s and 1970s, several countries started providing incentives to their farmers to take over the management of operations and maintenance, leaving the government organizations to concentrate on the main system. During 1970s and 1980s, the work on irrigation systems' management in participatory mode involving farmers began in many countries, like United States of America, Japan, Italy, Mexico, Columbia, South Korea, Vietnam, Indonesia, China, Nepal, Pakistan, Sri Lanka, The Philippines, Turkey, Egypt, Sudan, Jordan, Morocco, Senegal, and. Tanzania, with varying degrees of success. The United States of America, Philippines, Mexico, and Turkey have developed their own models of involving farmers' organizations in the operation, fee collection, maintenance, and management of either complete irrigation projects or part of the systems commensurate with their political and socio-economic environment; (Singh, 1998). The review on irrigation development gives an account of the efforts to accelerate people's participation in irrigation system development and management.

DESIGN OF IRRIGATION WELLS

Generally, wells are designed for the purpose of irrigation, drainage, and domestic and industrial uses. The design of each type of well for each purpose requires particular attention, taking into account its purpose. Designing a water well involves selecting the proper dimensional factors for the well structure and choosing the materials to be used in its construction. Good design aims to assure an optimum, combination of performance, long service life and reasonable cost.

The hydraulic and (hydro-geological characteristics of aquifers vary greatly. Irrigation wells should be designed and constructed to take advantage of the natural conditions at a given location. When this is done, and the materials of construction are properly selected an economical and efficient well structure of long life can be achieved. Irrigation wells are usually designed to obtain the highest yield available from the aquifer, and the highest efficiency in terms of specific capacity. These factors bear directly upon operating costs. Two general principles influencing the design of both Open wells and tube wells are the water requirement of the crops to be irrigated and the location of well.

Water requirement: The yield potential of a well is evaluated on the basis of hydrological conditions of the area rainfall, runoff and recharge. When the yield potential of an area is not a limiting factor, a properly designed irrigation well should provide the required quantity of water to irrigate the entire area owned by the farmer, keeping in View the area to be irrigated, cropping pattern, and other needs such as domestic and livestock requirements.

When the yield potential of a well is more than the field water requirement of the farm under the control of the well owner, it is always economical to sell the additional yield to the neighbouring farmer, rather than limiting the capacity of the well to the requirement of a small or medium size folding. Often, it is not economical to have separate wells for individual small holdings. In an area with predominantly small holdings, it is essential that high discharge irrigation wells are owned jointly by a group of farmers or by a suitable public agency like the State Tube well Corporation.

Location of the well: The topography of the farm, the type of water conveyance system, the method of irrigation to be used, recharge possibilities of the area, safety of the pumping set and other installations are the main factors to be considered in deciding the location of the well. If the general slope of the farm does not exceed, 1 per cent and surface methods of water conveyance and distribution are to be employed, the well should be constructed at a higher elevation. If underground pipeline water distribution is to be used or sprinkler or drip irrigation are used. It is not necessary to locate the well at a high point. The location at a lower point of the farm usually facilitates better well yield.

Design of irrigation wells may be broadly divided into:

(i) design of open wells in hard rock areas (consolidated formations) and unconsolidated, formations and

(ii) design of tube wells in unconsolidated formations.

References

Bagchi, K.S. (1995). Irrigation in India - History and Potentials of Social Management. Uplabhi Trust for Developmental Initiatives, Delhi, P. 149.

Central Ground Water Board (1982). Ground water Resources and Development Potential in India. Ministry of Irrigation, Govt. of India, New Delhi, P. 112.

Dhawan B.D. (1989), Water Resource Management in India, Vol.44, No.-3, pp. 233-282.

Directorate of Economics and Statistics, Ministry of Agriculture and Co-operation, Govt. of India, Land use Statistics at a Glance Part-I.:

Indian Council of Agricultural Research (2003). Vision of Water Research - Vision 2020. ICAR., New Delhi, P.45

Ministry of Water Resource, Govt. of India (2003). Vision of Integrated Water Resources Development and Management, Feb. 2003, P.53.

National Bureau of Soil Survey and Land use Planning, Nagpur 1997.

Chapter - 2

Irrigation Methods in India : An Overview

Dr. S.B. Verma and Dr. S.K. Arya

The objective of irrigation is to distribute water over the land in such a way that it will enter the soil and be stored uniformly within the potential root zone. It is impossible and impracticable in actual farming conditions to accomplish fully this objective. The uniformity of water application along an irrigation run offers a number of problems and many of the irrigation systems are poorly designed. Either too little or too much water is applied, water is wasted, drainage problems are created, saline conditions develop, and excessive soil erosion occurs. Further, the rate at which water is to be applied, the frequency and period of irrigation depends upon the properties of the soil, the crop and the climate. While preparing an irrigation schedule, all these factors must be taken into account.

The pressure of population and need for additional food supplies are necessitating a rapid expansion of irrigation throughout the world. Due to limited supplies available, more effective use of water is becoming essential. Both the design and the operation of irrigation systems can be most intelligently done if proper correlation amongst various design parameters is known.

Irrigation water can be applied below the surface of the soil, at the surface of the soil or over the surface of the soil. The terms subsurface, surface, and overhead or sprinkler are used to describe these three types of irrigations respectively. Sprinkler irrigation involves heavy initial expenditure and high evaporation losses. Subsurface irrigation is also very expensive and laying out of pipes present a number of problems. Surface irrigation is by far the most popular method in our country. Indian farmers are more acquainted with surface irrigation methods and can be easily trained to use water efficiently and economically.

1. Types of Irrigation Methods

There are seven basic techniques or methods of irrigation, most of which have several variations. Each technique and variation has characteristics that are adaptable for different locations and crops. The basic components and operation for each of the seven techniques are:

1.1 Basin

A level area of any size or shape bounded by borders or ridges retains all the applied water until it infiltrates. Any loss of water results from either deep percolation or surface evaporation.

1.2. Basin-check

A fairly level area of any size or shape bounded by borders and with no depressions which cannot be readily drained. The borders (or ridges) retain all the applied water for a sufficient time to obtain a relatively uniform depth of infiltration over the area and then the remaining water is drained off the surface and used to irrigate an adjacent border-check. Water is lost chiefly by deep percolation and evaporation[1]

1.3. Border strip

A sloping area, usually rectangular, is bounded by borders or ridges that guide a moving sheet of water as it flows down the bordered strip. There should be little or no slope at right angles to the direction of flow. The inflow of water is usually cut off when the advancing sheet has flowed six to nine tenths of the distance down the strip. Water is lost chiefly by deep percolation and runoff.

1.4. Furrow or Corrugation

A small sloping channel is scraped out of or pressed into soil surface. For high uniformity of wetting, the irrigation stream should reach the end of the channel in about one fourth of the time allotted for the irrigation; but the stream is not shut off until the root zone soil at the lower end of the furrow is adequately irrigated. Water in the soil moves both laterally and downward from the channel. Water is lost chiefly by deep percolation and runoff.

1.5 Sprinkler

Any or numerous devices for spraying water over the soil surface are called sprinklers. Water discharged from a sprinkler into the air should infiltrate the soil where it falls but it should not saturate the soil surface. For high uniformity of wetting, the. spray patterns from adjacent sprinklers must be properly overlapped. Evaporation, wind drift, and deep percolation are the chief causes of loss of water.

1.6. Drip (or Trickle)

A device used in trickle (or drip) irrigation for discharging water at some very low rate (less than 12 Ips) through small holes in tubing placed near the soil surface. Water moves through the soil both sideways and downward away from the point of application to form a bulb of wet soil. Typically, only a portion of the soil mass is kept quite moist by very frequent or continuous application. Water loss is mainly by deep percolation.

1.7 Water Table

In certain areas the water table can be adequately controlled and periodically raised to sub-irrigate the crop's root zone. Precise control of the water table requires certain natural conditions-pervious soil, level soil surface, naturally high water table, and low salinity of water.

Table 1 summarizes and compares the major physical characteristics that affect the adaptability of each of the seven basic irrigation techniques.

It also evaluates the probable Potential Application Efficiency of Low Quarter of a well designed and properly used system, employing each technique where appropriate. Most systems can be mechanized or even automated in order to reduce labour. This labour leaves no allowance for such items as salinity and control of microclimate and taken no account of costs or personal preferences of the irrigator.

Table 1. Major physical requirements and potential application efficiencies of the low quarter for the basic irrigation techniques

Irrigation method	Physical requirements at site					
	Soil uniformity	**Infiltration rate**	**Ground slope**	**Water supply**	**Labour intensity**	**PELQ %**
Basin	Uniform within each basin	Any	Level or graded to level	Large intermittent	High at infrequent intervals	60 to 85
Basin check	Uniform within each basin	All but extremes	Fairly smooth with no	Large intermittent	High at infrequent intervals	60 to80'
Border strip	Uniform within each strip	All but extremes	Mild and smooth	Large intermittent	High at infrequent inten/als	70 to 851
Furrow' or corrugation	Uniform along each furrow	All but very rapid	Mild or contour	Medium to large intermittent	High at infrequent intervals	70 to 75[1]

Sprinkler	Soil may be intermixed	All but very slow[2]	Any farmable slope	Small continuous	High to very low Daily[3]	65 to-85[2]
Drip or Trickle	Soil may be intermixed	Any	Any farmable Slope	Small continuous	Very low daily	75 to 90
Water table control	Uniform within each field		Level or graded to level	Large relative to area	Very low	50 to 90

1. Value of 90% can be gained under ideal conditions if runoff water is reused.
2. Except for center pivot and traveling sprinklers, which are best suited to use on soils that have medium and high infiltration rates.
3. Labour inputs range from high intensity for hand move, moderate for mechanical move, to low for automatic sprinkler irrigation systems.
4. Surface soils with medium capillarity must be undertaken with very pervious sub-soils.

2. Role of Irrigation Methods in Setter Use of Irrigation Water

Considerable amount of irrigation water is not beneficially used for plant growth, after it is diverted to the individual land holdings from the turnouts. It is estimated that 30 to 40 per cent water made available at the field boundary is lost" as deep percolation and / or runoff, while it is being applied in the field for irrigating the standing crops. This loss can be minimized to a great extend by following proper methods of irrigation. A method which is appropriate for the existing field conditions, i.e., soil topography, crop, available quantity of water, climate, field dimensions, and farm machinery being used, local customs etc., will greatly reduce the application losses resulting in better plant growth, improved drainage due to less percolation losses, besides making it possible to irrigate more area with the same quantity of water, and will, thus, increase production per unit of water. Together with the method of application, the quantity and time of application of water are also very crucial for achieving better water management at the field level.

Improvements in surface irrigation are most likely to result from a skilful combination of experience and thorough understanding of the processes involved.

Surface irrigation is accomplished by one of several application methods, including borders, furrows, checks and basins. In each case water moves over the land surface in open channel flow. The water may be directed in small earth channels called furrows or corrugations or it may move as shallow overland flow over a carefully smoothed soil surface as in border irrigation.

Surface irrigation methods of applying water are widely used in many parts of the world and will likely remain important as competing demands for energy discourage the use of energy intensive alternative methods. Competing demands for both energy and water are already encouraging efforts to improve the water use efficiency obtainable with surface irrigation.

Water is lost in furrow irrigation in two ways-it runs off and it goes too deep. Stopping or reducing theses losses could irrigate more land and saves labour and energy. In areas where more land is available than water to irrigate it, the value of this water is measured by how much crop it can produce and the cost of it and the labour and capital to apply it ceases to be of dominant economic importance. Efficient irrigation under such conditions is of great value. For furrow, runoff and uniformity of the depth of water infiltrated along the furrow are related to the speed of water entering the lower end relative to know how long it needs to be there to do a job of soaking in enough water.

This is conveniently expressed as the Advance Ratio (AR), the ratio of the Time of Advance to the Time of Irrigation.

Border strip irrigation has high potential application efficiency, 80% plus, but is usually operated at about 50% actual application efficiency. Further more, it has the dubious distinction seldom realized by its users, of being the most sophisticated, complicated, least adjustable method. But when border strip irrigation is used correctly, Actual Application Efficiency can go above 90% and labour and power requirements are very low.

The value of water is not its cost, nor the labour to apply it. It is measured by what the water and labour will produce when water is in short supply.

Sprinkler and drip (or trickle) irrigation methods are unique in contrast to surface methods in that they are independent of soil uniformity and topography in their adaptability. They are also compatible to a small steady stream of water whereas surface methods work best with large flows. However regarding adaptability and contrary to popular opinion, sprinkler has the lowest potential efficiency of any normal irrigation method. Also it is difficult to modify for drought conditions. The basic reason for the good reputation it does have is that most systems are fairly well designed and the design efficiency is presumed to be the operating efficiency. However, in general the pre nozzle losses are ignored and the coefficient of uniformity is frequently incorrectly thought of as being the efficiency of the method. It requires two yardsticks - Management Allowed Deficiency (MAD) and Soil Moisture Deficiency (SMD). MAD is first expressed as the per cent of the available moisture in the root zone that can desirably be used and correlated with the climate condition. This per cent is often taken as 50% but desirably should often be 40% to 80% depending on conditions. The second yardstick is how dry is it the root zone.

3 Factors Influencing Selection of Irrigation Methods

Since the objective is to apply water efficiently to provide favourable environment for the plants, it is therefore, essential to consider the factors, which have significant effect in the process of water application. The important factors affecting this phenomenon are as under:

3.1 Soil

The soil texture, structure, porosity etc. have a great influence on the water intake (infiltration) and water holding properties of the soil. Light textured soils with limitations of depth and hard pan will allow moisture freely to deeper section in a given time period as compared to heavy textured soils. Likewise, soils having single grain structure will allow water freely to move downward compared to platy and other structures. The movement of water over the land surface (in surface irrigation methods) is also affected by the soil properties. Soil depth is another important factor, as it affects the land leveling / shaping operations, so vital for the success of the surface irrigation methods. In shallow soils, it may not be possible to undertake deep cuts required for leveling fields, as it might expose the subsoil and render the field infertile for a long time.

3.2 Land Topography

Out of the total volume of water applied at a given point on the soil surface, water in excess of soil intake rate will remain or flow over the land surface, depending on the topography or slope of land. More the slope, more rapid will be the movement of water towards the lower end of the field. Undulating topography also has its effect on the selection of irrigation methods. In an undulating field water will stagnate at low points and at the same time high spots will not receive adequate water. If this situation cannot be remedied by leveling / shaping because of shallow soils, it will not be possible to adopt gravity or surface irrigation methods in such fields.

3.3 Climate

Rainfall, temperature, humidity, wind velocity, radiation, etc., also have *an* effect on selection of irrigation methods, because of their effect on evapotranspiration. Wind velocity and direction as well as temperature greatly affect *the* performance of sprinkler irrigation method.

3.4 Water supply

Water supply in terms of quantity and quality as well as the interval are important criteria for consideration while selecting appropriate method of irrigation. If available flow is top small to be used for gravity methods, pressure methods may be the only solution. Similarly, presence of salt in water is other factor of importance for selection of appropriate irrigation methods, since, the accumulation of salts in the root zone will adversely affect the production. In such a case a method, which will push the salt out of the root zone should be preferable.

3.5 Crops

Type of crops has very significant role to play in the selection of irrigation method. The flow of water on the field surface is greatly influenced by the resistance offered to the flow by the crop (Manning's n) which in turn depends on crop spread horizontally and vertically, as well as slowing/planting methods such as broadcasted, drilled, row planted, etc. The value of crops is also very important. High value crops may provide enough returns to justify the use of pressure irrigation methods, such as drip and sprinkler, but the use of these methods for low value crops may not be economical.

3.6 Farm machinery

In mechanized cultivation the ease of machine operation and movement should be taken into account, while considering the selection of method of irrigation. The movement of machinery for operations such as inter-culture, application of chemical and harvesting, should not get restricted on account of the method of irrigation.

Other factors of importance which should be given due weightage while deciding the selection of irrigation method, are field dimensions, local customs and preference of the farmers.

4 Factors Influencing Planning

The factors required to be considered in planning are soils, topography, crops to be irrigated, water supply existing facilities and available construction and farm equipment.

4.1 Preliminary plan

- Locate the high points in a field and determine the direction of irrigation and drainage.
- Determine soil boundaries, probable crop rotations and feasibility of land leveling.

- Locate field boundaries and farm roads
- From this preliminary plan it should be possible to determine the best delivery point for the water.

4.2 Planning procedure

After a preliminary plan has been made, studied and discussed with the farmer, detailed plans for any area on the farm can be prepared. First select a method of water application for each field and prepare a layout. Then design the delivery, application, and disposal facilities as well as the necessary roads.

4.3 Layout

Planning a general layout for subdividing and irrigating the area in the units of suitable dimensions is the next step. Area delineated according to slope and soil characteristics provide a basis for selecting the best field arrangement and for locating field ditches. Consider alternate layout.

4.4 Application

Design the application facilities. Determine (1) the amount of water that must be applied in a normal irrigation, (2) the time allowed for applying it and (3) the rate at which it can be applied. Then determine the amount of water that must be delivered to field. Plan for land leveling if it is needed. Locate and design the head ditch or pipe line to fit the method of irrigation used. Locate and design ditches, pipes, levees and the other structures needed to apply water to field in the amount and rate required by the crop and soil.

4.5 Delivery

Plan the delivery facilities so that they permit delivery of water to the different fields in the volume and rate required by the method of application-previously selected. Select and design the method of conveyance either ditch or pipe line. Locate and design all the necessary grade control or distribution-control structures, including measuring devices.

4.6 Disposal

Plan for the disposal of any irrigation wastewater and excess rainfall promptly and safely. Consider recovery of wastewater for reuse. Include all necessary disposal facilities-channels, pipe lines, tiles, structures and pumps.

5 Design of Surface Irrigation Systems

Surface methods are the following types: (i) Border (ii) Basin (iii) Furrow or corrugation (iv) Contour ditch (v) Contour levee irrigation.

5.1 Phases of Surface Irrigation

(a) Advance phase: That portion of the total irrigation time during which water advances in overland flow from the upper field boundary toward the lower field boundary.

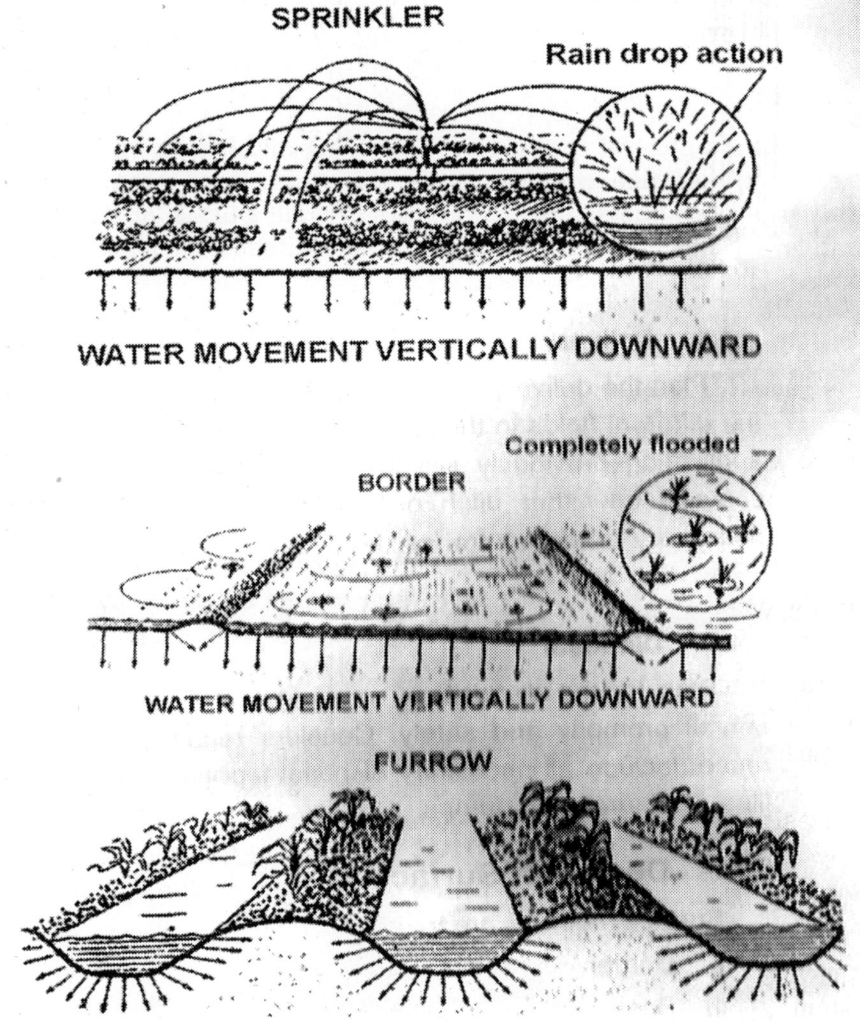

Fig. 1. Water movement vertically downward.

(b) Storage (continuing) phase: That portion of the total irrigation time between the end of advance and inflow shutoff. If shutoff occurs first, this phase is of zero duration.

(c) Depletion phase : That portion of the total irrigation time between inflow shutoff and the beginning of recession of the upper field boundary. This is also known as the recession lag time.

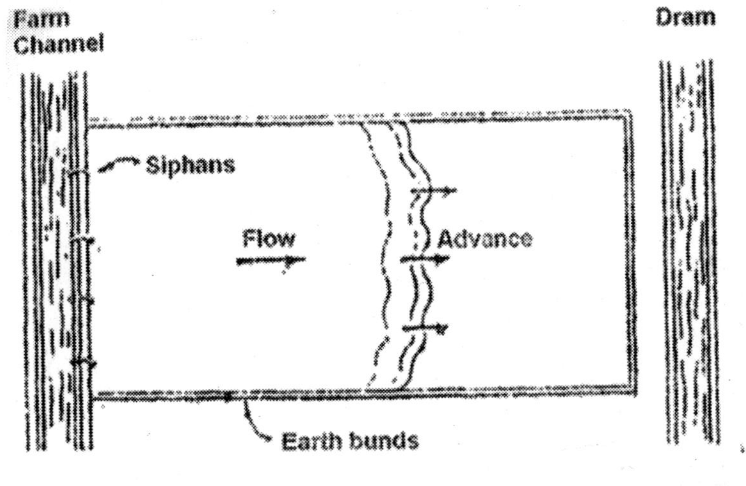

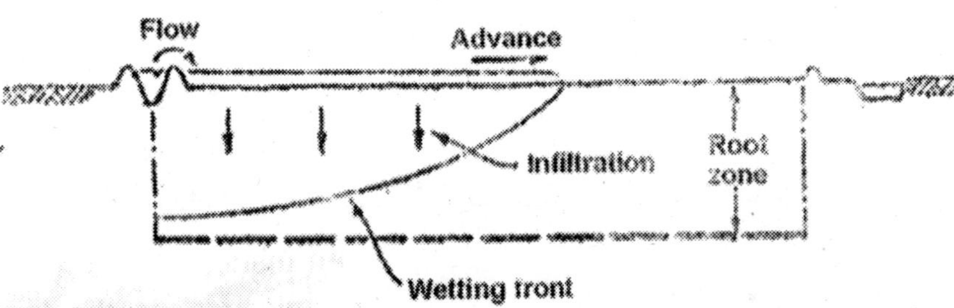

Fig. 2. Water movement in the soil (a) Water moving on soil surface (b) water moving in soil profiles

(d) Recession phase: That portion of the total irrigation time between the beginning of recession at the upper field boundary and the disappearance of the last water from the field surface.

5.2 Design objectives

A surface irrigation system should be designed rather than merely built in order to assure satisfactory adoption to the soils, topography and crops and to guarantee uniform irrigations and high water application efficiencies using the available stream size and water supply. Ideally, the system should be capable of repeatedly replenishing the root zone reservoir uniformly before the soil water has been depleted beyond specified limits. The available stream size, and the length and grade of the land units must be combine to achieve these results without excessive labour, waste of water erosion, and inconvenience to other farming operations.

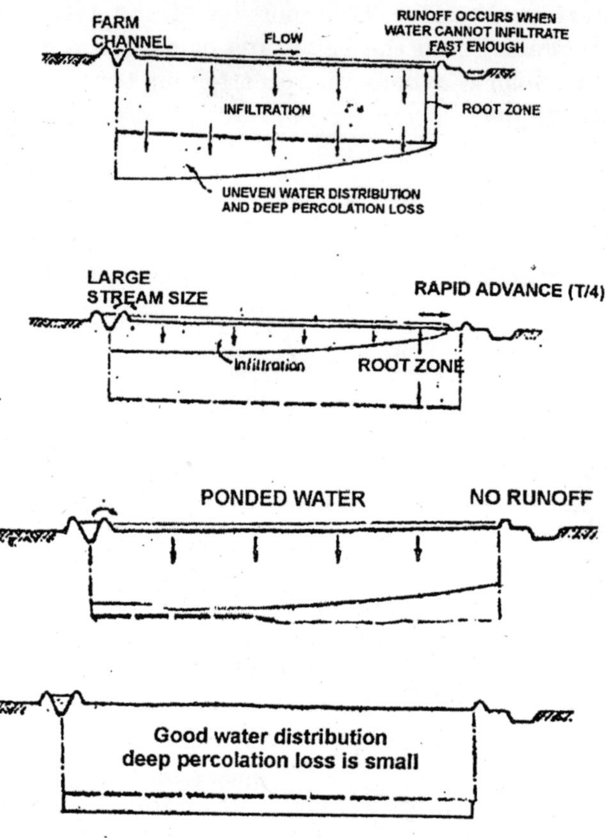

Fig. 3. Surface irrigation problems.

Designing a system implies that the behaviour or performance of the system can be predicted satisfactorily without trial and error process in the field. If the intake characteristics are known, the designer then predicts two major occurrences

(i) the advance of water sheet or furrow flow over the soil surface, and' the recession of this water sheet or furrow flow the surface. The water should remain on the surface sufficiently long (required contact time tc) to allow just the desired amount of water to infiltrate the soil. The required contact time is obtained using the cumulative intake time relationship for the soil in question. For maximum water application efficiency the design objective is to have the actual contact time as nearly equal to the required contact time as practical. The designer accomplishes this by adjusting the size of stream, length of run and other variables that can be manipulated until a satisfactory agreement is reached.

In surface irrigation methods water is applied, to the sloping or level field surface and it infiltrate into soil while it is advancing or held on the field surface respectively. The amount of infiltrated water at any point depends on the duration of time at the point, i.e., the time between the advance of water and its disappearance (recession) at that point, which is known as opportunity time. If the opportunity time is more or less same through out the field, and the soil is quite uniform, very high distribution uniformity and application efficiency can be attained. Fig. 3 describes this phenomenon and various terms associated with surface irrigation.

5.3 Design principles

The design of a surface irrigation system first involves evaluating the general topographic conditions, soils, crops, farming practices anticipated and farm operator's desires and finances for the field or farm in question. Information collected during the preliminary analysis should be sufficient to permit selecting one or more surface methods that will be most suitable. Then the basic information that will be needed to design the selected systems must be secured.

5.4 Design Data

The basic data needed to design a system can be grouped into five general categories:

Water: Annual allotment, method of delivery (continuous flow, rotation or demand system, pumped etc.) stream size available at any time and during peak water use period, quality of irrigation water expected amount and distribution of rainfall, and irrigation water requirement including leaching requirement.

Topography: Major land slopes, field sizes and shapes, uniformity of grades, minor topographic undulations, point of water delivery and surface drainage characteristics.

Soils: Feasibility of constructing canals and ditches without excessive seepage losses, structural stability for canals and ditches, maximum root zone depth, available water holding capacity, effects of surface flooding such as crusting and cracking cumulative intake as a function of time and expected variability between irrigations, erodibility, salt content and internal drainage capacity.

Crops: Types and proportions of each crop to be grown, rooting depths and allowable soil water deficits at various stages of growth, anticipated germination problems, relative sensitivity to inundation, harvesting procedures required, crop rotation systems, and grazing needs.

Other: Availability and cost of labour, financial resources available, local customs, degree of maintenance anticipated and maintenance equipment available and construction equipment available to the operator or through the local contractors.

All of the above items have some bearing on the system selected and its final design. Overlooking or neglecting to consider any one of them can impair the effectiveness of the surface method selected.

At least ten principle criteria are important in the design of a surface irrigation system. All ten are in turn governed by the overall economics of the entire farming operation. Thus the design of an irrigation system is complex and not readily subject to quantitative analysis.

- Store the required water in the root zone of the soil.
- Obtain reasonably uniform application of water.
- Minimize soil erosion.
- Minimize runoff of irrigation water from the field. Provide for beneficial use of runoff water.
- Minimize labour requirement for irrigation.

Minimize land used for ditches and other control to distribute the water.

- Fit irrigation system to field boundaries.
- Adapt system to soil and topographic changes.
- Facilitate use of machinery for land preparation, cultivating, furrowing, harvesting etc.

The irrigation system designer must consider many soil and topographic factors in selecting the method of irrigation. The suitability of each method (surface, sprinkler, and trickle) must be considered for specific site conditions. Soil and topographic factors that the designer considers in selecting the systems include:

- Topographic features such as surface irregularities, steepness of slope, changes in slope direction.
- Soil features such as water holding capacity, intake rate and depth.
- Geographic features such as field shape, natural drains, buildings, utilities or obstructions. Related factors that may influence selecting a surface system and land shaping include:

Water supply, quantity, delivery flow rate, delivery schedule and location.

Labour requirement and availability of labour

- Energy requirement, cost and availability
- Cost of system installation, operation and maintenance.
- Available farm equipment.
- The farmer's preference.

If surface irrigation method is selected, the field surface must be given detailed consideration. Rarely is it possible to establish a satisfactory surface irrigation system without some land grading to change the surface topography of a field to a planned grade.

6 Sub Irrigation

Inherent advantages make controlled sub irrigation an attractive proposition to the irrigator if he can devise the means of execution. The advantages are the avoidance of evaporative losses of open water or wet soil surfaces and the elimination of the impedance caused to cultivation by pipes and ditches.

Natural sub-irrigation

Natural sub-irrigation is so called because the conditions, which make it possible, are geographical and topographical. These are near-level terrain and a deep topsoil of very high lateral permeability at 2 m to 7 m depth by an impermeable stratum.

Sub irrigation is limited usually to areas where the soils are relatively permeable for a considerable depth, where surface slopes are gentle and where natural sub drainage is restricted. It must be practical to hold the water table. Thus a first analysis is to determine the possible lateral flow from the area when the water table is at its desired elevation.

From the Darcy's equation of continuity

$$q = k.\ i.\ a = k\ f/I\ a$$

where

q = flow in m^3 / s

k = hydraulic conductivity, m^3 m^2 = m / s

a = the cross sectional area in square meters through which flow occurs at right angles to it in the direction of flow, and

$i = f/1$

Example 1: It is desired to maintain the water table at some predetermined depth below ground surface in a highly permeable soil underlain by a restricting layer. The water table roughly parallels the ground surface, slopes about 1 in 176, and intersects the bottom of a nearby stream which is the only natural drainage way. The average depth to shale or tight clay was found to be about 15 m. Assume k for the soils is 0.3×10^4 m/s. Calculate the outflow.

Solution:

Cross sectional area from 1 km long section paralleling the creek and through which natural drainage from the irrigated area would have to flow would be 1000 m x 15 in =15,000 m^2.

$$a = \frac{0.3 \times 15{,}00}{10^4 \times 176 \times 1} = \frac{0.45}{176} m^3/s = 2.56 \text{ IPS}$$

If this is not considered adequate water removal then some drainage system must be installed.

The total water requirement for the area is the drainage plus the consumptive use requirement of the crops grown.

Chapter - 3

Irrigation Requirements of Common Crops

Dr. A.M. Michael

Water requirement of crops vary with their type and variety and stage of growth, properties of the soil and the climate. The tolerance of the crop to soil moisture stress is influenced by the genetic make up of the crop, the rooting characteristics and the properties of the stomata. Water management for efficient water use is profoundly influenced by the selection of suitable crops and their varieties based on the amount of water available in the area during different periods of the crop growing season.

Adequate soil moisture, temperature and air are the essential requirements for the germination of seeds. The seeds are to be planted at shallow depth (usually 5 to 7 cm) below the ground surface on firm moist soil to ensure proper germination. When the soil moisture at the time of sowing is inadequate a pre-sowing irrigation is necessary. Additional water over and above the crop water requirements may be necessary for leaching of excessive salts in order to maintain proper salt balance in the crop root zone. In wetland rice cultivation, puddling of the seed bed is necessary to minimize water loss due to deep percolation below the root zone.

Rice. Rice is the most important cereal crop grown. It is the staple food for over half of the population of the world. It occupies about 40% of the irrigated area in India and is grown extensively in other south Asian countries. It is a semi-aquatic plant and hence its water requirements are many times more than most other food crops. It is, therefore, a major consumer of the water and thus needs careful management in order to increase its efficiency of water use.

Rice is grown under varied soil and climatic conditions and one to three crops per annum are taken in India and other countries in monsoon Asia. Though rice could be grown on a variety of soils, it grows best on clay loams to clays since these are retentive of moisture and have low percolation rates (1-5 mm/day). It can also grow on soils with relatively high lime or alkali content. The crop thrives best under conditions of high temperature and humidity in North India as well as temperate regions. At high attitudes, the crop is grown only in the warm part of the year.

The cultural practices of rice vary widely, depending upon the variety and the local soil and climatic conditions. The conditions under which rice is grown could broadly be grouped into two, namely, wetland or lowland rice and upland rice. *Floating rice* is a particular variety of wetland rice which is grown in areas subject to deep flooding, with water depths ranging from 60 cm to 6 m. It is a long-stemmed variety which is sown before the onset of rains and starts growing quickly with the early rains. When flooding commences it grows rapidly and the plants float on the water surface.

Upland rice or dryland rice is directly sown by drilling or by broadcasting the seeds and it mostly depends on rain as the source of water. It constitutes only a small percentage (5%) of the total area under rice. Lowland rice culture is a widely accepted practice for high production. Under lowland conditions rice is generally transplanted on puddled soils and land is kept under submerged conditions, covering most of its growth period by rain and/or irrigation. Puddling reduces percolation losses, .checks weed growth and regulates soil and water temperatures.

Wetland rice cultivation has many advantages. Crop growth in the nursery, which occupies only 5 to 10% of the normal crop area, saves considerable amount of water and provides the time required to prepare the seedbed for transplanting. Weed problem is low in a transplanted crop field. The transplanted crop under predominantly submerged condition during their growth provides the required environment for growth.

Wetland rice production comprises of the following operations: *(i)* Preparation of the rice nursery and sowing good quality seeds, *(ii)* Preparation of rice fields (usually level basins in rectangular or contour benches), *(iii)* Transplantation of the seedlings when it is about one month old in the nursery, *(iv)* Water control, *(v)* Weeding, *(vi)* Fertilization, *(vii)* Pest control, *(viii)* Harvesting, threshing and storage.

On germination of the rice seed, the radical (the rudimentary root of the embryo) develops from the base of the grain, quickly followed by two additional roots. Subsequently, short lateral roots develop from the nodes of the stem below the ground level. Tillers are produced at the nodes and further adventitious roots are produced from the lower nodes and the plant develops a mass of adventitious roots. The tiller produced directly from the stem is the 'primary tiller', which is followed by further tillers. The roots tend to develop horizontally more than in the vertical direction. Hence, the nutrient uptake of the rice crop is mainly from the top layer of the root zone. A remarkable feature of the root system of rice plant is that it does not suffer from lack of oxygen under submerged condition. Using large basins is most suitable in wetland rice cultivation. For the distribution of water between basins, field channels are constructed between basins to deliver water to individual fields on either side, in case of gentle slopes (keeping field plots level), and to the down-slope fields in case of contour channels. The traditional practice of 'field to field' irrigation, in which water is turned to the adjoining lower basins after those at higher elevations are filled reduces the water distribution efficiency substantially and results in loss of fertilizer and nutrients.

In case of individual field plots, the traditional practice has been to keep the field submerged with water to a depth of about 8 to 10 cm or more. Field experiments conducted at various locations in India and other countries on wetland rice reveal that the practice of keeping the soil under shallow depth of submergence (about 5 ± 2 cm water) throughout the crop growth period is conducive to higher yields. The practice of shallow submergence directly saves considerable amount of water, as compared to deep submergence. These results, in general, hold good under different climatic conditions. The practice of continuous shallow submergence, however, is possible only when the water supplies are adequate and assured. Land needs to be scrupulously levelled to facilitate uniform spreading of water. Weeds, especially the grassy types, need to be controlled.

Experimental results are available to show that it is not always necessary to follow the usual practice of continuous submergence. The practice of intermittent submergence during the critical stages of initial tillering and /or flowering and maintenance of saturation to field capacity during the rest of the stages give yields comparable to those obtained under continuous shallow submergence. The water supplies, if limited, could be curtailed during the non-critical stages of crop growth. The shortage of water during initial tillering and flowering reduces the yield considerably, while the stages of tillering, grain formation and maturity tolerate water stress to a great extent. The water is drained completely in about 7 days before harvesting to facilitate the use of mechanical equipment in, harvesting.

Drainage needs of rice: When continuous land submergence by ponding is practised, it is necessary to drain the soil either through horizontal (surface) or through vertical (sub-surface) drainage, once or twice during the growth period, especially on poorly drained clayey soils having the rate of percolation less than 2.5 mm/day. This practice is beneficial in removing the toxic substances such as sulphides and in supplying oxygen to the root system. The drainage period could last from four to eight days, depending upon the type of the soil. In case of sandy soils, the drainage period is about three days while in clay soils it is about five to seven days. The following growth periods can be identified in a wetland rice crop:

(i) Growth period in the nursery (usually one month),

(ii) *Vegetative stage.* From transplanting to panicle initiation (initiation of the inflorescence with its main stem and sub-divided branches). This period may range from 45 days to 90 days, depending on the variety of the crop.

(iii) Mid-season stage. From panicle initiation to flowering, and it includes stem elongation, panicle extension and flowering.

(iv) Late season or ripening stage. Period from flowering to full maturity (duration about one month).

Transplanted rice may be of short duration, medium duration or long duration, depending on their growing period in the field after transplanting. The growth period of short duration rice usually ranges from 90 to 100 days, medium duration crops 100 to 130 days and long duration crops 130 to 160 days after transplanting.

Components of water loss in rice culture: Considerable amount of water delivered to the rice field in wetland cultivation is lost through deep percolation, evaporation from the water surface in the field and field channels, seepage from the bunds and field channel embankments, overtopping of the bunds, breaches and rat holes in the bunds, and runoff to drains. Vamadevan and Dastane (1968) observed that out of 1960 mm of water needed by rice, 1200 mm were lost through deep percolation (about 70%) on the sandy loam soils of Delhi and only 480 mm were actually used consumptively.

Great economy in water use by rice crop could be achieved if suitable measures are adopted to reduce the deep percolation losses. Some of the practices to improve water use in lowland rice are as follows (Dakshinamurthi, Michael and Shri Mohan, 1973).

1. Selecting heavy soils or those with hard pans in subsurface depths to reduce permeability.
2. Growing crop in, large blocks instead of in isolated small holdings
3. Scrupulous land levelling.
4. Puddling to reduce permeability.
5. Compacting soil, embedding polythene, sheet, applying chemicals such as bitumen, asphalt, etc. (these practices, however, are expensive and are not usually followed.
6. Treating fields, by, adding one to two per cent bentonite with small quantities of sodium chloride and sodium carbonate.

Wheat : Wheat *(Triticcum* spp.*)* is the most widely grown of all the cereal crops in the world. Wheat is grown in almost all temperate and most of the sub-tropical zones worldwide. It is also grown in some tropical countries at higher elevations. Wheat varieties can be broadly grouped into *winter wheat* and *spring wheat.* Winter wheat requires a cold period during early growth. Winter wheat has a growing period of 120 to 160 days. For winter and spring wheat, the minimum daily temperature for growth is about 5°C. The mean daily temperature for optimum growth and tillering is in the range of 15°C and 20°C. In regions of severe winter and frost winter, wheat requires a period of dormancy which may extend up to about 90 days, depending on the temperature conditions.

In India, wheat is grown on an area of about 19-20 million hectares and nearly 54 per cent of this area is irrigated. The best wheat producing region in India is the Indo-Genetic plain where cool and dry climate prevails during the growing season *(rabi)*. Areas that are always moist and warm are not suited to wheat. Well drained clay loams, loams and sandy loams are the best soils for wheat. Wheat is commonly sown around the middle of October under rainfed conditions and around the middle of November under irrigated conditions. The crop takes 4 months to mature in the northern region, 3 months in the central, and 3 in the southern regions.

Wheat farming in India has been almost revolutionised with the introduction of the high yielding dwarf wheat varieties in mid-1960s. Thereafter, a number of dwarf varieties have been evolved in the country which are highly responsive to irrigation and fertilizer applications. A short, stiff, strong stem is the main advantage a dwarf wheat has over the taller varieties. The shorter stem allows the plant to utilize irrigation water without lodging, and results in higher yields. Wheat is grown as an irrigated crop or rainfed crop, depending on the occurrence of rains during the growing period. In many temperate regions it is grown as a rainfed crop. The need for pre-sowing irrigation depends on the adequacy of soil moisture for germination of the wheat seed.

Considerable research information on the water management of wheat, using various criteria for irrigation scheduling based on soil, plant and climatic factors, has been obtained in wheat growing areas worldwide. The most practical criterion commonly adopted for scheduling of irrigations to wheat is the one based on the physiological growth stages critical in demand for water. In wheat, different growth stages such as crown root initiation, tillering, jointing, booting, flowering, milk, and dough could be well delineated. Experiments conducted to study the important stages critical in their demand for water have clearly indicated that some stages can tolerate moisture stress' to a certain extent. Most of the workers have observed that in case of dwarf varieties of wheat, irrigation at the crown root initiation stage (20-25 days after sowing) resulted in the maximum production per unit of water applied and, therefore, this stages was considered as the most critical stage for irrigation. It has been observed that regardless of the depth of planting, the crown root develops about 2 cm below the surface of the soil. If the soil zone around the crown is dry at the time of the crown root initiation, crown roots do not develop properly and only a few tillers are produced. An irrigation at the crown root initiation stage stimulates root development and tillering. The second important stage is that of flowering which takes place 80-85 days after sowing. Among the other stages, jointing and milk rank third, followed by tillering stage, and then dough stage. Irrigation is applied at the initiation of the stage so that the moisture in the soil is available during the stage in question. In general, wheat requires 3 to 5 irrigations in North India.

Fig. 1 Seedling of dwarf wheat showing crown root. Irrigation is critical at the crown root initiation stage.

There has been substantial contribution by the Plant Breeders of the Indian Agricultural Research Institute (IARI) and the Agricultural Universities in Indian since 1980s in developing stress tolerant wheat varieties which could be grown successfully under limited irrigation and rainfed conditions. The 'Kundan' variety of wheat developed at IARI in 1980s is recommended for cultivation under limited irrigation conditions in the northern plains zone of India. It is resistant to soil moisture stress and is maintaining its resistance since the past 25 years (Pandey, H.N., 2005). It can be grown successfully with 2 to 3 post-imergence irrigations.

Predominant drought situation prevails in many parts of Central India where wheat is cultivated under rainfed or limited irrigation availability. For such situations the IARI Regional Station, Indore has developed a number of *Aestiuum* and *Durum* varieties of wheat.

The *Aestiuum* varieties can be cultivated under rainfed as well as under one or two supplementary irrigations after germination and yield 30 to 40 quintals/ ha. The *Durum* varieties are semi-dwarf and under 2 support irrigations can yield 45 to 50 quintals/ha under October sowing.

Maize: Maize, also called as corn, is an important cereal crop grown in climates ranging from temperate to tropical during the period when mean daily temperatures are above 15°C and frost-free. About half of maize production in the world is from USA.

Maize is grown almost all over India under varied soil and climatic conditions. However, it is essentially a warm and humid season crop and in areas of mild climate it can be grown throughout the year. It grows best on fertile, well drained loamy soils. In North India maize is grown mostly during the monsoon period *(kharif)* of June-October as rainfed or irrigated and on a smaller scale in the summer season (Feb., Mar.-May/ June) under irrigated conditions. Three crops of maize could be grown in a year (June-September, October-January and February-May) in southern India because of favourable weather conditions. Most of the improved varieties of maize require 100-120 days to mature, except in the northern hills where they take 10-20 days more.

When maize seed germinates, four seminal roots appear. They are soon superseded by adventitious roots which develop from the lowest nod of the stem, below ground level. They begin to spread out laterally in the upper layers of the root zone soil. Subsequently, they grow downward as the plant grows. In about 10 to 12 weeks after sowing the entire root zone of 75 to 120 cm is filled with a mass of fibrous roots. However, a major part of the maize roots lie within 75 cm from the ground surface.

The irrigation requirements of maize vary with the type of soil and the season in which they are grown, depending upon the rainfall received. In regions where rainfall is inadequate maize can be grown with supplemental irrigation. In northern India, the crop is normally sown after a pre-sowing irrigations as the, monsoon arrives late. During the monsoon, supplemental irrigation may be required whenever soil moisture falls below the desired level. The early vegetative stage (20-40 days after sowing) and tasselling and silking stage (45-60 days after sowing) have been observed as critical in the demand for water by maize. Maize is very sensitive to excess water and hence, it is advisable to plant it on ridges or make ridges in the field after its establishment.

Sorghum. Sorghum *(jowar / cholam) is* next in area and importance to rice and wheat/India. It is grown for grain and fodder in the relatively dry tracts of central and southern India. Sorghum grows best in semi-arid areas with well distributed rainfall of about 300 to 1000, mm. The temperature in the growing season should be within the range of 16 to 40°C. Optimum temperature for growth is around 25°C. Sorghum has many features which make it a drought resistant crop. This is mainly due to the extensive root system, effective control of ET and stomata with an ability to recover rapidly after a period of water stress. In dry areas with low and/or erratic rainfall the crop can respond well to supplementary irrigation. Under conditions of deficient water supply it is advantageous to spread the water application to larger areas of sorghum and provide for irrigation at the critical periods of growth.

In India sorghum crop is grown mostly as a monsoon season crop (*kharif,* June-November). It is also grown in the October-March season on the moisture retentive deep black soils of South Gujarat and in those parts of Andhra Pradesh, Karnataka and Tamil Nadu which benefit from the North-East monsoon rains. The crop takes 110-130 days for maturity, depending upon the variety.

Sorghum is grown on a variety of soils ranging from clay to sandy loam. When grown on clay to clay loams the irrigation requirement is low and on lighter soils, irrigation requirement is higher. Initial seeding, pre-flowering, flowering, and grain formation stages coinciding with 2-4, 12-14, 14-16 and 17 weeks after sowing, respectively have been found critical for water demand of sorghum.

Pearlmillet: Pearl millet *(bajra)* is well suited to warm areas of low rainfall (200-500 mm). In India it is commonly grown in Rajasthan, Haryana, Punjab, Uttar Pradesh, Gujarat, Karnataka and Tamil Nadu. It is mostly grown as a rainy season crop *(kharif)* under rainfed conditions and it takes 85-90 days for maturity. Pearl millet is preferred when the rainfall is inadequate for maize and sorghum, as it is more tolerant to drought. Observations, However, have revealed that pearl millet *(bajra)* responds well to the applications of supplemental irrigations.

The irrigation requirement of the crop vary from 150 to 200 mm in most parts of India, Experiments have revealed that the crop tolerates about 75% soil depletion from 0-30 cm depth of soil on heavy clay soils. On sandy loam soils, however, best yields are obtained with irrigations at 50% depletion level.

Barley: Barley is commonly grown as a *rabi* crop (October-March) mainly in the States of Uttar Pradesh, Rajasthan and Haryana. It needs less water than wheat for its growth. It is grown as a rainfed crop in areas having rainfall of 400-500 mm in the growing season. In drier areas the crop needs irrigation for good growth. It is generally grown on sandy loam soils of medium fertility level. The crop has a moderate level of tolerance on saline and alkaline soils. The moisture sensitive period occurs at the end of the shooting stage and during earing. During this period, drought conditions have their maximum effect in reducing yield, and irrigations have the greatest effect in increasing yield.

Pulses: Pulses, or grain legumes, by virtue of their deep tap root system, utilize soil moisture efficiently and require less number of irrigations as compared to cereals. These crops, therefore, have been introduced in the cropping pattern in areas with limited water resources. Legumes in monsoon season *(kharif)* are normally grown without any irrigation in most parts of India, unless the rainfall distribution is very unfavourable. However, legumes grown in winter *(rabi)* and summer seasons depend on irrigation. Most of the legumes do not require irrigation at the early vegetative stage, as it may do more harm than good by interfering with the nodulation and restrict oxygen requirement of roots at this stage. These crops profit from irrigations during flowering and pudding,

Gram: Gram *(Cicer arietinum* L.), also known as chick pea *(chana),* is extensively grown in northern India in the winter season *(rabi)* on alluvial sandy loam to loam soils. In central and, southern India, the crop is grown on water retentive clay loam soils and hence is not normally irrigated. The crop is best suited for low rainfall areas. However, it has been found to respond to irrigation in the absence of winter rains, especially on the lighter soils of northern India. The crop generally requires only one or two irrigations when grown in *rabi* in northern India. Pre-flowering (45 days after sowing) and flowering (70 days after sowing) stages have been found to be most responsive to irrigation.

Pigeon Pea; Pigeon Pea *[Cajanus cojan* (L). Mill sp.], commonly known as red gram *(arhar),* is grown in practically every state of India. The crop can be grown on almost all types of soils, ranging from alluvial sandy loam to heavy black clay soils. It is mostly grown in the rainy season (June-November). It is grown as a *rabi* crop in the south where mild winter prevails, in which case the sowing is done in September-October, and harvesting in March-April. The crop benefits from irrigation if dry spells prevail for long periods. Flower initiation (about 75 days after sowing) and pod filling (about 100 days after sowing) are found to be most responsive to irrigation for short duration red gram.

Green gram: Green gram *(mung)* or mungbean *(Vigna radiata),* is cultivated almost all over India. The major producing states being Uttar Pradesh, Madhya Pradesh, Punjab, Andhra Pradesh, Rajasthan, Karnataka and Tamil Nadu. It is largely grown as a rainfed *kharif crop.* In some parts of Maharashtra, Andhra Pradesh, Karnataka and Tamil Nadu it is also grown in the *rabi* season as a second crop. The crop takes two and half to three months for maturity. In *rabi* and summer, the crop needs irrigation for higher production. The crop when grown in *rabi* or summer generally requires two to four irrigations for high yields. A pre-sowing irrigation is generally found to be beneficial.

Oilseeds: Oilseeds occupy approximately 25 per cent of the total cropped area of India. Like grain legumes pulses, oilseeds are also efficient in water use and hence require less water for their growth. These crops are mostly taken as rainfed but their yields could be boosted substantially if grown under limited irrigation conditions, especially in the arid and semi-arid areas and on lighter type of soils.

Irrigation requirement is mainly based on soil moisture depletion ' levels, depending on the rainfall during the growing season. Flowering and fruiting stages are observed to be most responsive to irrigation.

Mustard and Rapeseed. These are edible oilseed crops mainly grown in North India. These are crops of tropical as well as temperature zones and require cool weather for satisfactory growth. In India, these are grown in the winter season *(rabi),* mostly on sandy loam to clay loam soils. The optimum sowing time for the common varieties is the last week of September. Sowing after September end will delay flowering and result in attack by insects. Being a crop with well developed tap root system, mustard and rapeseed can utilize soil moisture from deep layers in the root zone and can produce high yields without any irrigation after germination. The crop has been found to be highly successful in the flood-prone areas of the Indo-Gangetic plains (covering an area of about 2 million ha) when sown in moist soils in the last week of September, with yields ranging from 20 to 30 quintals/ha. The technology sponsored by the IARI, New Delhi, revolutionized mustard cultivation in India in late 1980s. These crops benefit from irrigation if the winter rains are inadequate. Pre-flowering (about 40 days after sowing) and pod filling (about 80 days after sowing) stages have been observed to be the most responsive to irrigation.

Groundnut: Groundnut *(Arachis hypogaea),* also called peanut, has high oil content and nutritive value. The oil content varies from 48 to 51% and the protein content is about 25%. The crop is highly adapted to well-drained, loose, friable medium textured soils. The growing period of groundnut ranges from 90 to 115 days for sequential branched varieties and 120 to 140 days for alternately branched varieties. For optimum growth of groundnut the mean daily temperature ranges from 22 to 28°C. The crop yield gets reduced at temperatures above 33°C and below 16°C. The germination gets delayed at temperatures below 20°C. The crop is planted with a row spacing of 75 cm to 1 m. Groundnut is moderately sensitive to soil salinity. Being a legume, groundnut can fix nitrogen from the atmosphere.

Groundnut is commonly grown in semi-arid regions both under rainfed and irrigated conditions. It has a well developed tap root system extending to about 1.8 m depth. The major part of the root system is within the top 50 to 60 cm soil layer. The flowering period is most sensitive to water stress. Water deficiency during flowering causes flower drop and improper pollination. Next is the yield formation period. Water deficit during this stage results in reduced pod weight. In general, water deficiency during the vegetative period causes delayed flowering and harvest, accompanied by reduced growth and yield. To obtain higher values of total production per unit of water applied it is better to cover larger areas under the crop and reduce the quantity of water applied, taking care of the growth periods sensitive to water stress.

Sunflower; Sunflower is an important oilseed crop widely grown as a rainfed crop and under supplemental irrigation. The crop is photo-insensitive and can be grown-in almost any season. It takes about 3 to 4 months for its maturity. Sunflower is sensitive to frost. Mean daily temperature for good growth is about 18 to 25°C. It is grown on a wide range of soils. Optimum soil pH is in the range of 6.0 to 7.5. At lower values liming may be required. Sunflower is particularly sensitive to boron deficiency. Under erratic and low rainfall conditions, a deep soil with good water-holding capacity is required. Sunflower has deep root system which may extend up to 2 to 3 m in deep soils. Hence, the plant can extract moisture from deep layers.

The water requirement of sunflower is in the range of 600 to 1000 mm, which is high when compared to most crops. The ET of sunflower may be as high as 12 to 15 mm/day at the flowering stage. Despite the high water requirement the plant can withstand short periods of severe water stress. However, severe water stress during the early vegetative stage results in reduced plant height, but deeper root depth. Adequate soil moisture during the later part of the vegetative period is necessary for proper bud development. Flowering period is most sensitive to soil moisture stress which may result in considerable yield reduction, as only fewer flowers come to full development. Yield formation stage is the next most sensitive to water deficit, causing severe reduction in yield and oil content. Supplemental irrigation is highly desirable during periods of failure of rainfall at these stages.

Safflower. Safflower *(Carthamus tinctorius)* is an important oilseed crop in India, the Mediterranean basin, China, Japan, Mexico and USA. In India, safflower is grown largely in Maharashtra, Madhya Pradesh, Chhattisgarh, Andhra Pradesh and Karnataka. The crop is grown in the winter season *(rabi)* as a rainfed crop, especially on water retentive clay soils.

Safflower requires a fertile, fairly deep and well drained soil. The crop is not suited to lowland humid tropics. It is a draught resistant deep rooted crop. In deep soils, the roots may grow as deep as 3 m or more. The crop can withdraw water from depths as deep as their root system. Safflower is moderately resistant to soil salinity. However, high yields are produced on soils with neutral reaction. When the pH is lower than 6, liming may be required. The row-to-row spacing of the crop varies from 50 to 80 cm, with 15 to 35 plants per metre length on each row.

Safflower is susceptible to diseases under excess soil moisture conditions. Excessive humidity may cause head rot. Heavy rains during flowering may adversely influence pollination and seed filling. Soil moisture deficits during the early and late vegetative periods cause reduction in growth and prolong the growing period. Safflower. is fairly tolerant to soil moisture stress. However, flowering and yield formation stages are sensitive to water shortage. The water uptake gets reduced when about 60% of the available soil moisture has been abstracted. Under conditions of limited water availability, the overall production is increased by increasing the area under the crop and partially meeting the crop water requirements rather than meeting the full water requirements over a limited area.

Cotton: Cotton is the most important fibre plant in agriculture. Most of the cotton grown in the world is rainfed, except in regions with low annual rainfall where the crop is grown under irrigated conditions. Cotton crop falls into two groups:

(i) *Gossypium arboreumum,* and *Gossypium herbaceum:* These have short fibres which have low commercial value and are being progressively replaced,

(a) *Gossypium hirsutum:* This has medium length and strong fibres. Varieties of this type are most widespread in the world. *Gossypium barbadense* is a cotton variety of this group with long fibres which are smooth and fine and of high commercial value. The Egyptian cotton belongs to this variety.

In northern India, the crop is sown in the month of May after pre-sowing irrigation and at least two to three irrigations are required prior to the start of monsoon. During the rainy season, normally no irrigation is required, unless rainfall distribution is very erratic. In ' central and south India, sowing is coincided with the onset of monsoon in the month of June. In parts of Tamil Nadu and Karnataka, the crop is also sown in August-October as these regions receive rainfall during this period and the temperature remains high and favourable, unlike the cold temperatures in northern India.

Cotton is very susceptible to frost. The total growing period ranges from 150 to 180 days. Depending on the variety and temperature conditions, bud formation takes about 50 to 60 days from planting, 25 to 30 days for flower formation and 50 to 60 days from the opening of flowers to maturity. The optimum temperature for germination ranges from 18 to 30°C. Temperatures between 27 and 30°C are optimum for boll formation. Above 38°C the yield gets reduced. Cold and strong winds harm the young seedlings. Such winds at maturity will blow away fibres from open bolls.

The cotton plant has a deep and extensive root system when grown on deep, well drained, and fertile soil. The root system develops rapidly, and within 70'to 80 days may extend to a depth of about 1.8 m and laterally to the same extent. Rooting depth is affected by soil and drainage conditions, as well as by irrigation frequency. In general, rooting depths are shallower on clay soils, or with frequent irrigations. Cotton can extract most of the soil moisture to the depth of rooting and this characteristic has an important influence on irrigation of cotton. '

Pre-planting irrigation is important unless rainfall has brought the soil to field capacity to a depth of about 1 m. This will ensure good germination and rapid plant development. Under extremely unfavourable conditions, a light irrigation after planting may be needed to ensure germination.

Sufficient water should be added to the soil, either from rainfall or irrigation, to ensure that the entire root zone is brought to field capacity early in the growing season; say within 40 days after planting. Cotton is relatively tolerant to short periods of waterlogging. However, waterlogging and heavy rains may lead to lodging of the plants. The crop is relatively tolerant to soil salinity.

Adequate soil moisture is needed for rapid growth and good budding and formation of healthy bolls. However, excess water early in the growing period will restrict root-and crop development. Cotton requires adequate water supply, particularly just prior and during bud formation. Severe water deficits during flowering may restrict the growth, but with subsequent water application the crop growth recovers. In the vegetative period, irrigation is scheduled when about 60% of the available moisture in the top 75 cm of the root zone has been taken up by the crop. During flowering the soil moisture depletion of about 70% could be tolerated without adversely affecting the yield.

Irrigation after flowering and fruit setting period is needed only to supply sufficient soil moisture for fibre and seed development. The soil moisture levels maintained during the flowering and fruit setting period are usually adequate for this period. Particular attention should be given to avoid over-irrigation during this period as this may encourage delay in opening of bolls. The timing of the last irrigation will depend on the water-holding capacity of the soil. In general, it is desirable to allow the cotton crop to mature on stored soil moisture, and extract almost all available soil moisture at the end of the season. Thus, in north India, irrigation of cotton on deep loamy soils is not generally needed after late September, while on drier soils, it may be necessary to irrigate till mid-October.

Jute: Jute is an important fibre crop of India and Bangladesh. It is grown on a large scale is West Bengal and to some extent in Bihar. Uttar Pradesh, Orissa and Andhra Pradesh. The crop requires warm and humid climate, with temperature ranging from 24°C to 35°C and relative humidity of the order of 90 percent. The crop requires irrigation if rains are untimely and not well distributed. Sowing of jute in lowland is done is February, on midlands in March-April and on uplands in May-June. In some parts of Bihar and Uttar Pradesh, the sowing may continue upto mid-July. Irrigation schedules based on the maintenance of a soil moisture deficiency level of 50 to 60 per cent have been reported to be most favourable for jute.

Sugarcane. Sugarcane *(Saccharum officinarum)* is the most important source of sugar in the world. The crop requires a long, warm growing season with a high incidence of radiation and adequate soil moisture, followed by dry, sunny and fairly cool but frost-free ripening and harvesting periods. Nearly 70% of the area under sugarcane in India is irrigated but this constitutes only about five per cent of the irrigated area of the country. Sugarcane usually occupies the land for about 10 to 18 months and thus necessitates adequate irrigation for realising its potential yields.

In North India, planting is mostly done with the commencement of spring season (February-March). In the States of Tamil Nadu and Andhra Pradesh in the south, planting is done during December-February. In Maharashtra and parts of Karnataka, the crop is planted in December-February, October-November and July-August for 12, 15 and 18 months crops, respectively.

Sugarcane is propagated by planting stem cuttings from mature plants. Optimum temperature for sprouting (germination) is 32 to 38°C. Roots emerge quickly from the original cutting. After a time, secondary shoots emerge quickly from the original cutting, and they send down shoot roots which eventually develop into a mature root system.

Optimum growth in sugarcane is achieved between 22 and 30°C. Minimum temperature for active growth is about 20°C. However, for ripening a relatively lower temperature in the range of 20 to 10°C, which has a significant influence on the reduction of vegetative growth and enrichment of sucrose in the cane, is preferable. The growth period ranges from 9 months with harvest before winter frost to 24 months in Hawaii, USA. In general, the growing period is between 15 to 16 months. The first crop is usually followed by 2 to 4 ratoon crops, each taking about a year to mature. Adequate soil moisture throughout the growing period is important to obtain high yields.

Sugarcane has an extensive fibrous root system, with roots most extensive in the upper 60 to 90 cm of soil. Some roots may extend to depths' as much as 240 cm in well drained, deep loamly soils. Moisture extraction is greater in 4he upper 120 cm, decreasing rapidly below that depth. Rooting patterns will vary with soil type and drainage conditions.

Basically, three methods of water application are adopted: furrow, sprinkler and drip. The frequency and depth of irrigation vary with the growth periods of the crop. During the emergence and establishment of young seedlings, light and frequent irrigations are required. During the early vegetative period, the tillering is in direct proportion to the adequacy of soil moisture in the root zone, which requires frequent irrigations. During stem elongation the irrigation interval can be extended, but the depth of irrigation is increased. The response of sugarcane to irrigation is greater during the vegetative and yield formation stages.

Average soil moisture in the first 60 cm should generally be kept at a level of above 66 per cent of total available moisture. However, during winter time in North India, temperatures are too low for growth, and available soil moisture levels need only to be maintained at about 50 per cent level in the root' zone. Just before the harvest of the crop, irrigations may be somewhat delayed as the practice results in the accumulation of sucrose in the cane.

Sugarbeet. Sugarbeet *(Beta vulgaris) It* is a sugar enriched root crop, which provides about 40% of sugar production in the world. Sugarbeet needs a relatively long growth period, ranging from 140 to 200 days. The crop is grown under a variety of climatic conditions. The minimum temperature for seed germination is 7 to 10°C. During the vegetative growth period a higher temperature is preferred. High sugar yields are obtained when the night temperatures are in the range of 15 and 20°C and daytime temperatures between 20 and 25°C. During tried later part of the growing period, day temperatures above 30°C decrease the sugar yield substantially. Sugarbeet is grown in a variety of well-drained soils with medium to slightly heavy textures. Seeds are planted 1 to 2 cm deep in single or double rows. The row-to-row spacing of the crop range from 50 to 70 cm. Crust formation on the soil surface during seed germination can lead to poor crop stand. When the plant attains 4 to 8 leaves, thinning is done either manually or by machines.

Sugarbeet has a relatively high tolerance to soil salinity. However, soil pH below 5.5 is unfavourable for growth. The process of photosynthesis results in the formation of sugar in the leaves, a part of which is utilized in the metabolic processes in the plant and the rest is stored in the fleshy part of the roots, particularly during the growing period when the vegetative growth slows down. The sugar yield of the crop is determined by the size of the roots and their sugar concentration. With the rapid growth of the storage roots, sugar concentration reaches a high value, depending on the climatic factors, soil moisture content and the nitrogen level in the soil. Nitrogen is given in split applications, a small part at the time of planting and the rest after the thinning of the crop. However, excess application of nitrogen as well as application during the late growing season will reduce the sugar yield. Sugar content is also influenced by the crop variety and plant spacing. Often the sugar content in the root is greater than 15% of the fresh root weight. When grown for sugar, flowering and seed formation stages are avoided. When the crop is grown for seeds several weeks at low temperatures near 4°C are required to induce flowering.

Three distinct growth stages are recognized in sugarbeet. The first, ranging from 25 to 35 days is from the time of sowing to the emergence of seedlings. The second, lasting 75 to 115 days is from the emergence of the seedling to the point of constant growth. During this period the leaf surface develops faster than the root. In the third period, which extends from 40 to 50 days, the rate of growth is more constant and strong root develops along with a large canopy of leaves. A well-developed plant has a deep tap root and a set of well-branched lateral roots which fill the upper 35 cm of the soil, extending to a radius of about 50 cm.

Sugarbeet is grown both under rainfed and irrigated conditions, depending on the availability of rains. Sugarbeet is particularly sensitive to soil moisture deficits during the emergence of the seedlings and a period of about one month after emergence. Frequent light irrigations are desirable during these periods. On the other hand, over-irrigation will retard leaf development and can encourage flowering in the first year (bolting) which is harmful when the crop is grown for sugar. Water deficits during the mid-growing period (vegetative and yield formation periods) will adversely influence the sugar yield more than during the ripening stage. Soil moisture deficits, together with deficiency in nitrogen towards the end of the growing period, may reduce the growth of roots but increase the sugar concentration. Thus, except during the emergence and early growth periods, the crop is less sensitive to soil moisture deficits.

Tobacco: Tobacco *(Nicotiana tabacum)* is grown under a wide range of climates. The crop requires a frost-free period of 90 to 120 days from transplanting to last harvest of leaves. The optimum temperature range for growth is 20 to 30°C. A dry period is required for ripening and harvest of leaves. Tobacco is raised in nurseries (soil beds) and is transplanted 40 to 60 days after sowing when the plant is about 15 cm tall.

Tobacco is grown under a variety of soils and climatic conditions. Cigarette tobacco is normally grown on light soils and mainly in *rabi* (winter season) in the southern parts of India. *Bidi* tobacco is grown as a rainfed crop during *kharif* and *rabi* seasons, mostly in Gujarat, Karnataka and Maharashtra. *Hookah* tobacco is mostly grown in the alluvial soils of Uttar Pradesh, Bihar, West Bengal, Haryana and Punjab.

Tobacco when grown on heavy black clay soils normally does not require irrigation especially in the monsoon season *(kharif)*, but the crop grown on light soils and during winter season *(rabi)* requires irrigation. Furrow irrigation is more efficient than other surface methods of irrigation.

Tobacco has a well developed tap root with extensive and well developed secondary roots. However, about 75% of the moisture uptake is from the top 30 cm of the root zone. When grown under irrigated conditions tobacco requires careful scheduling of irrigation applications. In the nursery the seedlings are given light frequent irrigations. After 30 to 40 days of sowing the seedlings require less water in order to obtain robust plants. The period of maximum water requirement is 50 to 70 days after transplanting. This is followed by a decrease in water requirements. Moderate water deficits during the early vegetative period may enhance root development. However, water deficits during the mid-vegetative period (period of rapid growth) may result in reduced growth and smaller leaves. Severe water deficits during the yield formation and ripening periods adversely affect leaf weight and chemical composition which will influence the fire-holding capacity. However, a mild water deficit during ripening may be favourable as it will restrict the growth of new young leaves. On the other hand, excess water due to rains or irrigation will result in leaves of low quality. Waterlogging for two or more days will severely damage the plant.

Fodder Crops

Important fodder crops of the monsoon season *(kharif)* are maize, cowpea, pearl millet and sorghum. The important *rabi* fodders are oats and berseem or Egyptian clover. Some fodder crops like napier and lucerne are grown as perennial crops.

The objective in the irrigation of fodder crops generally is optimum vegetative growth of a certain quality. Since seed production is not generally the objective, the timing of irrigation with respect to grain yields does not apply. For example, in maize and sorghum *(jowar)* grain yields are not appreciably affected by early season stress (prior to bloom) if there is no severe wilting. Such a stress will have **a** marked affect on vegetative growth and fodder yield. Lucerne, for example, generally produces good seed yields only under moisture stress. As a general rule, optimum forage production is possible only with a continuous high moisture level. A safe rule to follow is to irrigate so as to keep the available soil moisture level above 50 per cent at all times.

Lucerne (Alfalfa): Lucerne *(Medicago sativa),* also known as alfalfa, is an important fodder crop of high nutritive value. It is a perennial crop planted from seed and reaches its maturity in the first year. The highest yields are obtained in the second year of growth. It is a legume which grows well on a variety of well-drained soils with pH values above 7. Lucerne is moderately sensitive to soil salinity. When properly innoculated it develops nitrogen fixing bacteria in the root nodes. It is a dense crop grown in closely spaced rows. It is cut for fodder several times during each growing season and may remain productive for 3 to 7 years or more. Lucerne is also grown as a short season annual crop.

An important feature of the crop from the point of view of irrigation is the well-developed tap root system. The tap root penetrates up to 4 m in deep soils. To stimulate growth, lucerne is to be irrigated frequently. Root development is adversely affected by dryness. To promote growth, irrigation is normally applied just after each cutting. Adequate soil moisture is maintained in the root zone throughout the growing period by irrigation when the rainfall is inadequate: However, excess irrigation may cause reduced soil aeration which is harmful to the crop.

Vegetable Crops

Vegetables form an important component of human diet, supplying major amounts of vitamins A and C and numerous nutrients, protein and fat. Vegetable crops are grown over a wide variety of soils and climatic conditions. One or the other vegetable crop is grown at any time of the year, depending upon the availability of irrigation water. The important monsoon season *(kharif)* and summer vegetable crops are lady finger, tomato, brinjal, clusterbean, sweet potato, and cucurbitaceous crops. The important winter *(rabi)* vegetables are cabbage, cauliflower, kholkhol, beetroot, radish, turnip, carrot, potato, onion and leafy vegetables like lettuce and spinach. Most of the vegetable crops are of short duration and quick growing in habit and need frequent irrigations for their maximum production. The irrigation requirements will vary, depending upon the duration of the crop and the season when grown.

Vegetable crops include perennials such as asparagus *(Asparagus officialis)* which have extensive and deep root system and hence are draught resistant. Another perennial vegetable crop is globe artichoke *(Cynara scolymus),* which also has extensive root system but requires frequent irrigations throughout the year. Soil moisture deficiency, high temperature particularly during the time buds are forming, results in loose buts which are not marketable. The crop does not tolerate waterlogging. Almost all other vegetable crops are seasonal. They include potherbs or greens, including spinach, and kole which are shallow rooted. The salad crops include lettuce, celery, endive, and parsley which are relatively long season crops with shallow root system's. Soil crust on the soil surface adversely affects the emergence of seedlings of lettuce. The other group of important vegetable crops are the crucifers, namely cabbage, cauliflower, broccoli and brussels sprouts. Important root crops amongst the vegetable crops are beets, carrot, radish and turnip which are rapid growing crops which readily respond to irrigation. Bulb crops include onions and garlic. Beans and green peas are the prominent legumes amongst the vegetable crops. Solanaceous vegetable crops include tomatoes which are deep rooted with high water requirements and high yields. Cucurbits include cucumbers, muskmelon, watermelon, pumpkin and squash which are crops with medium or deep root systems.

Adequate soil moisture availability in the root zone is of prime importance in the cultivation of vegetable crops. In arid and semi-arid climates irrigation is a basic requirement in the cultivation of vegetable crops. Supplemental irrigation is common in most climatic regions, including humid and sub-humid regions. In surface irrigated crops, furrow irrigation is usually preferred. Under conditions of limited water availability both the drip and sprinkler methods of water applications are suitable, though during the recent times drip irrigation is preferred in many situations.

The water requirement of vegetable crops and the frequency of irrigation vary with the type of crop, soil and climatic conditions. However, as a general rule, applicable to most of the vegetable crops, the available soil moisture in the crop root zone should be replenished when 50% off it has been depleted. Specific characteristics of some of the common vegetable crops influencing irrigation are presented in the following pages.

Cabbage: Cabbage *(Brassica oleragea)* requires a cool climate for growth. The total growing period depends on the growing season. Spring season crop usually has a growing period of 90 days, while autumn crop has an extended growth period up to 200 days, depending on climatic conditions, the variety of the crop and the date of planting. Normally, the growing period ranges from 120 to 140 days. Most varieties can tolerate short periods of frost. Optimum growth occurs at mean day time temperatures around 27°C. Mean relative humidity may be in the range of 60 to 90%. In general, heavy loam soils are preferred. Under heavy rainfall conditions, sandy and sandy loam soils are preferable as they have better drainage properties.

Cabbage has an extensive shallow root system. The majority of the roots are in the top 40 to 50 cm of the root zone. The transpiration of the crop increases with the increase in the growth, with the peak towards the end of the growing season. During the vegetative period, the development of the crop is slow and the water requirements low. During the yield formation stage, which is the period of rapid growth, the availability of adequate moisture in the root zone is important. Depending on the climate, stage of development of the crop and soil type, the frequency of irrigation varies from 3 to 12 days.

Carrots: Carrots grow best in deep sandy and sandy loam soils. Heavy, compacted and cloddy soils hinder smooth development of carrots. The crop requires adequate soil moisture throughout the growing season.

Pea: Pea *(Pesum sativum)* is grown both for fresh and dried seed. The crop varieties range from tall, climbing plants to small bush types. Bush type plants have shorter growings-period. Pea prefers cool climate with Optimum mean daily temperatures around 17°C, with a minimum of 10°C and a maximum of 23°C. Young plants can tolerate light frost, but flowers and green pods are injured by frost. Peas are grown on well-drained soils with pH in the range of about 5.5 to 6.5. Peas have tap root systems with many lateral roots. Root depths range from 1 to 1.5 m. However, moisture uptake is usually limited to the 60 cm to 1 m depth of the rootzone. The sensitive periods for soil moisture deficits are flowering and yield formation. For high yields soil moisture depletion should not exceed 60% of the total available moisture and 40% during flowering and yield formation periods.

Radish: Radishes are very short season crops, maturing in 4 to 6 weeks. They are. grown on a variety of soils. Since the crop is very shallow rooted, frequent irrigations are required. The top 15 cm of root zone should be maintained at about 0.3 atmosphere tension.

Onion: Onion is the most important bulb crop grown under a variety of climates from temperate to tropical. The crop flourishes in mild climates without extremes of temperature and without excessive rainfall. During the initial growth period, cool weather and adequate soil moisture are advantageous for the proper establishment of the crop. During the crop ripening period, warm, dry weather is desirable for high yields of good quality bulbs. The mean daily temperature for optimum growth of the crop ranges from 15 to 20°C. The length of the growing period varies with the climate, but usually ranges from 150 to 175 days, from sowing to harvest. The crop is usually grown in the nursery and transplanted after 30 to 35 days. Direct seeding is also practiced. The optimum soil temperature for germination is 15 to 25°C. For bulb production the plant should not flower, as it adversely affects the yield. For the initiation of flowering, temperatures lower than 14 to 16°C and low humidity are desirable. Onion is sensitive to soil salinity. Optimum pH is in the range of 6 to 7. Onion, like most other vegetable crops, is sensitive to soil moisture deficits. The crop is shallow rooted with roots concentrated in the top 30 cm of the soil, requiring frequent and light irrigations. When the soil is kept relatively wet, root growth is reduced and this favours the enlargement of the bulb. Irrigation is discontinued when the crop approaches maturity.

The growth periods of onion crop are the *establishment period,* reckoned from sowing to transplanting which takes 30 to 35 days, the *vegetative* period comprising 25 to 30 days, the *yield formation* (bulb enlargement) period of 50 to 80 days and the *ripening* period of 25 to 30 days. Onion is most sensitive to soil moisture deficit during the yield formation period, particularly rapid bulb growth which occurs about 60 days after transplanting. It is also equally sensitive during the planting. For a seed crop, the flowering period is very sensitive to soil moisture deficit. However, over-irrigation will reduce growth and yield. Frequent irrigations are required to prevent cracking of the bulb and to ensure good quality bulbs.

Potato: Potato *(Solanum tuberosum)* is grown in most regions throughout the world, but is particularly important in the temperate climates. The optimum mean temperatures are around 18 to 20°C.

Night temperatures below 15°C is desirable for tuber initiation. Temperatures below 10°C and above 30°C adversely affect growth. Potato varieties can be grouped into *early* (90 to 120 days), *medium* (120 to 150 days and *late* (150 to 180 days). Early varieties are bred for temperate climates with long day lengths of 15 to 17 hrs, while late varieties result in good yields under long and short day conditions. For tropical climates varieties tolerant to short days are required.

Potato is grown in 3 or more year rotations with other crops such as maize, bean and lucerne to maintain soil productivity, to check weed growth and prevent crop damage due to insects, pests and diseases, particularly soil-borne diseases. Potato requires well drained soils with adequate aeration, with pH ranging from 5 to 6. Under irrigations, the crop is grown on ridges. The crop is moderately sensitive to soil salinity.

Potato requires adequate soil moisture for optimum growth. The available soil moisture should not be depleted beyond 30 to 50%. Soil moisture depletion more than 50% reduces the yield of the crop. Soil moisture deficits during the period of stolonization, tuber initiation and yield formation are particularly sensitive to potato. Early vegetative and ripening stages are relatively tolerant to potato. However, there is considerable variation between varieties in crop response to soil moisture availability. Potato has a shallow root system and about 70% of the total moisture uptake is from the top 30 cm of the root zone and nearly 100% from the top 40 to 60 cm of the root zone. Under irrigated conditions, irrigation scheduling should be based on avoiding soil moisture deficits during the periods of stolonization and yield formation. To obtain water savings greater level of soil moisture depletion can be allowed during the ripening period. This will also help in advancing the maturity of the crop.

Tomato: Tomato *(Lycopersicon esculentum)* is the second most important vegetable, next only to potato. It is a rapidly growing crop with a growing period of 90 to 150 days. The crop is neutral to day length. Optimum mean daily temperature is in the range of 18 to 25°C with night temperatures between 10 and 20°C. The crop is sensitive to frost. The crop can be grown on a wide range of soils but prefers well-drained light loam soils with pH from 5 to 7. Waterlogging induces diseases, particularly bacterial wilt.

The seed is usually grown in nursery plots and seedlings are transplanted after 25 to 35 days. Tomato is moderately sensitive to soil salinity. The most sensitive period to salinity is during germination and early vegetative periods. The plant produces flowers from bottom to top.

The growth period of tomatoes are recognized as follows: *nursery stage* (25 to 35 days), *flowering stage* (20 to 30 days), *yield formation stage* (20 to 30 days) and *ripening stage* (15 to 20 days). Highest yields are obtained by frequent, light irrigations. Tomato is most sensitive to soil moisture stress during and immediately after transplanting, during flowering and yield formation. Soil; moisture deficit during the flowering, period causes flower drop. Moderate moisture deficit during the vegetative period enhances root growth. Tomato has fairly deep root system. In deep soils roots may penetrate up to 1.5 m. About 80% of the soil moisture uptake is from the top 50 to 60 cm of the root zone soil. For high yields of good quality, tomato needs controlled supply of water throughout the growing period. However, excessive irrigation during the flowering period induces flower drop and reduces fruit set. It may also lead to excessive vegetative growth and delay in ripening. Highest demand for water is during flowering. Irrigation during and after fruit set should be limited to the extent that it will not induce stimulation of new vegetative growth at the cost of fruit development. Furrow irrigation is the most commonly adopted method of surface water application. The crop is highly, smutted to drip irrigation. With drip irrigation even moderately saline water could be used to irrigate the crop. However, tomato is not suited to sprinkler irrigation. Under sprinkler irrigation fruit set may be reduced with an increase in fruit rotting.

Fruit Crops

Fruit crops may be of evergreen type or deciduous type. The deciduous trees shed their leaves during winter and remain dormant for 3-4 months. A typical example is the grapevine. Amongst the evergreen fruit trees are citrus, mango, banana, olive, pineapple, papaya and date palm. All fruit crops require adequate soil moisture during their establishment period of three to four years. Soil moisture should be maintained above 50% availability in. the surface 1 m depth of soil. Later, on full development, the crops may be irrigated when two-thirds of available soil moisture is depleted during blossoming, fruit setting and fruit enlargement periods. At other periods and during dormancy in the case of deciduous trees, irrigations may be applied when the soil moisture depletion reaches below 30% of the available soil moisture in the root zone. Papaya and banana are shallow rooted and comparatively short lived species. They need to be irrigated when the available moisture in the top 30 cm soil layer is depleted to about 75 to 80 per cent of availability. This will amount to irrigations at 8 to 12 day intervals during rainless periods in a tropical climate. Date palm can withstand high temperature and low humidity but needs irrigations during flowering and fruiting stages to obtain good yields.

Banana: Banana *(Musa* spp.) is an important tropical fruit crop. A mean daily temperature of about 27°C is optimum for banana. The growth gets restricted when temperatures fall below 16°C. Banana plant is multiplied by planting suckers which originate from the underground stem (corm or rhizome) which bears several sprouts or suckers which are removed for planting, except for one or two which provide the ratoon crop. Nutrients must necessarily be applied at the time of planting and at the start of the ratoon crop. The plant has a sparse, shallow root system. The feeding roots are mostly spread laterally near the surface. The root depth is usually limited to about 75 cm. In general, the soil moisture uptake is from the top 80 cm of the root zone.

The growth of the banana plant can be distinguished into *vegetative stage* (from planting to shooting) which takes about 7 to 9 months and *flowering and yield formation stages* (from shooting to harvest) which take about 2 to 3 months. The number of ratoons varies with the variety. The average life of a ratoon plantation is over 3 years. Some varieties are replanted after each harvest. In irrigated plantations ample and frequent supply of water is necessary. Water deficits in excess of 35% adversely affect growth and yield. Establishment period, early phase of the vegetative period and flowering period are vital in terms of water requirement. Water deficit during the yield formation period adversely affect the fruit size and quality. Under limited water availability, the total production will be higher when full water requirements are met over a limited area than when it is met partially over a large area. This is unlike many other crops described earlier. Furrow and basin methods are the commonly adopted methods of surface irrigation in bananas. Both drip and sprinkler methods of irrigation are suitable in bananas. Drip irrigation results in considerable savings in water.

Citrus: Citrus is a widely cultivated perennial fruit tree of various species, including orange (about 70% of the total citrus crop), lemon, grape fruit and mandarin. Citrus is cultivated up to 1800 m altitude in the tropics and 750 m in the sub-tropics. For large scale cultivation the crop is usually not suited to the humid tropics. Optimum daily temperatures for growth are in the range of 23 to 30°C. There is a great reduction in growth at temperatures above 38°C and below 13°C. Strong wind is harmful to citrus, as they cause drop of flowers and young fruits.

Citrus is grown in well-drained deep soils which allow the tap root to penetrate to depths of 1 to 2 m. Most citrus species develop a single tap root. The lateral roots are spread laterally and form the feeding roots. Light, and medium soils without impervious layers and not subject to waterlogging are preferable. The soil pH in the range of 5 to 8 is preferred. Citrus is sensitive to high concentration of salts in the soil. The plant is propagated by bud grafting.

Adequate soil moisture is necessary for unrestricted growth of citrus. Water deficit retards growth and induces leaf curl and drop; young fruits fall and mature fruits are of inferior quality. Water deficits are avoided during periods of vigorous vegetative growth. However, prior to flowering, and fruit set too vigorous and luxuriant growth may impair production of high quality fruits. The flowering period is highly sensitive to soil moisture stress.

Citrus plant requires well planned irrigation scheduling. The plant requires good aeration and over-irrigation is harmful, particularly in young trees. Too frequent and heavy irrigations will affect root development and yield as well as leaching of nutrients. Surface irrigation of citrus are by furrow and basin methods. Both sprinkler and drip methods of irrigation are suitable and will result in uniform distribution of water and savings in water. Sprinkler irrigation is helpful in protecting the crop from frost.

Grape: Grape is an important fruit crop requiring a long, warm, dry summer and cool winter. Grapes are adapted to a wide range of well-drained soils. Largest vines and high yields are obtained in deep, fertile soil. Grapes are moderately sensitive to soil salinity.

Most grape varieties are propagated by cuttings grown in nurseries to produce roots. Grape vines are deep rooted with roots penetrating to depths of 2 to 3 m. Grapes can adjust moderately to limited water supply. Prior to and during flowering adequate soil moisture is essential for flower development. During the vegetative period, flowering and early yield formation stages soil moisture depletion should not exceed 0.35 to 0.45 when the ET_m is 5 to 6 mm/day. Later in the growth period, moisture depletion can be at a higher level, while towards and after harvest a high level of soil moisture depletion is required.

When the rainfall during the growth period is insufficient, irrigation should be scheduled to fill the root zone to field capacity at regular intervals, particularly during the periods when the crop is sensitive to moisture stress. In such situations irrigation should be applied before the vegetative growth starts. In shallow and light soils properly scheduled irrigations are required till harvest. Furrow method of surface irrigation is common in grape cultivation. Drip irrigation is the most suitable irrigation method under limited water availability. Sprinkler irrigation may also help in protecting the crop against frost. However, sprinkling during the ripening period increases the chance of bunch rot.

Pineapple: Pineapple *(Ananas comosus)* is grown mainly in regions with high relative humidity. For good growth pineapple requires a mean daily temperature of 22 to 26°C. Mean daily maximum and minimum temperatures are 30 and 20°C, respectively for the whole growing period. Pineapple can grow on a wide range of soils, but well-drained sandy loam soils are most suitable. The lime content of the soil should be low. The crop is sensitive to waterlogging.

Pineapple is usually grown in double rows on raised beds. The plant population is about 50,000/ha. Shading is sometimes used to protect the crop from scorching when temperatures and radiation are high. The growing period, from planting to harvest, ranges from 1 to 2 years and in ratoon crops 9 to 18 months, depending on the planting material and climatic conditions. The root system of pineapple is shallow and sparse. In deep soils the maximum root depth may be about 1-m, but roots are generally concentrated in the top 30 to 60 cm from where the full water requirements are extracted by the plant. Flower initiation is induced by low temperature, water deficit or hormone sprays. Hormone spray results in uniform fruiting and harvest periods.

Pineapple can tolerate long dry periods due to its ability to retain water in the leaves, which is used by the plant during periods of water scarcity. However, pineapple is sensitive to soil moisture stress, particularly during the vegetative period. Soil moisture deficits retard growth, flowering and fruiting. However, soil moisture deficits during flowering is less harmful. On the other hand, it may hasten fruiting. Frequent irrigations or rains at the time of harvest may cause poor quality of fruits, and make the crop susceptible to fungal attack, causing heart rot. Under deficient water availability, irrigation scheduling is based on preventing water deficits during the period of vegetative growth. Water supply can be restricted during the period of ripening. Soil moisture depletion up to 75% of the available level can normally be allowed, except during the vegetative growth period. About a month prior to harvest, irrigation is discontinued. Furrow, sprinkler and drip methods of water application are suitable in pineapple cultivation. Pineapple fruit contains about 80 to 85% water and 10 to 14% sugar. Frequent irrigations prior to harvest decrease the sugar content.

Spices and Condiments

Important crops in this group are turmeric, ginger, cardamom, cumin, coriander, betal vine, black pepper and chillies. Only limited information is available on the water management practices of these crops. Turmeric, ginger and chillies should be grown on broad ridges and irrigated to maintain over 50 per cent of available moisture in the zone of maximum root spread which is the top 60 cm layer of soil. Coriander and Onion are winter crops and need irrigation at intervals of 10 to 12 days on light soils and 15 to 20 days on heavy soils. The water requirements of these crops are around 500 to 600 mm.

Lawns and Gardens

In parks and gardens, seasonal flowering plants need to be irrigated frequently to maintain the surface 30 cm soil layer moist. These plants should receive irrigations at an interval of five to eight days depending upon the severity of the climate. A lawn may be irrigated at an interval of eight to ten days in summer and 15 to 20 days during winter to maintain its good growth. In lawns, one has often to resort to surface irrigation methods and hence proper consideration should be given to levelling at the time of planting to ensure that water will spread uniformly. The young shrubs and live-fences should be irrigated along with the lawns. Fully developed ornamental trees may be able to tap deep percolated water and hence no special care may be necessary for their growth.

Plantation Crops

Apple, coconut, date palm, arecanut, black pepper, cocoa, coffee, tea and rubber are amongst the major plantation crops grown in many regions of the world. There are many other fruit crops and spices which are important in specific regions. Plantation crops are usually high price crops. Many of the plantation crops are developed in favourable environments so that irrigation may not be a major limiting factor in their production at the global level. However, most plantation crops respond well to irrigation during periods of water scarcity at specific growth periods.

Coffee: Coffee crops are normally of two species, namely, *Coffee Arabica* and *Coffee robusta.* Only Coffee Arabica is irrigated on a significant scale. Coffee Robusta commands a much lower price in the world market and is usually grown in hotter and wetter areas. Coffee Arabica provides the highly aromatic variety, forming the bulk of the commercial blends of coffee, which is usually grown in moderately high altitudes. Mean monthly minimum temperature for coffee plantations is 10°C. .

Coffee varieties have a vigorously developed root system and can withstand soil moisture stress to a great extent. Properly controlled irrigations, however, increase the yield and bean quality of Coffee Arabica. Limited soil moisture availability controls flowering and hence better period of Picking. Continuous replacement of soil moisture through irrigation is not advisable in the monsoon regions. Amongst the surface methods of water application, furrow irrigation (usually contour furrows) is preferred. Both sprinkler and drip irrigation are suitable. Drip irrigation is preferred in. regions with limited water availability.

Tea: Tea *(Thea sinensis)* is one of the important crops of India and Sri Lanka which together produces over half of the world production. It is also grown in many regions of the tropics and sub-tropics under cool and high altitude conditions.

The major root system of tea plant is in the top 70 cm of the soil. Irrigation is highly desirable in regions with prolonged drought periods. Sprinkler and drip methods of water application are suitable in tea plantations. Furrow irrigation, particularly contour furrows is also suitable.

Cocoa: Cocoa *(Theobroma cocoa)* is a tree crop grown in wet, tropical lands around the world. The climatic requirements are high humidity and comparatively high temperature. Cocoa is a shallow rooted plant with most of its active roots spread in the top 60 cm of the soil. The crop is sensitive to soil moisture stress. Soil moisture stress below 60% of the field capacity moisture level affects the plant growth. Furrow irrigation is the most commonly used surface irrigation method. Drip irrigation is suitable and will result in substantial savings in water.

References

Parihar, S.S., Sandhu, B.S. (1987), Irrigation of Field crops : Principles and Practices. ICAR, Krishi Anusandhan Bhavan Pusa, New Delhi, P. 142.

Saxena, K.K. (1987), Design and Operation of Irrigation projects, Indian Institute of Technology, New Delhi, P. 372.

Sri Ramulu, US (1998), Management of Water Resource in Agriculture, New Age International (P) Ltd. New Delhi, P. 159.

Water Management Division, Ministry of Agriculture, Govt. of India, New Delhi (1991), Scheduling Irrigation to meet Crop Needs III, 1-6.

Vamdevan, V.K. and Pastane, N.G. (1968) Drainage Needs crops, IL RISO, Anne, XVI, 3 P. 243-50.

Chapter - 4

Crop Water Requirements and Irrigation

Dr. S.B. Verma & Dr. Jibu Jha

The estimation of the water requirements (WR) of crops is one of the basic needs of crop planning on the farm and for the planning of any irrigation project. Water requirement may be defined as the quantity of water, regardless of its source, required by a crop or diversified pattern of crops in a given period of time for its normal growth under field conditions at a place. Water requirement includes the losses due to evapotranspiration *(ET)* or consumptive use (C_u) plus the losses during the application of water (unavoidable losses) and the quantity of water required for special operations like land preparation, pre-sowing irrigation and transplanting. It may thus we expressed as follows:

$$WR = ET \text{ or } C_u + \text{application losses} + \text{special needs.}$$

Water requirement is, therefore, a 'demand' and the 'supply' would consist of contributions from any of the sources of water, the major source being the irrigation water *(IR)*, effective rainfall *(ER)* and soil profile contributions (S) including that from shallow water tables. Numerically, therefore, water requirement is given as:

$$WR = IR + ER + S \tag{1}$$

Irrigation Requirement

The irrigation requirement of a crop is the amount of water to be supplied by irrigation to a disease-free crop growing in a large field with adequate soil water and fertility and achieving full production potential under the prevailing soil and climatic conditions. Irrigation requirement includes the evapotranspiration of the crop *(ETc)* (or consumptive use *Cu)*, maintaining a favorable salt balance within the root zone of the crop. Irrigation requirement does not include the water which is available to a crop through precipitation, soil moisture 'storage and soil moisture accretion through capillary movement of groundwater or seepage from such surrounding areas.

With the estimated value of *ETc* the irrigation requirement of a crop can be computed, adopting the following relationship (using identical units of depth, i.e., mm/cm/inches):

$$1 = ETc - P + RO + DP + L + D_r (Q_f - 9) \quad (2)$$

in which,

I = Irrigation water requirement

ETc = Evapotranspiration of the crop

RO = Runoff from the crop field from irrigation and/or rainfall

DP= Deep percolation in the field from irrigation and/or rainfall

L = Leaching requirement

D_r = Depth of the crop root system of

Q_f Root zone soil moisture content at field capacity level

Q_f = Root zone soil moisture content prior to irrigation.

The field irrigation requirement of a crop, therefore, refers to the water requirement of crops, exclusive of effective rainfall and contribution from soil profile, and it may be given as:

$$IR=WR-(ER + S) \quad .(3)$$

The farm irrigation requirement depends on the irrigation needs of individual crops, their area and the losses in the farm water distribution systems, mainly by way of seepage. The irrigation requirement of an outlet command area includes the irrigation requirement of individual farm holdings and the losses in the conveyance and distribution system.

Field Water Balance

The water balance of a field is an itemized statement of all gains, losses, and changes of storage of water occurring in a given field within specified boundaries during a specified period of time. The task of monitoring and controlling the field water balance is vital to the efficient management of water and soil. A knowledge of the water balance is necessary to evaluate the possible methods to minimize loss and to maximize utilization of water which is so often the limiting factor in crop production.

Gains of water in the field are generally due to precipitation and irrigation. Occasionally, there may be gains due to accumulation of runoff from higher tracts of land, or to capillary rise from below (especially where a water table is present at some shallow depth). Losses of water include surface run-off from the field, deep percolation out of the root zone (drainage), evaporation from the soil surface, and transpiration from the crop canopy. The change in storage of water in the field can occur in the soil as well as in the plants. The total change in storage must equal the difference between the sum of all grains and the sum of all losses.

Accordingly, the water balance equation may be stated as follows:

(Gains) - (Losses) = (Change in storage)

$$(P + I)-(R + D + E + T) = (\Delta S + \Delta V) \qquad ..(4)$$

In which, *P* is precipitation, I= irrigation, *R* runoff from the field, *D* *do*wnward drainage out of the root zone, *E* evaporation from the soil, *T* transpiration by the crop canopy, Δ S the change in soil water content of the root zone, and Δ V the change in plant water content. All of these quantities are usually expressed in terms of water depth per unit of land area (ha-cm) or units of depth (cm).

An important consideration in water balance studies is the period, or time interval, for which the balance is made. Too short a period might be impractical, while too long a period might mask the occurrence of short-term critical stages. At such critical stages as flowering and fruit-set even temporary imbalance in crop-water status can be have a lasting effect. It is necessary to ensure that the crop water balance is maintained positive continuously during the growing season.

Measurements of field water balance is usually conducted by the use of weighing or floating lysimeters. However, they cannot be used when the mixed nature of the species composition, the spatial distribution of the vegetation, the characteristics of the root systems, the size of plants, or other factors make it impossible to simulate the natural environment in the lysimeter itself. In such cases,, determination of soil water storage at different points in the community provide the only means for evaluating the water balance.

Effective Rainfall

In its simplest sense, effective rainfall means useful or utilizable rainfall. Rainfall is not necessarily useful or desirable at the time, rate, or amount in which it is received. Some of it may be unavoidably wasted while some may even be destructive. Just as total rainfall varies, so does the amount of effective rainfall.

The term effective rainfall has been interpreted differently by specialists in different fields. To a canal irrigation engineer, the rain which reaches the storage reservoir directly and by surface runoff from the surrounding area is the effective portion. Geo-hydrologists would define as effective rainfall that portion of rainfall which contributes to groundwater storage. Agriculturists consider as effective that portion of the total rainfall which directly satisfies crop -water needs. In the field of dry land agriculture, when the land is left fallow, effective rainfall is that which can be conserved for the following crop.

Thus, it may be seen that even though the concept is the same in all the above cases, the values of effective rainfall are different for different agencies for the same total rainfall. Since there are such varied-

interpretations of what may be regarded as effective rainfall, it is difficult to develop a definition to suit all the interested disciplines. Rainfall which is ineffective according to one may be effective according to another. From the point of view of the water requirement of crops, the Food and Agricultural Organisation of the United Nations (Dastane, 1974) has defined the annual or seasonal effective rainfall as that part of the total annual or seasonal rainfall which is useful directly and/ or indirectly for crop production at the site where it falls, but without pumping. It therefore includes water intercepted by living or dry vegetation, that is lost by evaporation from the soil surface, the precipitation lost by evapotranspiration during growth, that fraction which contributes to leaching, percolation or facilitates other cultural operations either before or after sowing without any harm to yield and quality of the principal crops. Consequently, ineffective rainfall is that portion which is lost by surface runoff, unnecessary deep percolation losses, the moisture remaining in the soil after the harvest of the crop and which is not useful for next season's crop. The growing season is counted from the start of the first tillage operation until the harvest.

The above concept of effective rainfall is suggested for use is planning and operation of irrigation projects. The irrigation water supply in a given year should be planned to complement rainfall. Since annual rainfall varies from year to year, an irrigation project is to be planned based on data, over a long period to calculate the effective rainfall on the basis of probability of occurrence.

Factors Influencing Effective Rainfall

(i) Rainfall characteristics: A soil has a definite and limited infiltration and moisture holding capacity. Hence, greater amount and intensities of rainfall normally reduce the effective fraction of rainfall and increase runoff. A well distributed rainfall in frequent light showers increases the effective rainfall and is conducive to crop growth than heavy downpours. In India the intensity, frequency and amount of rainfall are high during July and August and hence the effective fraction is low. From November to April most of the rainfall is effective due to its low intensity, frequency and amount.

(ii) Land slope: The slope of the land has a profound influence on the time available for the rain water to infiltrate into the soil (infiltration opportunity time). Water stays longer on flat and levelled land and thus has a longer opportunity time than on slopping land where there is a rapid runoff.

(iii) Characteristics of the soil: The soil properties influencing infiltration and moisture retention influence the effective fraction of rainfall. High values of infiltration rates- and hydraulic conductivity of the soil enhance the infiltration and reduce runoff. The moisture content of the soil at the time of occurrence of the rain (initial moisture content), affects the effective rainfall considerably. The higher the moisture content, the lower the infiltration rate and higher the surface runoff which reduces the effective rainfall.

(iv) Management practices: Any management practice which influences runoff, infiltration, hydraulic conductivity or evapotranspiration also influences the degree of effective rainfall. Bunding, terracing, contour tillage, ridging and mulching reduce runoff and increase effective rainfall.

(v) Crop characteristics: Crops with high water consumption create greater deficits of moisture in the soil. The effective rainfall is directly proportional to the rate of water uptake by the plant. Crop characteristics influencing the rate of water uptake are the degree of ground cover, rooting depth and stage of growth. Soil moisture stored in deeper layers can be tapped only when roots penetrate to these depths. Deep-rooted crops, therefore, increase the proportion of effective rainfall in a given area.

(vi) Carry-over soil moisture: It is the moisture stored in the crop root zone between cropping seasons or before the crop is planted. This moisture is available to meet the consumptive water needs of the crop. The contribution of a fairly good rain occurring just prior to sowing may be equivalent to one full pre-sowing irrigation. An additional benefit of excess pre-sowing season rain is the leaching of salts accumulated in the root zone in the summer season. Carry-over of soil water is of great importance in rainfed farming in water scarcity areas.

(vii) Groundwater contribution: Soil moisture contribution from the groundwater table is determined by the depth of groundwater below the root zone and the capillary and conductive properties of the soil.

Measurement of effective rainfall: Evaluation of effective rainfall involves the measurement of rainfall and/or irrigation losses by surface runoff, percolation beyond root zone and soil moisture use by crops. Precise measurements are often made by weighing type lysimeters.

Groundwater contribution: The contribution to the root zone soil moisture from the groundwater table is determined by the depth of groundwater below the root zone, the capillary and conductive properties of the soil and the soil moisture content in the root zone. Both the rate and the distance of water movement are important criteria. For heavy soil the distance of movement is high and the rate low, while for light textured soil the distance of movement is low and the rate high.

Surface and sub-surface in- and out-flows: Computation of surface inflow normally does not apply, except for areas subject to occasional flooding. Under efficient irrigation practices surface outflow is small. Management losses and waste of water due to technical faults are normally accounted for in irrigation efficiency. Sub-surface inflow is only of local significance in areas where there is upward movement of water from deeper subsoil caused by seepage from reservoirs and canals. Sub-surface inflow may also occur locally on or near the toe of sloping lands.

Deep percolation: Deep percolation is that amount of water which passes below the root zone of the crop and can continue for a long time after field capacity has been reached in the crop root zone, following irrigation or heavy rain. Total water loss by deep percolation in irrigated conditions can account for 20% or more of the total amount of water applied. However, soil water movement in arid below the root zone, after an initial downward outflow, can later be reversed to an upward inflow from the wet sub-soil to the drying root zone above.

Net Irrigation Requirement

The net irrigation requirement is the depth of irrigation water, exclusive of precipitation, carry-over soil moisture or groundwater contribution or other gains in soil moisture, that is required consumptively for crop production. It is the amount of irrigation water required to bring the soil moisture level in the effective root zone to field capacity (Fig. 1). Thus, it is the difference between the field capacity and the soil moisture content in the crop root zone before starting irrigation.

This may be obtained by the relationship given below:

$$d = \sum_{i=1}^{n} \frac{(M_{fci} - M_{bi})}{100} . A_i . D_i \qquad (5)$$

In which,

d = net amount of water to be applied during an irrigation, cm

M_{fi} = field capacity moisture content in the ith layer of the soil, per cent

M_{bi}= moisture content before irrigation in the ith layer of the soil, per cent

A_i= bulk density of the soil in the ith layer

D_l= depth of the ith soil layer within the root zone cm, and

n = number of soil layers in the root zone D.

In drawing up the seasonal or monthly net irrigation requirements for a given crop or cropping pattern the main variables composing the field water balance include:

(i) crop water requirements as determined by climate and crop characteristics, ETc

(ii) contribution from precipitation, and groundwater, and

(iii) carry-over of soil moisture. The deficit in the soil water balance is met by the net irrigation requirement.

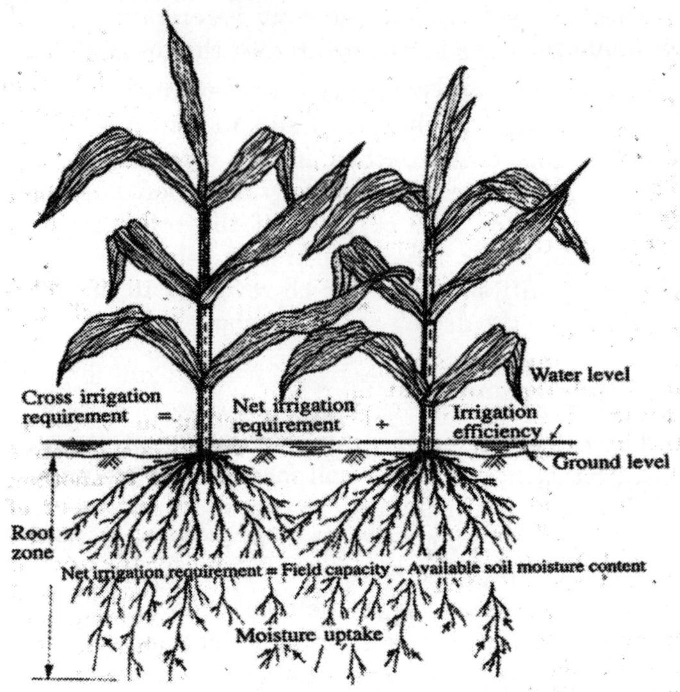

Fig. 1. Concept of gross and net irrigation requirements.

Gross Irrigation Requirement

The total amount of water applied through irrigation is termed as 'gross irrigation requirement'. In other words, it is the net irrigation requirement plus losses in water application and other losses. The gross irrigation requirement can be determined for a field, for a farm, for an outlet command area or for an irrigation project, depending on the need, by considering the appropriate losses at various stages of growth of the crop.

Gross irrigation requirement (in field)

$$\frac{\text{Net irrigation requirement}}{\text{Field efficiency of system}}$$

For example, if the net amount of irrigation is 8 cm and the field efficiency is 75 per cent, the gross amount of water to be applied to the field is

$$\frac{8}{0.75} = 10.66 cm$$

The gross irrigation requirement *(IR)* at the field head, for instance, can be determined as follows:

$$IR = \sum_{I}^{n} \frac{d}{E\ (application)} \tag{6}$$

IR = seasonal gross irrigation requirement at the field head (cm)

d = net amount of water to be applied in each irrigation (cm)

$E_{(application)}$ = water application efficiency, and

n = number of irrigations in a season.

Irrespective of the method of irrigation, no system is 100 per cent efficient and not all the water applied during an irrigation enters the soil and is held in the root zone. Unavoidable losses are caused through seepage and leaks in the water conveyance system, non-uniform distribution of water over a field, percolation below crop root zone and wastage due to surface runoff at the end of borders and furrows. In case of sprinkler irrigation, additional losses may be caused by evaporation from the spray and by the retention of water on the plant foliage.

Irrigation frequency. Irrigation frequency refers to the number of days between irrigations during periods without rainfall. It depends on the consumptive use rate of a crop and on the amount of available moisture in the crop root zone. It is a function of crop, soil and climate. Sandy soils must be irrigated more often than fine-textured deep soils. Moisture-use rate varies with the kind of crop and climatic conditions and increases as the crop grows larger and the days become longer and hotter. In general, irrigation should start when about 50 per cent and not over 60 per cent of the available moisture has been used from the zone in which most of the roots are concentrated. The stage of growth of the crop with reference to the critical periods of growth is also kept in view while designing irrigation frequency.

In designing irrigation systems, the irrigation frequency to be used is the time (in days) between two irrigations in the period of highest consumptive use of the crops grown. Irrigation frequency depends on how fast soil moisture is extracted when a crop is transpiring at its maximum rate. The average moisture-use rate during this period is used to plan irrigation systems. For an irrigation system to be adequate, it must have sufficient capacity to supply the water required during this period. The design irrigation frequency may be computed as follows:

Design frequency (days)

Field capacity of the soil in the effective crop root zone.

Moisture content of the same zone

$$= \frac{\text{at the time of starting of irrigation}}{\text{Peak period moisture use rate of crop}}$$

Irrigation period. Irrigation period is the number of days that can be allowed for applying one irrigation to a given design area during the peak consumptive use period of the crop being irrigated. It is the basis for irrigation system capacity and equipment design. The irrigation system must be so designed that the irrigation period is not greater than the irrigation frequency.

Irrigation period

Net amount of moisture in soil between start of

$$= \frac{\text{Irrigation and lower limits of moisutre depletion}}{\text{Peak period moisture use rate of crop}}$$

IRRIGATION SCHEDULING

Irrigation scheduling is the process of determining the time to irrigate and how much water is to be applied (irrigation depth) in each irrigation. Proper scheduling is essential for the efficient use of water and other inputs in crop production. Irrigation schedules are planned to either fully or partially provide the estimated water requirement of the crop.

Amount of water to apply per irrigation. With full irrigation the amount of water required per irrigation is computed as follows:

$$1 = D_r(fc - f_m) \text{ irrigation efficiency} \tag{7}$$

in which,

D_r = Depth of root zone

f_c = Soil moisture content at field capacity

f_m = Soil moisture content prior to irrigation.

Full irrigation: This provides adequate water to meet the entire irrigation requirement and is aimed at achieving the maximum production potential of the crop. Excess irrigation, however, reduces crop yields by adversely influencing soil physical properties like soil aeration, soil temperature and the microbial activities in the soil. Full irrigation is justified when there is no scarcity of water and the cost of irrigation is low.

Deficit irrigation: This meets the water requirement of crops only partially. It is economically justified when reducing the amount of irrigation supply below the full level causes production costs to decrease faster than the decline in the value of crop harvest. Deficit irrigation is practiced when there is water scarcity or when the irrigation system capacity is limited. With 'deficit' irrigation the crop root zone is not always filled to the field capacity moisture content level. In locations with appreciable amounts of precipitation during the irrigation season, it is possible to fill the root zone only partially so that some precipitation can be stored in the crop root zone. The recent concepts on the management of 'deficit irrigation', i.e., replacing only a part of the water that is used by the plant and achieving maximum possible production per unit quantity of water and/or land require careful analysis and planning. For example, it has been observed that the yield and fruit quality of matured apple trees are not affected by partial replenishment of the water used during the period prior to fruiting. Full replacement is required during' the period following fruiting. Many research studies indicate that with increasing water shortage globally it may be more economical to plan irrigation systems to obtain economical crop production per unit of land. However, care should be taken to ensure that there is no water shortage during the 'critical stages' in the growth of the crop.

Ideally, the irrigation depth and/or the irrigation interval vary with the stage of growth of the crop. At the beginning of the growing season the depth of irrigation is kept low, but the applications are frequent. This is due to the low value of *ETc* of the growing seedling and their shallow root system. During the mid-growth stage of the crop the depth of irrigation is increased and irrigations are given less frequently. Controlling irrigation depths is not difficult under sprinkler and drip methods of irrigation. However, with surface irrigation methods the scope for changing irrigation depths is limited. The amount of effective rainfall which may occur during the crop season is deducted from the amount of irrigation water estimated to be applied in the irrigation which follows. Alternatively, provision may be made to allow for the precipitation which is expected during the crop growing period.

Irrigation Indices

Scientific irrigation must be based on an understanding of the soil-water-plant-atmospheric relationship. Irrigation needs of crops depend on the evaporation demand of the ambient atmosphere, soil-water regime in the crop root zone and the plant foliage.

Visual plant symptoms: Visual symptoms in plants to decide on the time to irrigate are changes in the colour of plants, curling of the leaves and tendencies of wilting. These are observed by looking at the crop as a whole and not individual plants. When the crop comes under water stress the appearance changes from vigorous growth to retarded growth. Under conditions of water stress the number of emerging leaves get reduced, and often there is a marked change in the colour of the leaf which gets darker and sometimes grey. Successful interpretation of crop stress requires keen observation and experience in crop fields. However, symptoms may sometimes be misleading due to changes in crop varieties and nutritional disorders as well as insect and pest incidence. Another limitation in resorting to the appearance of crop foliage is that by the time the symptoms are evident irrigation has already been withheld for too long a period for most crops and the yield losses have become substantial.

Plant water content and water potential: The status of water in the plant is generally indexed from the measurement of leaf-water content and leaf-water potential. Their values at any time of the day will depend on the lag between the evaporative demand of the atmosphere and the moisture uptake rate by the crop. When their values fall below certain critical limits specific to the plant species and their growth stages, important physiological and growth factors are adversely affected. Hence, these values can serve as reliable indicators for irrigating crops.

Plant temperature: Almost the entire radiation energy received on the leaf surface during the daylight hours is utilized for evapotranspiration *(ET)*. The remaining energy, if any, is used to heat the leaf tissues and the ambient air. When *ET* reduces due to water deficit in the plant, the energy saved in the process is partly used to raise the leaf temperature. Many investigations have shown that the leaf canopy temperature is a sensitive index in crops like soybean, oats, barley, lucerne (alfalfa), wheat, sorghum and maize (Prihar, S.S. and Sandhu, B.S., 1987).

Soil-water regime: A commonly recommended method to decide on irrigation schedules is soil moisture measurements in the field. When the soil moisture content has dropped to a certain critical level, say about 50% of field capacity level in the crop root zone, irrigation is applied. Normally irrigation is not delayed beyond 60% field capacity moisture content.

Irrigation schedules based on prediction of *ET*: Irrigations can be suitably scheduled on a farm if the allowable maximum water depletion in the crop root zone and the evapotranspiration *(ET)* for short periods during the growing period are known.

Irrigation scheduling based on pan evaporation: Prihar *et al.* (1974) suggested a 'practical approach' relating irrigation requirement to the cumulative evaporation during a time period as measured in a standard open pan. The relationship is expressed as IWE_{pan} in which IW is the depth of irrigation water and E_{pan} is the cumulative open pan evaporation. The rainfall during the period, if any, is subtracted from the value of E_{pan}. The validity of the procedure will depend on the proper installation of the evaporation pan and the rain gauge and the precision in measurements of pan evaporation and rainfall. Further, the suitability of the method is situation-specific arid limited to the particular variety of a crops.

General Guidelines on Planning Irrigation Schedules

The characteristics of different soil groups and soil depth influencing irrigation scheduling are as follows:

(i) Shallow and/or sandy soil: In sandy soils in general and soils with limited depth (due to the presence of hard pan close to the ground surface) the soil moisture holding capacity is low. These soils require frequent irrigations and the amount of water applied in each irrigation is kept low.

(ii) *Loamy soils:* Loamy soils can store more water than sandy soils or shallow soils. Hence, irrigation is applied less frequently with larger depth of water in an irrigation.

(Hi) Clayey soils: The moisture holding capacity of clayey soils is more than loamy soils. Hence, irrigations are less frequent and the quantity of water in an irrigation is higher.

References

Prihar. S.S. and Sandhu, B.S. (1997). Irrigation of Field crops : Principles and Practices. I.C.A.R., Krishi Anusandhan Bhavan, Pusa, New Delhi, P. 142

Saxena, K.K. (1997), Design and Operation of Irrigation Projects, Indian Institute of Technology, New Delhi, P. 372

Sri Ramulu US (2000), Management of Water Resources in Agriculture, New Age International Pvt. Ltd., New Delhi, P. 159;

Water Management Division, Ministry of Agriculture, Govt. of India, New Delhi (1991). Handloom of Irrigation, Scheduling Irrigation to Meet Crop Needs. 111-16.

Chapter - 5

Micro-Irrigation for Sustainable Water Management in Agriculture

D.K. Singh and R.M. Singh

Agriculture is the largest user of fresh water for irrigation to fulfill the increasing demand of food, feed, fiber, fuel and other need of ever-increasing population. Water availability to agriculture is characterized by declining in share with time, over exploitation, improper mismanagement, deterioration in quality as well as environment. This could be addressed using sustainable management of irrigation water. As per the Brundtland Commission report of United Nations, sustainable development is the development that meets the needs of the present without compromising the ability of future generations to meet their own needs. Sustainable water management aims to improve water productivity, availability and quality over an area. Improved and efficient irrigation techniques and practices enhance water application efficiencies, and thereby reduction in associated damage to environment, which include water logging, soil salinity, runoff and leaching of nutrients and other chemicals polluting water sources, and over exploitation of ground water. Micro irrigation systems have been characterized to be the most efficient irrigation techniques. These have efficiencies even more than 90%, doubled water productivity with improved quality of produce, reduced runoff and leaching of nutrients and chemicals to water sources. Application of fertilizer through drip irrigation called furtigation could be adopted for efficient use and saving of fertilizer, fertilizer savings of 20-60% and 8-41% increase in yields of horticulture and vegetable crops has been realized through furtigation. Micro irrigation is getting popularity having large potential due to its being economic feasible, widely accepted in society and environment friendly. Therefore, micro irrigation systems could be used for sustainable water management.

Introduction

National Water Policy of India (2002) emphasizes that water is a prime natural resource, a basic human need and a national asset. The planning and management of this resource and its optimal, economical and equitable use has become a matter of the utmost urgency. The fresh water is essential for sustaining all forms of life. Agriculture is the largest user of fresh water for irrigation. Water is being over exploited to fulfill the increasing demand of food, feed, fiber, fuel and other need of ever-increasing population. Also, conventional water management has been characterized to pose problems of water logging, increased salinity and alkalinity of soil, depletion of ground water table, salt water intrusion in ground water due to poor drainage and irrigation water management.

Improper use and mismanagement of available water, climate change, urbanizations and increasing population are considered as the main causes of increasing water scarcity. The increasing population and urbanization force the greater need of water for domestic consumption. Access to water is so crucial to human well-being and development that has now become a main concern for the global community. To emphasize the need for immediate action, year 2003 was designated as the International Year of Freshwater by the United Nations to provide an opportunity to raise awareness, motivate people, and mobilizes resources in order to manage water in a sustainable way. This paper presents prospects of micro irrigation for sustainable management of water.

1. Water Associated Crisis

The four main factors of population growth, urbanization, increased consumption and climate change are responsible for aggravated water scarcity (IPCC, 2007). Water scarcity is defined as per capita supplies less than 1700 m^3/year (IPCC 2007). It has been assessed that one in three people are already facing water shortages. About 1.2 billion, people which is around one-fifth of the world's population live in areas of physical scarcity, while another one quarter of the world's population which is around 1.6 billion people live in countries lacking the infrastructure to take underground and rivers water, thus facing an economic water shortage. The Water use in the world (2005) presented in Table 1 indicates two-third of total water is utilized in agriculture (Comprehensive Assessment of Water Management in Agriculture, 2007).

Table 1. Water use in the world (2005)

Activities	Water
Agriculture	67
Household	9
Water supply	8
Electricity and gas	7
Manufacturing	*4*
Mining	2
Others	3

The nature of rainfall in India is characterized as erratic with uneven distribution over space and time. Only about 62 mha, which is around 44% of the cropped area is irrigated. There is a need to bring more cropped area under assured irrigation to increase agriculture productivity and production. The ultimate irrigation potential of the country has been estimated to be about 140 mha out of which about 76 mha could be from surface water and about 64 million hectare from ground water sources. The per capita water availability, in terms of annual average utilizable water resource of India, which was 3450 m in the year 1951, was 1250 m in 2005 and would reduce to 760 m''' in the year 2050, placing it in the category of water stressed country. The water demand will increase in future (Figure I).

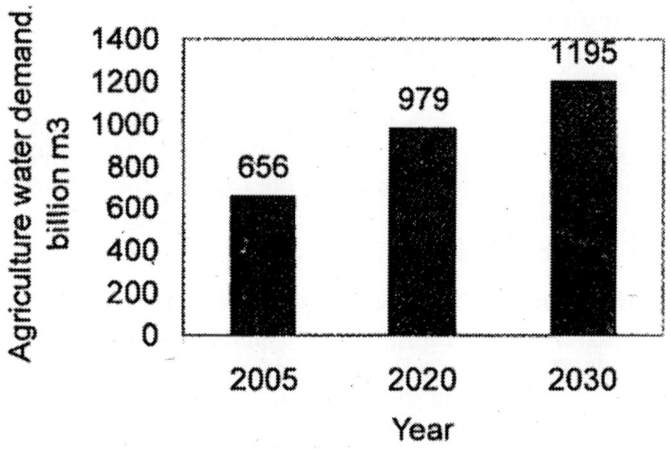

Figure 1: The annual agriculture water demand in India is projected to increase at rate of 2.4% during 2005 to 2030. The water demand will be double by 2030.

The National Commission for Integrated Water Resources Development (CIWRD) has assessed that about 83% of water is used for irrigation and remaining for domestic, industrial and other purposes. It has assessed the projected demand as 1180 billion cubic meters for the high demand scenario for the year 2050. The share of irrigation water in the overall demand has been estimated to reduce from the present level of about 83% to about 69% by the year 2050.

The run off from farm, untreated municipal sewage, domestic wastewater, agricultural run off, industrial effluents, fertilizer runoff, manure form livestock, all introduce pesticides, sewage and nutrient in the water. Also, some chemicals such as fluoride, arsenic, lead, nitrate, chlorinated solvents, petrochemicals contribute to water contamination. The use of contaminated water for drinking purpose leads to several diseases. The poor quality water causes water borne diseases such as, typhoid, cholera, and dysentery due to bacterial infections. These cause heavy economic loss to the individual and nation as a whole.

2. Sustainable Water Management

More and more water is required to fulfill the needs of food, feed, fiber, and energy through agriculture for increasing population. Its availability to agriculture is declining day by day. Also, over exploitation, improper use and mismanagement have led to deterioration of quantity and quality of water as well as environment. It is causing many problems to the society and responsible to loss of man, money and quality nature. This issue could be addressed with proper understanding of the problem and posing the best solution and planning through sustainable management to prevent irrigation water crises for present and future generations.

2.1 Sustainability

Sustainability is the capacity to endure. Sustainability interfaces with economics through the social and ecological consequences of economic activity. It is the potential for long-term maintenance of well being, which has environmental, economic, and social dimensions. Sustainability was one of eight goals in Millennium Development Goals agreed at the United Nations Millennium Summit in September 2000.

2.2 Sustainable development

The use of resources in a way to meet human needs for the present as well as future generations while preserving the environment may be referred as the Sustainable development. The Brundtland Commission report of United Nations in 1987, defines sustainable development as the development that meets the needs of the present without compromising the ability of future generations to meet their own needs. It is environmentally friendly forms of economic growth activities (agriculture, logging, manufacturing, etc.) that allow the continued production of a commodity without damage to the ecosystem (soil, water supplies, biodiversity or other surrounding resources) (www.ecokids.ca/pub/eco_info/glossary/index.cfm).

Sustainable development requires balance between the economy, ecology and society for present and future generations. It could be broken into three main parts viz. (i) environmental sustainability, (ii) economic sustainability, and (iii) social sustainability.

2.3 Sustainable Development Models

In concentric circles model of sustainable development, three constituent parts of economy, society and environment could be represented as concentric circles (Ott, 2003). If society and environment development are considered then it is only bearable development. It is equitable if together social and economic development is taken care of. However, it is viable if environment is considered along with economic development. Sustainable development is achieved if all three developments are in balance, emphasizing sustainable development: at the union of three constituent parts (Adams 2006 and IUCN 2006).

The economic development, social development, and environmental protection are interdependent and mutually reinforcing pillars of sustainable development. Strong foundation is required to balance three pillars for stability of sustainable development. Ethics and strong commitments should be treated as the foundation to the pillars of the sustainable development (Singh 2011).

3. Sustainable Water Management Model

The sustainable water management can be achieved by using appropriate techniques of water management to improve water productivity, and availability over an area for irrigation. A conceptualized model for sustainable water management could be visualized with its three constituent parts, the users, available water and environment all represented by circle and interacting each other (Singh *et al.*, 2009).

The constituent parts interact in such a way that users are affected and affecting the available water in a closed system of environment which is affecting both the users and available water and vice versa. The circle represented for user is expanding because of increasing population and the other circles remain same. The circle represented for available water remains same if assumed that quantity of available fresh water remains same in hydrologic cycle over a long period of duration. The circle represented for environment remains same under sustainable condition.

4. Strategies for Sustainable Water Management

The available water is utilized by the users, the ever increasing population. However, the water resources of country have not been hardnessed adequately. On the one hand, most of the rainwater flows into sea without being harnessed and on the other hand, ground water is depleting due to its over-extraction. The following points need to be taken care of:

(i) Reduction of water losses involved during the course of water use or consumption to enhance the availability and productivity of water. It could be achieved through:

  (a) Use of improved irrigation technology/systems

  (b) Improving irrigation efficiencies, water use efficiencies and water food print.

(c) Optimum utilization of water

(d) Adoption of precision agriculture using GIS, RS and GPS.

(e) Adequate drainage to improve soil health and soil water plant interaction.

(ii) Enhancing the availability of both surface stored water and the ground water storage through the process of water harvesting, and conservation.

(iii) Practicing the artificial ground water recharge: it will enhance the availability of water to present and future generations without the compromising their ability to meet their own water demand.

5. Irrigation Methods

Several methods of irrigation from traditional surface flooding to modern drip irrigation systems have been evolved. Each irrigation method has its own advantages, disadvantages and suitability to various soil, crop, climatic and socio-economic situation.

The traditional surface irrigation poses numerous problems such as; soil salinity; seepage, conveyance and evaporative losses; higher energy cost; faster soil erosion; more wastage of fertilizer and other nutrients. Surface irrigation methods also result in higher weed population; increased operational difficulties and cost; uncontrolled, unmeasured and uneven water supply.

Surface-irrigation methods are supply driven rather and driven and cause mismatch between the need of the crop. These are also associated to increased disease and increased cost of cultivation due to low production.

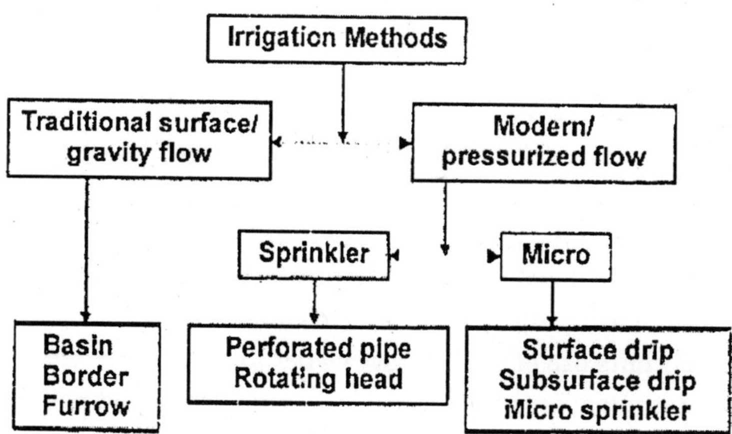

Figure 2. Methods of irrigation are classified as traditional which flow under gravity and modern, called pressurised irrigation, that consists of sprinkler and micro irrigation.

6. Micro irrigation system

Micro irrigation system refers to low-pressure irrigation system that spray, mist, sprinkle or drip. It is the precise, slow and frequent application of water to the plants in the form of discrete drops, continuous drops or tiny streams, through the devices called the drippers or emitters or applicators located at selected points along water delivery lines called as laterals. It provides irrigation with high frequency application of water in and around the root zone of plants. It consists of network of pipes along with appropriate emitting device, control system, fertigation and filtration system and other fittings.

Emitting devices are to dissipate pressure and discharge water in micro irrigation system. These allow a small, uniform flow of water at a constant rate. Drip irrigation uses drippers as water emitting device. These are drippers, micro sprinkler, sprayers and mist according to requirement of crops. The water is also applied below the soil surface if emitting devices are buried in soil, called subsurface drip. The drip irrigation is also termed as trickle irrigation. Use of micro sprinklers is also getting popularity. However, drip irrigation; lateral placed at soil surface consists of the maximum area under micro irrigation

6.1. Suitability of Drip irrigation

Drip irrigation is adopted extensively in areas of acute water scarcity and especially for crops such as Coconut, Grape, Banana, Ber, Citrus, Sugarcane, Cotton, Maize, Tomato, Brinjal and plantation crops. It is suitable to almost 80 crops in different soil and agro climatic conditions.

6.2. Advantage of drip irrigation

Drip irrigation offers several advantages over traditional irrigation methods.

1 Less weed growth: restricted weed growth to wetted areas only

2 Uniform and controlled water distribution closer to plant roots

3 Partial soil wetting: only to crop root zone

4 Moisture always at or near field capacity in the root zone

5 Less fertilizer/ nutrient loss due to localized application

6 High water distribution efficiency

7 Permits use of poor quality water

8 Recurrent irrigation is easily accomplished and has the effect of keeping soil moisture between field capacity and saturation

9 Soil factor plays less important role in frequency of irrigation

10 Levelling of the field not necessary

11 No soil erosion

12 Cultural operations are possible even during irrigation Low labour cost

13 Possibility of regulating variation in water supply.

14 Simultaneous of application water and fertilizer i.e. fertigation.

15 Low water delivery rate, low water pressure, precise placement of water.

16 Minimum application, field runoff and deep percolation losses.

6.3 The limitation of drip irrigation

The drip irrigation has the following limitation:

1 Higher initial cost can be more than overhead systems

2 Durability of the components

3 Plant Performance: Studies indicate that most plants grow better when leaves are wetted as well

4 Poor system and crop performance if system not designed properly

5 Skilled persons are required to operate and maintain the drip system

6 Regular maintenance is required

7 Power source is required for operation of the system and automation except for the systems operated with gravity

8 Susceptible to crop damage in case of system failure for more than 3-5 days

9 Micro climate can not be controlled.

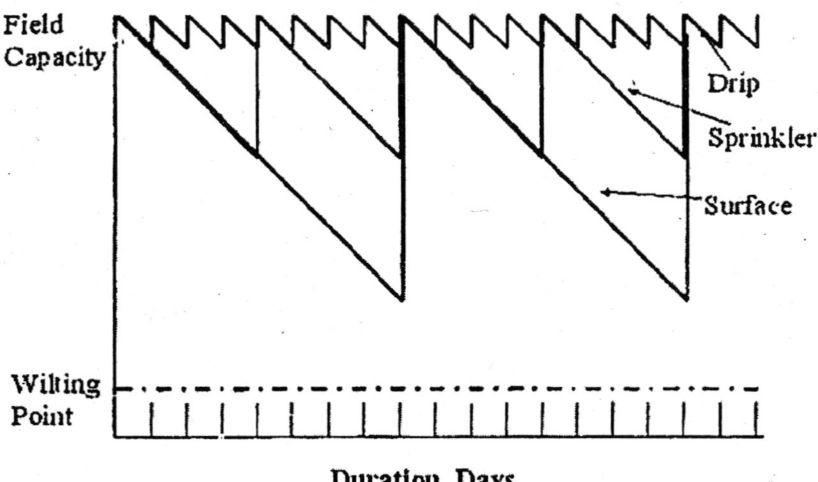

Figure 3 : Moisture availability to crop with different irrigation methods.

6.4. Moisture availability to crop

More than 80% of worlds irrigated land is under surface irrigation methods, yet its field level application efficiency is often only 40-50%. In contrast micro irrigation may achieve field level application efficiency of 80-90%, as surface runoff and deep percolation losses are minimized (Heerman et al., 1990 and Postel, 2000). The moisture availability to crop under different irrigation methods has been depicted in Figure 3.

Moisture availability to the crop through drip irrigation remains near to the field capacity and nearly uniform as compared with sprinkler and surface irrigation methods, which imposes less stress to the plants and results higher crop yield.

6.5. Enhanced productivity and water use efficiency

Using micro irrigation, water can be applied to crops efficiently with enhanced water use efficiency. Thus the micro irrigation may allow more crop cultivation per unit applied water, and allow crop cultivation in an area where available water is insufficient to irrigate through other irrigation methods. Fertilizers and chemicals could also be applied efficiently using this irrigation method. It has potential to save water and fertilizer with enhanced crop production.

Micro irrigation could utilize the marginal quality water also. The relative advantage of using MIS over conventional method of irrigation for some crops (INC ID, 1994) has been presented in Table 2.

Table 2: Advantage of micro over conventional irrigation of crops

Crop	Yield, t/ha		% Yield increase	% Water saving	Increase in water use efficiency, %
	Conventional	Drip			
Banana	57.5	87.5	52	45	176
Chilly	4.2	6.1	44	63	219
Grapes	26.4	32.5	23	48	136
Sweet Lime	100	150	.50	61	289
Pomegranate	55	109	98	45	167
Tomato	32	48	SO	31	119
Water Melon	24	45	88	36	195
Sugarcane	128	170	33	56	204

6.6. Potential of drip irrigation

Use of efficient pressurized irrigation systems such as drip and sprinklers irrigation can be used to improve water use efficiency including water saving and increased yield. In India, more than two million-hectare land under vegetables and high value crops is being irrigated through pressurized irrigation system with efficiencies of 60-95%. Pressurized irrigation includes sprinkler as well as micro irrigation consisting of micro sprinklers, subsurface drip and drip irrigation. Micro irrigation is growing fast in India. About one million-hectare land under vegetables and high value crops is being irrigated through drip irrigation in India during 2008 (Table 3).

Table 3. Area under Micro irrigation (ha)

State	Drip	Sprinkler	Total
Rajasthan	15248	684748	699996
Maharashtra	462240	207205	669445
Haryana	6243	512657	518900
Andhra Pradesh	317935	182260	500195
Agartalla	169795	216978	386773
Gujarat	158727	128942	287669
Tamilnadu	124951	26739	151689
West Bengal	123	150020	150143
Madhya Pradesh	12518	104049	116567
Chhattisgarh	2627	44763	47391
Orissa	3361	23187	26548
Uttar Pradesh	10577	10555	21132
Punjab	10427	10276	20702
Kerala	14119	2516	16635
Sikkim	80	10030	10110
Nagaland	0	3962	3962
Goa	762	332	1094
Himachal Pradesh	116	581	696
Arunachal Pradesh	613	0	613
Jharkhand	133	365	498
Bihar	107	180	287
Grand Total	**1310956**	**2320586**	**3631542**

(*Source :* NCPAH, New Delhi)

Government has planned to cover 14 M ha area under micro irrigation during XI Plan. The drip irrigation has vast potential in the country as depicted in Table 4. Net potential area for drip irrigation is estimated to be 21.27 mha for the country (Narayanamoorthy, 2004).

Table 4: Potential of drip and sprinkler irrigation (area in mha)

State	Drip	Sprinkler	Total
Utter Pradesh	3.35	10.53	13.88
Maharashtra	4.11	3.30	7.41
Madhya Pradesh	1.19	5.98	7.17
Rajasthan	1.86	4.01	5.87
Punjab	0.65	3.57	4.22
Others	0.84	3.04	3.88
Haryana	1.03	2.75	3.78
Karnataka	2.02	1.47	3.49
Bihar	0.84	2.44	3.28
Andhra Pradesh	2.16	0.99	3.15
Gujarat	2.04	0.89	2.93
West Bengal	1.12	0.99	2.11
Kerala	1.69	0.25	1.94
Tamil Nadu	1.37	0.21	1.58
Orissa	0.84	0.42	1.26 3.56
Others	1.89	1.67	
Total	27.00	42.51	69.51

To increasing the efficiency of water use and enhanced crop yield the area under micro irrigation is further likely to increase. The drip irrigation can be made more applicable for irrigating a wide range of agronomic, horticultural and fruit crops by installing the laterals below the soil surface, called subsurface drip irrigation.

6.7 Fertigation

Application of fertilizer through the drip irrigation system is called fertigation. It is the most advance and efficient practice of fertilization. Fertigation combines the two main factors in plant growth and development, water and nutrients. Fertigation is the most efficient method of fertilizer application, at it ensures application of the fertilizers directly to the plant roots (Patel & Rajput, 2001a). In fertigation, fertilizer application is made in small and frequent doses that fit within scheduled irrigation intervals matching the plant water use to avoid leaching. Fertilizer use efficiency upto 95% can be achieved through drip fertigation (Table 5).

Table 5: Fertilizer use efficiency

Nutrient	Fertilizer use efficiency (%)		
	Soil application	**Drip**	**Drip and fertigation**
N	30-50	65	95
P #	20	30	45
K	50	60	80

6.7.1. Advantages of fertigation

Fertigation has the following advantages:

- Less labour, equipment and energy needed for receiving, storing and fertilizer application
- Reduced soil compaction
- Prevents damage to crop during delivery
- No restrictions or limitation on application timing
- Accurate and uniform distribution for superior efficiency
- Application restricted to most active root zone which reduces waste
- Adaptability of nutrients supply to the growth curve resulting in better crop response
- Split applications for better control of run-off and leaching into groundwater.
- Extremely efficient method of accurately delivering uniform, minute quantities of minor elements
- Complete adaptability to automation.
- Can be used for other purposes, i.e. pestigation, soil amendments, maintenance
- Can overcome negative effects of saline/waste water.

6.7.2. Response of Fertigation to crops

Significant savings in the use of fertilizers and increase in yield of various crops have been reported by researchers (Table 6) (Anonymous, 2001) (Patel and Rajput, 2001a, 2001b, and 2002). The research conducted at various places on fertigation indicated that crop yields were substantially increased from 26-40% in pomegranate, 11-41% in straw berry and 8 - 31% in grape.

Table 6: Response of crops under fertigation

SI. No.	Crop	Saving in fertilizer, %	Increase in yield, %
1.	Okra	40	18
2.	Onion	40	16
3.	Banana	20	11
4.	Castor	60	32
5.	Cotton	30	20
6.	Potato	40	30
7.	Tomato	40	33
8.	Sugarcane	50	40

6.7.3. Potential of fertigation

Studies revealed significant fertilizer savings of 20-60% and 8-41% increase in yields of horticulture and vegetable crops. Fertigation could be adopted for increased yield and fertilizer saving. It is the essence of drip irrigation. Fertigation is a must in order to realize the full potential and benefits of drip irrigation system. With increasing area under drip irrigation in India, there is tremendous opportunity of using fertilizer through fertigation system for many crops.

7. Conclusions

The sustainable water management approach can be used to address the issues of irrigation water crises for present and future generations. It can be achieved by using appropriate techniques of water management to improve water productivity, and availability over an area for irrigation. Improved and efficient irrigation practices may be adopted to enhance water application efficiencies, and thereby reduction in associated damage to environment due to water logging, soil salinity, runoff and leaching of nutrients and other chemicals polluting to water sources, over exploitation of ground water. Micro irrigation systems have been characterized to be the most efficient irrigation techniques. These have efficiencies even more than 90%, doubled water productivity with improved quality of produce, reduced runoff and leaching of nutrients and chemicals to water sources. Fertigation could be adopted for efficient use and saving of fertilizer. Fertilizer savings of 20-60% and 8-41% increase in yields of horticulture and vegetable crops has been realized through fertigation. Micro irrigation is getting popularity with large potential area because of being economic feasible, widely accepted in society and environment friendly. Therefore, micro irrigation systems could be used for sustainable water management.

References

Adams. W.M. (2006). The Future of Sustainability: Re-thinking Environment and Development in the Twenty-first Century. *Report of the IUCN Renowned Thinkers Meeting,* 29-31 January 2006.

Comprehensive Assessment of Water Management in Agriculture. (2007). Water for Food, Water for Life: A Comprehensive Assessment of Water Management in Agriculture. London: Earthscan, and Colombo: International Water Management Institute.

Hecrmann, D.F., Wallcnder, W.W. and Bos, M.G. (1990). Irrigation efficiency and uniformity. In: Management of farm irrigation system. Hoffman G.S., Howell, T.A. and Soloman, K.H. (Eds.), ASAE, St. Joseph, M.I.: 125-149.

IPCC Fourth Assessment Report. (2007). Intergovernmental Panel on Climate Change. http: / /www, ipcc. ch / ipccreports / ar4-wglAtm.

IUCN. (2006). The Future of Sustainability: Re-thinking Environment and Development in the Twenty-first Century. Report of the IUCN Renowned Thinkers Meeting, 29-31 January 2006 http://cmsdata.iucn.org/downloads/iucn_future of susta nability.pdf.

INCID (1994). Drip irrigation in India. Indian National Committee on Irrigation and Drainage. New Delhi.

Millennium Development Goals. (2000). The United Nations Millennium Summit in September 2000.

Millennium Ecosystem Assessment. (2003). Ecosystems and Human Well-Being: A Framework for Assessment (Island Press, Washington, DC). http: / /en.wikipedia. org/wiki/ Millennium_Ecosystem_Assessment.

Narayanamoorthy, A. (2004). Drip irrigation in India: can it solve water scarcity. *Water Policy.* 6(2): 117-130.

NCPAH, (2001). National committee on plasticulture applications in Horticulture, progress report 2001, Department of Agriculture and Co-operation, MoA, Gol, New Delhi:44.

NCPAH. (2008). Area under Micro Irrigation. National Committee on Plasticulture Application in Agriculture, New Delhi.

National Commission for Integrated Water Development. (1999), Report of the National Commission for Integrated Water Resources Development Plan. Vol. I, Ministry of Water Resources, Government of India, New Delhi.

National Water Policy. (2002). Ministry of Water Resources, Government of India, New Delhi.

Ott, K. (2003). The case for strong sustainability. In: Greifswald's Environmental Ethics. Ott, K. and P. Thapa (cds.). Greifswald: Steinbecker Verlag Ulrich Rose.

Patel Neelam and Rajput, T.B.S. (2001a). Effect of fertigation on growth and yield of onion. CBI Publication. 282: 451-454

Patel Neelam and Rajput, T.B.S. (2001b). Fertigation of ok commercially available granular fertilizers. Proceedings of International symposium on importance of potassium in 270-273

Postel, S. (2000). Redesigning irrigated agriculture. In: State of the world 2000, Stark, L. (Eds). W.W. Norton and Co., New York: 39-58.

Rajput, T.B.S. and Patel, Neelam (2002). Fertigation: theory and practice. Indian Agricultural Research Institute Publication No. IAR/WTC/2002/2

Singh, D.K. Chandola, V.K. and Singh, R.M. (2009) Sustainable Management of irrigation water, An Approach to address impact of climate change in agriculture. Flash Book Publication, New Delhi.

Sing D.K. singh, R.M. and Chandola, V.K. (2011) Sustainable water management in agriculture : Issues and strategies to address. Proceeding of National Seminar on Sustainable Management of Water Resources Banaras Hindu University, Varanasi, 14-15 Jan, 2011, 59-74.

United Nations (1987). Report of the world commission on Environment and Development General Assembly Resolution 42/187, 11 Dec., 1987.

Chapter - 6

Water Management for Irrigation in Kerala

M. Lathika

Despite Kerala's bounty of natural wealth, the state finds itself increasingly under pressure in the two vital resources of land and water. Efficiency in their use is, therefore, very important.

Traditionally, the crop choice of the farmers in Kerala had a strong association with the physiographic nature of the land they cultivate and the suitability of the crop to the land. The water rich valleys (0 - 7.5 m altitude) were earmarked for moisture loving, submergence tolerant crops like paddy, sugar cane, etc. In the lower slopes (of 7.5-75.0 m altitude), where the water table is fairly high, but the soil is drained, annual/ seasonal crops like tapioca, plantains, vegetables, etc. and perennial crops like coconuts, rubber, pepper, cashew, arecanut, etc, are preferred. The perennial cash crops of tea, coffee (*Robusta* variety), pepper and even coconuts, rubber, etc, dominate the high land (75-1,000 m) regions. Hogh ranges (700-1,500 m) are devoted to crops like cardamom, coffee (*Arabia*) and tea (GOK 1998; GOK 2003; Rao 2003). Land utilisation in Kerala has undergone drastic changes, as much of the land hitherto engaged for water intensive foodgrain production has been either converted to suit human needs other than cultivation or is increasingly employed for crops demanding less water. This must have a direct repercussion on the intensity of water use for agriculture. A vast pool of the state's scarce financial resources is being used for improving irrigation. It is reported that the state has already (up to the Tenth Plan) spent more than Rs. 38 billion on irrigation, flood control, command area development (CAD) and anit sea erosion (GOK 2003; GOK 2007). A large chunk about 70%) of this sum had been allocated exclusively to major (with cultivable command area greater than 10,000 hectares) and medium (with command area greater than 2,000 hectares) irrigation systems.

This paper seeks primarily to assess the water resources of the state and the total water requirement of crops. It them examines critically, with the continuing shift of crop area in favour of cash crops, if the strategy of pumping more funds into the irrigation sector still stands to reason. The subject of the paper is confined to the water needs of only agriculture.

The paper is organised into four sections. Section 1 is an appraisal of the water resources of the state. The cropping pattern in the state practised over different periods has been analysed in Section 2. Assessment of the water requirement of various crops cultivated in Kerala, across the Plan periods since 1980-81, is made in Section 3. Section 4 seeks to ascertain, in the light of the changed cropping pattern, whether the irrigation infrastructure built up over the years yielded returns, especially with regard to the financial efficiency parameters. The study relies solely on secondary data available from various government agencies.

1. Rainfall and Water Resources

Owing to the high precipitation rate in Kerala, the region has generally been regarded as water surplus. The parameters that determine the availability of water in the state the presented in Table 1.

Table 1. Water Resources of Kerala as Compared to India

Parameter	Kerala	India
Geographical area (sq km)	38.864	32,87,732
Population 2001 Census (mn)	31.84	1,028,61
Population 2006 (mn, projected)	34.20	1,094.1
Total rainfall and snowfall (bcm, 2003)	88.45	4,057,35
Surface and replenishable groundwater (or, water potential of river basins, bcm)	70.17	1,869
Water that can be put to beneficial use	42.67	1,122
Of which : Surface flow	34.77	690
replenishable ground flow	7.90	432
% to total	60.82	60.03
Per Capita annual availability of utilisable water (m^3)	1.248	1.026

Source : Complied/computed from GoI 2006; GoK 2003; Tata 2001.

The total surface and replenishable groundwater in Kerala is estimated to be 70,165 million cubic metres (mcm) (GOK 1998; GOK 2007). But only about 60% of it could be put to beneficial use, as the rest of the available resources are lost as run off causing heavy floods many a times (GOK 1998; GOK 2001a; GOK 2003). The groundwater resource available in Kerala is estimated at 7,900 mcm (GOK 2005). The per capita utilisable water availability from both surface flow and replenishable ground flow comes to 1,340 m^3 (as on 2001), which is seen reduced to 1,248 m^3 by 2006. Thus, the region, according to international standards, is under ``water stress".

As per a rough estimate (as of 1974) of the projected demand for water, Kerala would require around a total of 49,700 mcm of water (GOK 1998). Agriculture requires 30,000 mcm and domestic requirements work out to another 7,500 mcm. Kerala with its long coastal belt stretching to over 590 km has to spend a significant proportion of its water resources for preventing salt water intrusion. It is estimated that this needs another 12,200 mcm. The utilisable water resources as per the earlier assessment are above 42,000 mcm. The surface water balance does not, therefore, suffice to meet the demand for water in Kerala.

Kerala receives about 3,000 mm of rainfall annually, which is almost three times that of India (1,170 mm). The 41 west flowing rivers (of more than 14 km long) run to a total distance of 3,092 km across the length and breadth of Kerala and each kilometre of these rivers, on an average, has a gross catchment area of 11.33 km^2 (computed from GOK 2003). The three other rivers draw from a total catchment area of another 2,866 sq. km of Kerala before they join the Cauvery in the east. As per the state public works department estimates (1974), of the total annual yield of 77,883 mcm of water flowing through these 44 rivers 70,165 mcm is available in Kerala with only 42,672 mcm that could be utilisable in the state alone (GOK, 2001a).

The crop season wise average rainfall experienced in the state as well as in India as a whole has been computed from the month-wise data for a long period of 135 years (from 1871 to 2005) and is presented in Table 2.

Table 2. Average Seasonal Rainfall (in mm) in MET Subdivisions in Kerala and India (Period : 1871-2005)

Season	**Seasonal Rainfall (N = 135)**		**Monthly Rainfall (mm)**	
	Kerala	**India**	**Kerala**	**India**
Kharif (May-Oct.)	2168	901		
Virippu	(18)	(10)	434	181
Rabi (Nov-Jan)	491	132		
Munadakan	(30)	(26)	123	33
Other (Feb-Apr)	164	54		
Punja	(42)	(29)	55	18
Total	2822 (14)	1088 (9)	236	91

Source : Indian Meteorology Department.
Values Within parenthesis are the respective coefficient of variation (N=135).
Seasons given in Italics are the season name in local language.

This table shows the richness of Kerala in rainfall. Kerala receives more than three-fourths of the total annual down-pour during virippu/ kharif season, which is, by its quantity, 2.41 times that of India. The mundakan/rabi season, as compared to the major season kharif, is dry. The variation of rainfall over the years is larger in Kerala, with a higher coefficient of variation (cv) that that of the whole of India for all the seasons. Over the 135 years for which data is available, Kerala experienced normal rainfall (±10% of the average of 135 years) in only 20 years. India as a whole also had normal rains in about 21 years.

The rainfall diversity has a spatial dimension as well. Table 3 shows the average monthly rainfall by districts, during the nine years up to 2004. Except for the district of Idukki, all other districts experienced a lower average rainfall in 1996-2004, when compared with the normal rainfall figures supplied by the meteorology centre, Thiruvananthapuram. The districts of Kozhikode (27%) and Wayland (35%) had huge deficits of rainfall during this period. The state, as a whole, had about 12% less rainfall than the normal during the period. Table 3 demonstrates not only that the rainfall during the period was far short of the normal, but also that the deviation varied vasty from region to region and also from month to month. This calls for an effective conservation of water during the water plenty days and its judicious use during the dry periods. The construction of dams would thus be justified.

Table 3. Spatial and Temporal Distribution of Rainfall (mm) in Kerala - Average of 1996-2004

District	**January**	**February**	**March**	**April**	**May**	**June**	**July**	**August**	**September**	**October**	**November**	**December**	**Total**	**Normal**	**Deviation from Normal**
Thiruvana-nthapuram	10	28	25	122	196	281	182	154	204	341	183	56	1,782	2.204	-20.46
Kollam	11	47	47	161	255	391	310	274	219	476	170	58	2,419	2.555	-6.65
Alappuzha	21	51	60	175	291	504	416	342	270	431	175	47	2,783	2,965	-7.62
Pathana-mthitta	20	43	43	156	300	552	445	362	273	428	165	53	2,840	3,134	-10.63
Kottayam	15	27	62	143	284	555	493	387	279	426	151	51	2,873	3,130	-9.27
Idukki	5	28	47	158	287	695	786	631	332	472	179	52	3,672	3,379	7.84
Ernakulam	8	25	24	107	321	662	610	432	298	430	159	54	3,130	3,274	-4.73
Thrissur	0	21	8	66	237	646	579	393	247	368	91	25	2,681	3.262	-17.14
Palakkad	1	16	24	79	153	445	457	310	144	279	118	22	2,048	2,390	-15.19

Kozhikode	2	8	14	74	237	743	664	396	198	322	91	29	2,778	3,668	-27.29
Malap-puram	3	8	13	80	271	713	613	386	204	366	132	37	2,826	2,906	-2.65
Wayanad	6	12	28	102	161	522	570	418	153	249	96	37	2,354	3,591	-35.14
Kannur	4	1	4	47	253	927	816	532	172	338	92	36	3,222	3,465	-7.36
Kasargod	4	0	3	35	300	1,023	874	563	176	312	58	23	3,371	3,581	-6.42
Mean	8	23	29	108	253	619	558	399	226	374	133	41	2,770	3,108	-11.74
Normal	15	17	40	113	263	697	765	439	252	297	166	43			
Deviation	-45	39	-27	-5	-4	-11	-27	-9	-10	26	-20	-5			
Subdiv normal	11	17	35	112	244	687	636	377	224	286	156	37	2,822		

Source : Compiled/computed from GoK, Statistics for Planning, various issues.

2. Crops of Kerala

As already noted, owing to the location of Kerala, the soil and climatic conditions are highly suitable for a multitude of crops, that too, round the year. The crops can attain their full yield potential only with supply of water in addition to what is rendered by rainfall or a gravity driven rainwater runoff, as even in Kerala, intolerably long dry spells are the norm rather than the exception in all seasons. Even so, despite the high potency for water conservation (due to high annual rainfall) and the longer wet spells (than the average Indian level) in Kerala, the area with irrigation facility still hovers around 14.7% (2002-03) of the gross cropped area of the state - a level far below the average for India (38.7%). The traditional farmers on their part, choose the cropping pattern to suit the rainfall rhythm experienced during various months of a year.

Water intensive crops enjoy increasingly less preference in Kerala. Unlike other Indian states, more than 70% of the total area under cultivation comprises ``orchards" (GOI, 2005), which demand much less water per day of the crop than that is required by wetland crops like rice, sugar cane, etc. Many of these crops flourish well under rainfed management itself, though watering them during dry spells of the year would enable them to attain their potential yield.

The suitability of a crop to a region depends on various soil and climate parameters of the region. The regions that are found to be ``most suitable (S1)", lees suitable (S2)" and ``least suitable (S3)" for cultivation of each of the major crops separately are reported to have been identified based on the parameters of the region, namely, elevation, rainfall temperature, humidity, physiography, slope, soil depth, soil texture, presence of stones and rocks and soil drainage. These variables are very crucial in determining the irrigation needs of crops as well. The yield potential of the crops in respect of these suitability regions are also seen estimated, as a ``first approximation" (KSLUB 1997). These are given in Table 4.

Table 4. Area and Yield Potential of Crops under Different Suitability Regimes

Crop	Area (ha) Suitable under			Current (2005-06) Yield Tonnes/ ha	Yield Potential (Tonnes/ha)			Water Use Efficiency 2005-06 (kg/ha as a Ratio of mm of Water)	
	S1	S2	Total		S1	S2	S3	min	max
Rice	1,25,496	2,76,742	4,02,238	2,285	6	4.5	3	1.43	2.29
Pepper	1,22,140	94,552	2,16,692	0.368	1.2	0.9	0.6	0.31	0.74
Ginger	14,567	11,395	25,962	4.604	20	15	10	3.84	6.58
Cardamom	78,325	39,483	1,17,808	0.236	0.5	0.35	0.3	0.13	0.19
Banana	1,26,325	30,835	1,57,160	8.010	40	30	20	3.64	6.68
Cashew	71,131	53,969	1,25,100	0.872	1.2	0.9	0.5	0.73	0.97
Pineapple	23,950	7,420	33,370	6.681	60	45	30	6.68	9.54
Tapioca	1,53,755	1,11,630	2,65.385	28.367	35	25	17.5	40.52	56.73
Vegetable	24,158	2,196	26,354	-	-	-		-	-
Coconut	3,68,230	5,72,230	9,40,460	7,046	14,000	10,500	7,000		
Tea	51,860	25,430	77,290	1.609	2	1.5	1	0.89	1.29
Coffee	73,0.50	4,215	77,625	0.711	0.8	0.6	0.4	0.40	0.57
Rubber	3,23,141	1,05,047	4,28,188	1.495	1.5	1.25	0.75	0.83	1.20
Arecanut	(not available)			1.099	1	0.75	0.5	0.61	0.88
Total	15,56.128	13,35,144	28,91,272						
% to GCA (2005-06)	52	45	97						

Coconut yield is in number of nuts per ha.

S : Computed by author. Minimum value corresponds to upper limit of water requirement of the crop and the minimum value to the lower limit.

Source : Compiled from KSLUB 1997. Data on productivity: Economic Review, 2007.

Table 4 shows the potential of agriculture in Kerala. More than half of the current gross cropped area is classified as highly suitable for crop growth. The yield levels of the crops in 2005-06 are also shown. It can be discerned that the yield levels in Kerala are very low, except for a couple of crops, even with regard to the national standards. While the yield of coconuts in Andhra Pradesh (in 2005-06) was 8,577 nuts and in Tamil Nadu 13,133 the yield in Kerala was as low as 7,046 nuts, against the national average of 7,608 and the highest yielding state (Maharahstra) with 15,189 nuts.

The yield of rice in Kerala, though slightly better than that of average Indian level (2.047 tonnes), has been stagnant and is range bound for some years now (GOK 2007a). The only crop that is seen steadily improving its performance is rubber. Table 4 also provides the yield potential of crops under the three crop suitability regimes. These estimates are perceived, as noted above, as approximates projected with reference to given and limited information available. But these estimates categorically assert that the yield performance, barring a few crops like rubber, tapioca and arecanut are staggeringly low.

Given the suitability conditions prevailing in the state for almost all these crops, an urgent evaluation of the factors that hampered the yield growth in Kerala is called for.

In this paper, we address the issues related to irrigation of crops in Kerala. The ``water efficiency" (WUE) of a crop which is defined as the ratio of the yield (kg/ha) and water requirement (mm) of the crop is also computed and presented in Table 7. Interestingly, the tapioca, with WUE of 56-73 kg/ha mm is found to be the most water efficient crop in Kerala, followed by pineapple (9.54 kg/ha mm) and banana (6.68 kg/ha mm). The spice crops like cardamom (0.19 kg/ha mm) and coffee (0.59 kg/ha mm), for which the economic part constitutes only a very small proportion of the total biological yield naturally turned out to be the least water efficient crops in the state.

At the outset, let us examine the effectiveness of irrigation of crops, as evident from the secondary data available. Trendata on yield for the irrigated and unirrigated crop is available for paddy. We have, therefore, analysed the yield of paddy in Kerala under irrigated and non-irrigated conditions to ascertain the relative yield efficiency of irrigation. The seasonwise data of paddy yield corresponding to the 10 years up to 2005-06 is used for this analysis (source of data : GOK 2007). Two factor analysis of variance was done on the data. As the yield performance in a season cannot be considered independent of the previous season's performance, the Analysis of Variance, ANOVA, in split plot fashion is attempted, with season in the sub-plot (Gomez and Gomez, 1984), following the linear fixed effect model for ANOVA (Das and Giri 2003).

$$y_{ijk} = \mu + a_i + b_j + \delta_{ij} + e_{ijk}, with\ \ e_{ijk} \sim$$

$N\left(o, \sigma^2\right)$ where,

p = number of irrigation status (here, ``irrigated" and ``un-irrigated"),

q = number of seasons (sub-plots),

r = number of years (replicates), and

i = 1, 2, . . . p; j = 1, 2, . . . , q; k = 1, 2, . . ., r

μ = general mean, a_i = effect due to i^{th} level of ``irrigation status",

b_j = effect due to j[th] season,

δ_{ij} = interaction between i[th] level of irrigation status and j[th] level of season, and

e_{ijk} = second error term. The two seasons for the same irrigation status (sub-plot in the same main plot) being correlated, say with a coefficient p, they are related with E $(e_{ijk}, e_{ijk}) = \rho\sigma^2$.

The results are presented in Table 5.

Table 5. Mean Paddy Yield (Kg ha^{-1}) under irrigated and unirrigated Conditions (10-years Till 2005-06). By seasons

	Autumn	Winter	Summer	Mean
Irrigated	3,508.7	3,241.1	3,748.9	3,499.6
Unirrigated	3,036.8	2.864.2	3,163.2	3,0214
Mean	3,272.8	3.052.7	3,456.1	

$F_{1.18}$ for irrigation (I) = 17.81**, $F_{2.36}$ for season (S) = 6.10** $F_{2.36}$ IxS = 0.04

$CD_{(05)}$ = Irrigation = 238.03; Season = 231.00

Source of data : GoK 2007.

The table shows that irrigation has a great effect on enhancing the yield levels ($F = 17.81$, $p < 0.01$). Irrigation pushed up the yield by about one-sixth (about 500 kg per hectare) of that of the unirrigated level. The effect of irrigation was visible, irrespective of the seasons. Irrigating in summer, fetches the highest marginal yield though summer paddy in Kerala extends to less than 30% of the acreage of that in the winter. Summer is the best yielding season for paddy and winter yields the least. Note that the Kharif season is the wettest season in Kerala, both in quantity of rainfall and in the number of rainy days, with more than half the days turning out to be wet.

The spatial extent of occupation of the crops and area under irrigation of major crops of Kerala are furnished in Table 6. To gauge the shift in cropping pattern, the area growth and the share of these crops in the gross cropped area (GCA) corresponding in 1980-81 and 2005-06 are also given in the table. Kerala utilises about four fifths of its non forest area for cultivation. The striking feature of the trend in cropping pattern (during the period 1990-91 to 2005-06) was that while some crops lost heavily in area, some other crops still managed to improve their acreage considerably. During the 15 years, paddy lost its area at an annual exponential rate of 5.21%. Area shrinkage was rapid for ginger (-3.26%) and tapioca (-2.81%) also. Tapioca and ginger are crops of high WUE. Paddy was the crop that suffered the most. This table also reveals that Kerala has experienced major shifts in its cropping pattern. A shift from foodgrains to certain plantation crops notably rubber, arecanut and banana is conspicuous. Tapioca being the crop of highest WUE, a crop that almost matches its potential yield level and a crop that proved to be a crop in need, especially for the poor when they struggled to meet their energy requirements at times of price escalation for foodgrains, this debacle of tapioca would spell disaster for a state like Kerala, which is perhaps the highest food insecure state in India.

Table 6. Area (ha) under Crops and Area irrigated@

Crop	Crop Area (ha) 5-year Average 2005-06	Crop Areas as % of GCA		Exponential Growth of Area (from 1990-91 to 2005-06)	Crop Area (ha) Irrigated (5-years Ending 2005-06)	Area Irrigated as % of Crop Area (5-year Average Till 2005-06)	Crop Area Irrigated as % of Gross Area Irrigated
		1980-81	2005-06				
(NSA)	21,68,196	75.56	72.57	-0.33	-	-	-
Paddy	2,97,187.6	27.79	9.08	-5.21	1.78.716.6	57.90	37.12
Pepper	2,15,698.8	3.75	7.92	1.69			
Ginger	9,499.2	0.44	0.31	-3.26			
Cardamom	40,798.2	1.87	1.28	-1.57			
Banana	1,11,144.2	1.71	3.87	4.35	28,473.8	25.56	5.91
Cashew	86,293.8	4.90	2.66	-2.19			
Tapioca	98.626.6	8.49	3.13	-2.81	1.631	1.65	0.34
Coconut	9,00,223.4	24.02	29.62	0.23	1,57,424.6	17.38	32.7
Arecanut	99,789.8	2.12	3.56	3.63	33.927.6	33.33	7.05
Tea	36,855.4	1.31	1.15	0.52			
Coffee	84,375.2	1.45	2.79	0.36			
Rubber	4,80,885.6	7.16	16.27	1.00	40.121.5	8.34	8.33
Vegetables	52.502	0.51	1.56	0.25	13,400.2	19.32	2.78
(Others)	4.73.879.6	14.35	16.81	1.40	277.29.6	8.33	5.76
(GCA)	29,87,759	100	100	-0.12	4,81,424.9	16.03	100

@Some cells in the table are left blank as the respective data is seen missing in the departmental publications, possibly because the values are negligible. However, others would include these also. Two major crops that miss the table are mango and pineapple which occupy about 2.9% and 0.4% of GCA, respectively.

Source : GoK 2005, Statistics for Planning 2005, Department of Economics and Statistics. GoK (2007), Economic Review (various years). State Planning Board.

The crop which claimed 8.49% of GCA in 1980-81 was prevailing only in 3.13% of the GCA in 2005-06. Ginger, cardamom, cashew, etc. had experienced a slump in acreage and share in GCA. However, banana, coconut and cash crops like tea, coffee and rubber took no retreat in their acreage over the period. Rubber, which takes a tally of 16.27% of GCA in 2005-06 had a share of only 7.16% in 1980-81.

Similarly, area under banana (4.35% of GCA) and arecanut (3.63% of GCA) expanded substantially during the period. With practically no growth in GCA in Kerala, it follows that the gain in acreage under one crop should be at the expense of other crops. There are already reports that some crops gained in area in Kerala really at the expense of cereals or cereal substitute crops (Lathika et al., 2004; Jeromi 2007), making the already fragile food security system of the state more vulnerable. The table also shows that the utilisation of irrigation facilities also varied widely across crops, both in area irrigated as percentage of crop area and the crop area irrigated as percentage of gross area irrigated in the state. Though paddy occupied only about 9% of the GCA (as of 2005-06) of the state, it claimed about 37.12% of the total area irrigated. As paddy is a highly water demanding crop, it is natural that the larger area (about 60%) of the crop could be brought under irrigation. Coconut which claims 30% of the total irrigated area of the state. Thus these two crops appropriate about 70% of the total irrigation. This seem quite puzzling for many reasons. In Kerala, homesteads are the most prevalent form of garden land use, where the households are surrounded by a multitude of crops of diverse utility and economic value interspersed without following any crop geometry of sorts. When as irrigation facility reaches the area of the homestead, it is well high impossible that the facility is used only by one or two crops in the mix. It might be that the irrigation status of only these limited crops is being reported. This calls for a thorough re estimation of the total area really brought under irrigation, under all crops. For a real evaluation of the irrigation projects in Kerala this deems highly inevitable. This might also be a limitation of this study.

3. Water Requirement of Crops

Water plays an important role in all plant functions, the main being the photosynthetic activity - the process of using solar energy that is absorbed by the chlorophyll of plant leaves to generate carbohydrates and sugars for its own consumption by splitting weather into hydrogen and oxygen.

The net assimilates in the plants are free for consumption by the higher species. Agriculture is all about optimising the quantity and quality of these balance assimilates of plants to suit the need regime of the respective farmers. The total quantity of water that is needed by the plants for its growth and other functions and the water that is wasted through evaporation from around the plant constitute what is called the Consumptive Use (Cu) of the plant. As the metabolic process needs very little water (less than 1% of the whole water that is passed through it), the other two water needs, namely, evaporation and transpiration - jointly called evapotranspiration (ET) - amount almost fully to the total requirement of water by the plant. Though consumptive use covers exclusive needs for the crop growth, some field operations like land preparation and desalination also demand water. The total volume of water that is required at the field level must account for the application losses also (ICAR, 1987; Reddi and Reddy 2006). Application losses vary between types of irrigation and texture and structure of soil. The total annual water requirement of various crops in Kerala has been computed using the ``per hectare water requirement" values collected from literature. Some assumptions and approximation are made for the purpose, which are listed under the respective table. As only the range of water requirement is available in the literature, considering the wide spectrum of situations for the crops, we have adopted both the limits (lower and upper) of the quantity of water per hectare that is required, for computation of the total water need for agriculture.

Table 7. Range of Water Requirement (Million M) of Crops in Kerala Per Season in the Initial Year of Various Plan Periods, by Crops

Crop	**Water Requirement (mm)**[s]	**Range of total Annual Water Requirement in the State (Million M) in the year**				
		1980-81	**1985-86**	**1992-93**	**1997-98**	**2005-06**
Paddy[a]	1,000 - 1,600	8,017-12,827	6.783-10,852	5,413-8.661	4,308-6.893	2,757-4,412
Pepper[b]	500-1,200	540-1,297	608-1,459	891-2,138	902-2,164	1,203-2,887
Ginger[b]	500-700	63-69	78-100	77-108	62-86	46-65
Cardamom	1,250-1,800	676-973	758-1,091	546-786	511-736	486-700
Banana and other plantains	1,200-2,200	591-1,084	636-1,166	781-1,432	968-1,774	1,411-2,587
Cashew[b]	900-1,200	1,271-1,695	1,240-1,653	1,009-1,345	852-1,136	727-969
Pineapple	700-1,000	35-46	40-53	44.58	82-109	115-153
Tapioca[b]	500-700	1,225-1,715	1,015-1,420	709-993	607-850	475-665
Vegetable[c]	500-700	73-103	143-201	247-346	188-264	236-331
Coconut[d]	1,250-1,800	8,662-12,473	8,809-12,684	10,788-15,535	11,054-15,918	11,249-16,199
Tea[d]	1,250-1,800	471-679	435-626	433-623	433-623	438-631
Coffee[d]	1,250-1,800	522-752	821-1,182	1,050-1,512	1,038-1,494	1,058-1,524

Rubber[d]	1,250-1,800	2,584-3,720	4,129-5,946	5.322-7,664	5,816-8,375	6,180-8,899
Arecanut	1,250-1,800	766-1,102	734-1,056	793-1,142	917-1,320	1,353-1,948
(Others)[e]	700-1,000	2,897-4,139	2,587-3,696	2,057-2,938	2,386-3,409	3,485-4,978
(All crops)		28,393-42,694	28,816-43,195	30,160-45,281	31,069-46,504	31,219-46,948

$: Compiled from various technical sources. Certain modifications made to suit Kerala conditions in consultation with irrigation experts.

a : For paddy, special needs on account of nursery and pudding/transplanting are also accounted.

For other crops, apart from ET values, 75 mm required for pre-sowing irrigation (medium soils).

b : As the water requirement of these crops was not directly available from data, the quantity required for related crops (in terms of the relevant parameters) is substituted.

c : Water requirement values of tomato, brinjal and bhindi assumed for this. Area data was not available for a couple of years; it would thus figure in only under others.

d : Values of plantation crops, as given in Reddi and Reddy 2006 are used.

e : Water requirement is assumed as applicable for a moderately water demanding crop.

Source : Computed from data of area from GoK, Statistics for Planning, various years. GoK, Economic Review, various years.

Table 7 provides the quantity of water requirement for various crops at the initial year of all five year plans since 1980-81. The most striking feature of the results is that while paddy which needed more than 28% of the minimum total water requirement during 1980-81, the need for it in 2005-06 was only 8-83% of the minimum total water requirement for the whole gross area under cultivation in Kerala. It may be noted that the area under paddy the crop that demanded maximum water for its growth and is sensitive to water stress is still on the decline. On the other hand, water requirement of coconut and almost all cash crops recoded a sharp rise. The water demand for banana and vegetables also shot up tremendously. Note that requirement increases in proportion to area and the economic net returns of a few crops like rubber is almost inelastic to irrigation. But since in Kerala, crop shifts are largely in favour of cash crops which demand less quantity of water than that demanded by water intensive crops like paddy, and since most of the farmers of cash crops are still in the habit of using the monsoon rainfall, the real use of the costly irrigation water might have come down considerably. This calls for a serious rethink of expenditure on irrigation systems in Kerala over the years.

4. Irrigation Infrastructure and Economic Efficiency

Table 8 provides the public expenditure at current prices on both types of irrigation - major and minor and other public demands of water like anti sea erosion, CAD, etc. Irrigation systems received very little of public money till 1980. Starting from a meagre amount of Rs 51 million in the First Five Year Plan, it skyrocketed to Rs. 3,020 million in the Sixth Plan. The Sixth Plan envisaged some major projects like Meenachil, Vamaanapuram, Kakkadavu, Chaliar, etc, with an original estimate of Rs. 350 million, Rs. 198.2 million, Rs. 133.5 million and Rs. 106.1 million, respectively.

Table 8. Expenditure Incurred on Irrigation and Allied Sectors during Various Five Year Plans (in Rs. Million)

Plan	Expenditure at Current Prices							Total Irrigation Projects (at 1993-94 Prices)	% of Expenditure of India (1993-94 Prices)
	Irrigation Projects			Flood Control	Anti Sea Erosion	CAD	Total Irrigation Projects		
	Major & Medium	Minor SWD	Minor GWD						
I (1951-56)	51.1	-	-	-	-	-	51.1	791.13	1.16
II (1956-61)	89.24	22.62	-	-	18.94	-	111.86	1,553.59	2.07
III (1961-66)	103.1	56.47	-	6.3	45.7	-	159.57	1,817.36	1.57
Annual Plans (1966-69)	101.5	65.3	-	10.98	12.37	-	166.8	1,309.9	1.69
IV (1969-74)	289.2	112.2	1.08	15.8	54.01	-	402.48	2,521.78	1.67
V (1974-78)	751.3	126.67	7.34	22.83	45.41	-	885.31	3,478.61	2.16
Annual Plans (1978-80)	723.5	105.53	5.25	31.3	36.91	-	834.28	2,886.78	2.55
VI (1980-85)	2,620.7	340.9	58.44	53.08	126.03	8.36	3,020.24	7,102.64	2.63
VII (1985-90)	3,019	357.22	89.08	81.32	98.23	147.58	3,465.3	5,921.55	1.93
Annual Plans (1990-92)	1,345.9	283.97	66.15	61.29	95.31	151.05	1,696.02	2,157.17	1.64
VIII (1992-97)	5,869.3	1286.49	313.99	406.01	725.94	477.8	7,469.77	6,745.32	2.33
IX (1997-02)	7,033	2258	328.92	338.42	500.89	328.06	9,619.92	6,536.74	1.58
X (2002-07)	6,000	2050	-		500	-	8,050.0		
Total (Up to X Plan)	27,996.84	5,220.36	870.25		2,837.06	1112.8	34,087.46		

Source: GoK Steering Committee reports on water resources, 1998 & 2003; CWC 2006
SWD : Surface water development, GWD : groundwater development.

The Eighth Plan witnessed a major revision of cost estimates of pending projects. Kallada project started with an original estimate of Rs. 132.8 million in 1961. This remains still an unfinished project. But, till the end of the Eighth Plan this project had taken a huge sum of Rs. 5,051.5 million (about 40 times as that of the original estimate). It envisaged a net command area of 92,800 ha (GOK 2003). But the actual area realised is still in question. Chaliar project cost was later revised to 6,450 million. Such unduly delayed projects not only put a heavy financial stress on the exchequer, but also cast a shadow on the efficacy of the already built up component of the project as it remains unutilised. Till date, the total expenditure for irrigation, flood control and anti sea erosion works and for CAD, comes to more than Rs. 38,000 million and the major and medium projects alone appropriated about three fourths of it. Yet, the minor irrigation projects could achieve a comparable irrigation level. Up to the fourth year of 10th Five Year Plan (that is, as on 2006), 47% of the total achieved irrigation in the state was solely from the minor irrigation projects (GOK 2007). However, the real cost incurred will be a more useful measure to compare the cost increase across various plans.

The table furnishes the cost at 1993-94 prices also. It could be seen that only one third of the total cost incurred up to March 2002 (worked out to about Rs. 43,000 million on irrigation projects alone) was spent during the first half of the study period (from 1951-52 up to 1978-80). There has been a steady growth in expenditure on irrigation projects ever since. During the 10 years since 1992-93, about one third of the total cost was spent. But the period since 1992-93 saw a steep fall in the area under the most water demanding crop, namely, paddy and a slight decline (-0.33%) in the net sown area (NSA) (Table 6). However, the expenditure by both the central and state governments on irrigation as a share of total plan outlay is reportedly on the decline (GOK 2007) over successive plans. Thus, this table brings two points to the fore. One, minor irrigation projects received a much smaller share of the pie though their achievement levels were comparable with that of the major and medium projects, and two, the expenditure pattern saw a sudden surge upward in the post 1980 phase of growth, during which the cropping pattern in the state took a tilt towards plantation/cash crops that demand very less water compared to water-intensive food crops like paddy.

In this context, the financial efficiency of irrigation projects in Kerala could be a very relevant issue that should be examined. The salient information regarding the efficiency of irrigation projects in Kerala, as compiled from the Central Water Commission publication (2005), is given in Table 9.

Certain other parameters which are computed from this data are also furnished. For a comparison, the data related to India is also provided. It could be seen from Table 9 that the potential created up to the end of the

Ninth Plan was substantial both in respect of the major projects and the minor projects. Till this period, 55.60% of the ultimate potential has been utilised by major and medium projects in Kerala. This is largely comparable with that of India (53.04%).

Table 9. Financial Efficiency of Irrigation Projects of Kerala and India

	Item	Kerala	India
1	Ultimate irrigation potential (000 ha)$	1,000	58.465
	1.1 Major and medium : surface water		
	1.2 Minor : Surface and groundwater	1,679	81,543
2	Potential created (000 ha) up to IXth Plan	609	37.046
	2.1 Major and medium : surface water		
	2.2 Minor : surface and groundwater	640	56.900
3	Potential utilised (000 ha) up to IXth Plan#	556	31,010
	3.1 Major and medium : surface water		
	3.2 Minor : surface and groundwater	603	49,408
4	Potential created as % of ultimate potential	60.90	63.36
	4.1 Major and medium : surface water		
	4.2 Minor : surface and groundwater	38.12	69.78
5	Potential utilised as % of ultimate potential	55.60	53.04
	5.1 Major and medium : surface water		
	5.2 Minor : surface and ground water	35.91	60.59
6	Profit/Loss	22,329	9,60,079
	6.1 Capital outlay (up to 2002-03) Rs Mn		
	6.2 Working expending and interests (year 2002-03)	833	1,48,152
	6.3 Gross receipts - year 2002-03	37	7834
7	Expenditure per hectare of potential created	17,878	10,219
	7.1 Capital outlay (upto 2002-03) (Rs.)		
	7.2 Working expenditure and interests (Rs)- year 2002-03	666.93	1,576.99
	7.3 Gross receipts (Rs) - year 2002-03	29.62	83.39
8	Recovery of working expense to gross receipts (%) - year 2002-03 (%)	4.44	5.29

$: Irrigated potential of Kerala estimated by the State Irrigation Department is around 1.5 million ha, out of which 0.90 Mn ha is projected for minor irrigation sources (GoK 1998).

Data on area brought under irrigation, as provided by the state agencies are much lower if these are accounted for, the per hectare achievement will further shoot up.

Source : Compiled/computed from CWC 2005.

But the potential created and utilised by minor projects, which is below 40% of the ultimate potential, is strikingly lower than that of India. The area created and utilised (in hectares) is higher for minor projects in Kerala, as it is so for India as a whole. However, when it comes to cost, the capital expenditure per hectare of irrigation potential created in Kerala is higher than that for all of India. While it is Rs 17,878 in Kerala, it is just Rs. 10,219 for India, as a whole. The gross recipes per hectare till this period is just Rs. 29.62 which is about one third of the national receipts in that account. The working expenses are also substantially lower in Kerala. The percentage recovery of working expenses to the gross receipts was 4.44%, against the national tally of 5.29%. In short, Kerala's efforts in utilising the surface water and groundwater are not gratifying, and whatever potential created in Kerala cost it heavily, in terms of capital costs, in comparison with the all India Level. Considering the fact that only about one fourth of the total cost was spent on minor irrigation projects and that the achievement rate of these projects was reported to be comparable as that of the major/medium ones, it could be safely concluded that the minor irrigation projects are cost effective. The social costs of major irrigation projects are also put at very high levels (Thukral 1992; IIT 2008). There is great potential yet to be created in Kerala for minor irrigation projects. All these warrant a serious relook at the socio-economic benefits these projects originally envisaged and that the proposals for minor projects merit greater consideration. However, the conclusions on the basis of an assessment of financial efficiency should be corroborated with field data on irrigation of crops. An attempt is made towards this end, in the next section.

The actual marginal cost incurred during a plan period per hectare of the irrigation area added during the period is estimated. In order to assess the real achievements of irrigation in terms of the total requirement of water, an index, named in this paper as ``water requirement realisation index (WRRI)", is computed, using the formula.

$$WRRI = \frac{\sum_k w_k\ A_k^r}{\sum_k w_k\ A_k^o} \times 100;$$ where w_k is the per hectare water requirement of k_{th} crop (Table 10), A_k^r, *and* A_k^o are area under irrigation actually realised under the crop and area put under cultivation.

Thus, WRRI, by definition, gives the ratio of total water actually required for the irrigated area under all crops to the total water actually needed if all the area under cultivation of those crops were irrigated. These estimates are presented in Table 10.

Table 10. Extent and Efficiency of Irrigation in Kerala over the Plan Period

Plan	Total Area Irriga-ted (ha)	Irrigation Area Added during the Period (ha)	Marginal Cost for Irrigation (at 1993-94 Prices) Rs/ha	Water Requirement Realisation Index (WRRI) %	
				Maximum	Minimum
Up to Annual Plan (1978-80)	3,80,926	3,80,926	37,695	15.65	14.97
VI (1980-85)	3,99,152	18,226	3,89,698	16.53	15.84
VII (1985-90)	3,84,561	-14,591	-	16.41	15.84
Annual Plan (1990-92)	3,87,411	2,850	7,56,900	15.47	14.87
VIII (1992-97)	4,58,569	71,158	94,794	18.47	17.93
IX (Exp : up to 3/2002) (Provisional)	4,32,217	-26.352	-	16.77	16.68

Source : Computed from data source as given in tables above.

The values of WRRI are computed, both with the minimum requirement level and with the maximum levels. These estimates are presented in Table 10. As was already noted, despite heavy expenditure already incurred on irrigation during various five year plans, the total area brought under irrigation does not appear to be satisfactory. At the end of the Seventh Plan and the Ninth Plan (up to March 2002), it recorded even a negative growth (as the reported area under irrigation declined during the period). The expenditure incurred up to the annual plans 1978-80 (from 1951-52) was modest (Rs. 37,695 per hectare). The marginal expenditure just took a vertical escalation thereafter. This was the period when the acreage of paddy, the most water demanding crop of the lot, commenced its slide steeply down. During the annual plans 1990-92, it ran to about Rs. 7,57,000 per hectare. We have already seen in an earlier section that there was an advantage of about 500 kg of paddy (Table 5) on irrigating the paddy crop. But considering the whopping expenditure per hectare of added irrigation area, it should be a straight conclusion that major irrigation projects are, in no way, cost-effective.

The change in cropping pattern made the scene still worse. Where the minimum rate of water requirement (per hectare) is concerned, the level of irrigation for the crops was met, up to 1978-80 plan years, to the extent of only 14.97%. Though it increased to 17.93% during the Eighth Plan, the next plan period witnessed a considerable decrease (to a level of 16.68%). In terms of the maximum rate of irrigation also, it recorded slight declines in each plan period under consideration, except for the Eighth Plan.

The whole scenario on the realisation of water requirement of crops in Kerala is captured in Figure 1.

Having spent such large sums on irrigation, it is quite disturbing that there is no great improvement in the achievement levels of irrigation. However, these estimates in the study did not consider the socio-economic benefits other than crop irrigation. It might even be that all the area under crops that have benefited has not been brought under the data coverage of the government publications. For instance, cardamom is a crop that is yield sensitive to irrigation. Knowing it well, the farmers do not, as a practice leave the crop fully to the benevolence of rain. But no government publication is seen to report irrigation under cardamom. It could be that the data on irrigation area reflect only the acreage under the command area of the projects.

In some of the areas of irrigation projects in Kollam and Thiruvananthapuram districts, it was observed that the residents in the command area of the projects, as a whole, appeared to be happy as they felt that the irrigation canals, when charged, have elevated the water table, saving the wells around from going dry. It can be concluded with sufficient confidence, that even after accounting for these benefits, that the irrigation facilities already in place in Kerala have not been adequately utilised and that a little increase in irrigation that was really achieved (as per the available data) is not commensurate with the heavy increase in the costs of the irrigation projects of Kerala.

5. Conclusions

Public investment in irrigation was very much needed in India, more so in the infancy of the centralised planning process. Kerala also embarked on a series of large budget irrigation projects right from the First Five Year Plan period. This was greatly justified even in a wet tropical region like Kerala, where irregularity in rainfall was the norm and the domestic production of food could hardly meet just one sixth of the requirement of the population and achieved yield levels fell extremely short of the potential levels projected. Irrigation systems in Kerala are becoming heavily expensive and water is now much more precious. However, the land brought newly under irrigation is too small, to enable the state to recover the huge cost already incurred. Therefore, the fact that the domestic production of foodgrains registered a sharp decline owing to the rapid shrinkage in area and that the irrigation infrastructure remained grossly underutilised even by other crops, warrants a fresh look at the whole agenda of development of major irrigation facilities in the state.

NOTES

1. coconutboard.nic.in, as on October 2008.
2. GoK, 2006; Rubber Board, various issues; GoK, Statistics for Planning, various issues.
3. Rainfall received during the first crop season (kharif) is more than sufficient to meet the water requirement. Thirty seven per cent yield variability in paddy during kharif can be explained, based on pre- monsoon and monsoon rainfall. However, the pre monsoon showers during April to May are very important as it may hamper proper germination of seeds. the wet spells are detrimental for the stage from flowering to maturity (Rao, 2003).
4. The crops in a crop mix atypical the homesteads of Kerala will share the common resources of land and atmosphere including the water that is used for the evaporation and the water that is wasted through seepage and application losses in the field. Therefore, the estimate on the total water requirement of all crops in Kerala, computed in the paper as the sum total of requirements of individual crops, may be on the higher side. However, with the kind of data on irrigation and crop acreage as is available from published sources, only the method of estimation that is adopted in the paper seems feasible.

REFERENCES

CWC (2005); Water Data Book (New Delhi : Central Water Commission (Information System Organisation).

CWC (2006); Water and Related Statistics (New Delhi; Central Water Commission, Information System Organization).

Das, M. N. and N.C. Giri (2003); Design and Analysis of Experiments (New Delhi : New Age International).

GoI (2005); ``Situation Assessment Survey of Farmers : Some Aspects of Farming" (New Delhi: National Sample Survey Organization, Government of India).

GoK (1993); ``Statistics for Planning 1993" (Thiruvananthapuram : Department of Economics and Statistics, Government of Kerala).

GoK (1998); ``Ninth Five Year Plan 1997-2002; Report of the Steering Committee on Water Resources" (Thiruvananthapuram : State Planning Board, Government of Kerala).

GoK (2001a) : ``Data Book on Agriculture 2000" (Thiruvananthapuram : Department of Economics and Statistics, Government of Kerala).

GoK (2003); ``Tenth Five Year Plan 2002-2007; Report of the Steering Committee on Water Resources" (Thiruvananthapuram : State Planning Board, Government of Kerala).

GoK (2005) : ``Statistics for Planning 2005" (Thiruvananthapuram : Department of Economics and Statistics : Government of Kerala).

GoK (2007a) : Economic Review 2006 (Thiruvananthapuram : State Planning Board, Government of Kerala).

Gomez, K.A. and A.A. Gomez (1984); Statistical Procedures for Agricultural Research (New York : John Wiley & Sons).

ICAR (1987) : Handbook of Agriculture, Indian Council of Agricultural Research, New Delhi.

IIT (2008) : ``Impact of Tehri Dam : Lessons Learnt. Water Welfare Secretation", Indian Institute of Technology, Roorkee.

Jeromi, P.D. (2007); ``Farmers Indebtedness and Suicides : Impact of Agricultural Trade Liberalisation in Kerala", Economic & Political Weekly, 4 August.

KSLUB (1997) : Kerala State Resource Based Perspective Plan 2020 AD (A First Approximation), (Thiruvananthapuram : Kerala State Land Use Board).

Lathika, M, Mathew Kurian and C.E. Ajith Kumar (2004) : ``Scope and Constraints of Growth in Productivity of Coconut in Kerala", Productivity 45(3).

Rao, G.S.L.H.V. Prasada (2003) : ``Agricultural Meteorology" (Thrissur : Kerala Agricultural University).

Reddi, G.H. Sankara and T. Yellamanda Reddy (2006) : Efficient Use of Irrigation Water (New Delhi : Kalyani Publishers).

Rubber Board (various issues) : Indian Rubber Statistics, Kottayam.

Tata (2001) : Statistical Outline of India 2001-02 (Bombay : Tata Services Limited).

Thukral, Enakshi Ganguly, ed. (1992) : Big Dams : Displaced People (New Delhi : Sage).

Chapter - 7

Need to Augment Irrigation Capacity in Agriculture

Dr. N.T. Somashekaraiah & Dr. Harendra Raj Gautam

The outstanding feature of Indian agriculture is its dependence on rainfall. Due to unequal distribution of rainfall during the year and its variation from year to year in respect of quantity, incidence and duration, the Indian agriculture gets affected. Hence the development of irrigation was accorded priority in the India's Five Year plans.

Considerable amount was spent for the development of irrigation in the Five year plans. Irrigation accounted for an expenditure of Rs. 456 crore during the first five year and the irrigation potential created was of the order of 3.66 million hectares.

The expenditure on major, medium and minor irrigation projects during second & Third Five Year plans was Rs. 522 crore and Rs. 909 crore respectively. An irrigation potential of 2.83 million hectares and 4.52 million hectares has been created during the second and third Five year plans respectively.

Total outlay for irrigation in the Eleventh Plan (2007-12) has been kept at Rs. 2,32,311 crore (at current prices). Substantial expenditure has gone in for developing the major and medium irrigation potential especially the major river valley projects like the Bhakra Nangal Project (Punjab), Beas Project (Punjab and Haryana), Hirakund Dam Project (Orissa), Damodar Valley Corporation (Bihar and West Bengal), Nagarjunasagar Project (Andhra Pradesh and Karnataka) etc. However, minor irrigation continues to occupy an important place as its share in the total irrigation potential of 102.8 million hectares created upto the end of Tenth Plan (2006-07) was 58.8 per cent (60.4 million hectares).

The expenditure in annual plans (1966-69) and Fourth Five Year Plan (1969-74) accounted for Rs. 760 crore and 1,750 crore respectively. A total irrigation potential of 3.49 million hectares and 7.10 million hectares has been created in the annual plans and the Fourth Five Year Plans.

The expenditure in the Fifth Five Year Plan was Rs. 3,073 crore and 7.92 million hectares of irrigation potential has been created. An account of Rs. 2,553 crore has been spent on irrigation in the annual plans of 1978-79 and 1979-80. A potential of 4.48 million hectares was created in the annual plans of 1978-1980. During the Sixth Five Year Plan, the total expenditure on irrigation was Rs. 9,318 crore and a potential of 11.30 million hectares has been created in the Sixth Plan.

During the Seventh Five Year Plan, the total expenditure on irrigation was 14,360 crore and the potential of 13 million hectares has been created in the eleventh Plan. The outlay on major, medium and minor irrigation projects during Eighth and Ninth Five Year Plans was Rs. 36,649 crore and Rs. 63,682 crore (including flood control) crore respectively. An irrigation potential of 95.40 million hectares has been created during 2000-2001. The outlay on irrigation and flood control in the Tenth Plan (2002-2007) was kept at Rs. 95,743 crore.

It has been estimated that India has around 166 million hectares of agricultural land out of which 140 million hectares of land would ultimately be irrigated. By 2000 only 95 million hectares were provided irrigation facilities. That leaves still 45 million hectares of potential irrigation for the successive five year plans. According to the irrigation Commission, the cost of exploiting this potential would workout at Rs. 30,000 crore (at 1980-81 prices); it would cost more than Rs. 75,000 crore now (in current prices). As the financial resources available to the Government are limited, the Seventh and Eighth Plans took up only a few major and medium projects but laid emphasis on the completion of ongoing projects.

In order to expand irrigation facilities, the Government of India has adopted the following programmes.

(a) NABARD has set up Rural Infrastructure Development Fund (RIDF) under which loans are granted to States for speedy completion of minor irrigation projects.

(b) The Government of India has launched, since 1996-97, a programme called ``Accelerated Irrigation Benefit Programme" (AIBP) under which the Centre is providing additional Central assistance by way of loans to the States for early completion of selected large irrigation and multi purpose projects. AIBP was specifically started because a large number of river valley projects both major and medium have spilled from plan to plan, mainly because of financial constraints of state government.

Some of these projects were in an advanced stage of construction and could provide irrigation benefits in four or five agricultural seasons. The completion of these projects, however, was beyond the resources capability of the State Governments.

Under AIBP, funds were provided to the States in the form of Centre loan assistance. The Central Government also introduced in February

2002 a Fast Track Programme for completion of some irrigation works in two or three working seasons with full Central modified, with the adoption of a grant component - i.e. 30 per cent grant for General Category States and 90 per cent grant for Special Category States.

Between 1997 and 2004, the Government of India has released Rs. 14,670 crore as loan assistance under AIBP for 181 irrigation works and 3,810 minor irrigation works. So far 28 projects have been completed and irrigation potential of 2.2 million hectares has been created up to March 2003.

Under Accelerated Irrigation Benefit Programme (AIBP), the State Governments were provided Rs. 24,867.4 crore as Central Loan Assistance (CLA)/grant for 229 major/medium irrigation projects and 6,205 surface minor irrigation Schemes up to January 29, 2008. So far 91 major/ medium and 4,605 surface MI Schemes have been completed. In the current year, as on January 29, 2008, Rs. 3,127.5 crore has been released for AIBP.

To cover a larger area under irrigation, the Government sanctioned a National Project for Repair, Renovation and Restoration of Water Bodies directly linked to agriculture, in January 2005 with an estimated cost of Rs. 300 crore to be shared by the Centre and States in 3:1 ration. The water bodies having cultivated command areas of more than one ha and up to 2,000 ha were included under the pilot scheme in one or two districts in each State. The Scheme was approved for 26 districts in 15 States. Central share of Rs. 179.3 crore has been released to the States till November 30, 2007, covering 1,098 water bodies. The physical work for restoration has been completed for 733 water bodies and the work is in progress in the remaining 365 water bodies.

Following the pilot scheme, restoration of water bodies has also been taken up in States having considerable number of water bodies with the World Bank assistance. The World Bank Loan Agreement has been signed with Tamil Nadu for Rs. 2,182 crore to restore 5,763 water bodies having a cultivated command area (CCA) of 4 lakh ha. The Rs. 835 crore Andhra Pradesh Community Based Tank Management Project was signed with the World Bank in June 2007 for restoration of 3,000 water bodies with a CCA of 2.5 lakh ha. The project of Karnataka was signed for Rs. 259 crore with World Bank in November 2007 for restoration of 1,225 water bodies involving a CCA of 0.52 lakh ha. The proposals from Orissa and West Bengal Governments have also been submitted to the World Bank.

Irrigation in the 11th Plan

The 11th Plan envisages creation of an additional potential of 16 million hectares at an estimated required outlay of about Rs. 2,10,000 crore. Since irrigation is a State subject, most of this (about Rs. 172,000 crore) has been earmarked for financing by States and an analysis of State's own preliminary 11th Plan allocations shows that this might actually be exceeded. Further, although financial resources appear adequate except in some poorer States, guidelines for the Accelerated Irrigation Benefits Programme (AIBP) have already been changed to expand its scope and to increase the Central share for selected areas.

The Government has also been creating irrigation potential through public funding and assisting farmers to create potential has been created through major, medium and minor irrigation schemes. The total irrigation potential in the country has increased from 81.1 million ha in 1991-92 to 102.8 million ha in 2006-07. The potential created so far is estimated to be 73.5 per cent of the ultimate irrigation potential (Table 1). Of the total potential created, however, only 87.2 million ha (84.9 per cent) is actually utilised.

Table 1. Irrigation Potential Created and Utilised 2002-2007 (Million hectares)

Source	Potential created	Potential utilised	Per cent utilisation
Major and Medium Irrigation	42.4	34.4	81.3
Minor Irrigation	60.4	52.8	87.4
Total Irrigation	102.8	87.2	84.9

Source : Ministry of Finance, Economic Survey 2007-08

Irrigation is one of the six components for the development of rural infrastructure under the Bharat Nirman and aims at creating the irrigation potential of 10 million ha by 2008-09. The target under Bharat Nirman is to be met largely through the completion of ongoing major and medium irrigated projects/schemes. During 2005-06 and 2006-07, 1.68 million ha and 1.94 million ha of irrigation potential, respectively, is reported to have been created.

Of the 182.7 million hectares of land used for cultivation, only about 62 million hectares is currently irrigated; the rest depend entirely on monsoon rains. Hence, from the agriculture sector's point of view enlarging the cropped area under assured irrigation is critical for the economy. India's agricultural sector currently uses about 90 per cent of total water resources. Irrigated agriculture has been fundamental to economic development, but unfortunately caused groundwater depletion. India draws 80 per cent of its irrigation water from groundwater. As water scarcity becomes a bigger and bigger problem, rural and farming areas will most likely be hit the hardest. India receives an average of 4,000 billion cubic meters of rainfall every year. Unfortunately, only 48 per cent of rainfall ends up in India's rivers. Due to lack of storage and crumbling infrastructure, only 18 per cent can be utilized. Rainfall is confined to the monsoon season mainly from June to September, when India gets, on an average, 75 per cent of its total annual precipitation. Once again, due to India's storage crunch the government is unable to store surplus water for the dry season. The latest official figures show that more than 60 per cent of India's 62 million irrigated hectares is fed by groundwater. According to two independent studies which used satellite data from GRACE (the Gravity Recovery and Climate Experiment) indicate that northern India was losing more groundwater than anywhere else in the world except for the Arctic ice sheets. One of the studies put the annual net groundwater loss in Punjab, Haryana and Rajasthan at 109 cubic km, which is roughly equivalent to 109 billion tonnes. The water table has dropped dramatically in many areas and this is one of the main problems Indian agriculture face today.

In order to augment the depleting ground water resources, it is essential that the surplus monsoon runoff that flows into the sea is conserved and recharged to augment ground water resources. Ground water storage that could be feasible has been estimated as 214 billion cubic meters (BCM) of which 160 BCM is considered retrievable. Central Ground Water Board has prepared a conceptual plan for artificial recharge to ground water for the country. Out of total geographical area of 3,28,7263 sq. km. of the country, an area of 4,48,760 sq. km. has been identified suitable for artificial recharge. The total quantity of surplus monsoon runoff to be recharged works out as 36.4 BCM.

Rainwater harvesting is of course a focus area for sustainable agricultural production but attention should also be paid for efficient management of the available water resources. Even if 5 per cent of annual rainfall were harvested properly, that will produce a substantial quantum of water to the tune of 900 million litres. There are two main practices of rainwater harvesting. In the traditional method, storage of rainwater is done on the surface for future use in the form of underground tanks, ponds, check dams, weirs etc. Other method is a new concept of rainwater harvesting where different type of structures like pits, trenches, dug wells, hand pumps, recharge wells, recharge shafts, lateral shafts with bore wells and spreading techniques are generally used. Rainwater harvesting saves water, improves soil moisture, improves the quality of water, allows drought proofing, prevents flooding, saves energy required to lift water, reduces soil erosion and above all it is an ideal solution of water problem in areas having inadequate water resources.

Several measures for development and management of water resources are undertaken by the respective State Governments with a view to optimally utilize the available resources which include creation of storages, restoration of water bodies, rain water harvesting, artificial recharge to ground water and adoption of better management practices etc. Storage capacity of about 225 billion Cubic Metre (BCM) has been created so far. The total estimated storage capacity of the various projects under construction is about 64 BCM as per present assessment. Further, the State Governments have identified various other schemes for investigation and planning and the estimated storage for such schemes is about 108 BCM.

The Central Government has implemented a scheme since 2008 in which 4,455 million wells will be dug up for recharging groundwater in 110 blocks of the identified states for the benefit of farmers. The scheme has been implemented before monsoon of 2008. The Artificial Recharge of Ground Water Through wells scheme was launched keeping in view the concerns of over exploitation of groundwater resources in the country as well as to ensure sustainable water resource management and irrigation facilities in the affected rural areas. About 80 per cent of these groundwater stressed areas over exploited, critical and semi critical areas are located in hard rock areas in the states of Andhra Pradesh, Gujarat, Madhya Pradesh, Maharashtra, Rajasthan and Tamil Nadu where rapid decline of groundwater levels have been observed on long term basis. It is a state sector scheme and will be implemented by the respective state government in association with the Panchayati Raj Institutions (PRIs), the Central Groundwater Board, the NABARD and NGOS. The farmer who has dug well in their agricultural land in the identified areas is beneficiary for the scheme.

The Artificial Recharge of Ground Water through wells scheme was launched keeping in view the concerns of over exploitation of groundwater resources in the country as well as to ensure sustainable water resource management and irrigation facilities in the affected areas. The scheme approved for a total cost of Rs. 1,798.71 crores with 100 per cent subsidy to marginal and small farmers and 50 per cent subsidy to other farmers. The Central Ground Water Board has prepared a Manual and subsequently a Guide on Artificial Recharge to Ground Water which provide guidelines on investigation techniques for selection of sites, planning & design of artificial recharge structures, economic evaluation & Monitoring of recharge facility. These are of immense use to States/U.T.s in planning and implementation of recharge schemes for augmentation of ground water in various parts of the country.

The Union Ministry of Water Resources constituted a standing sub committee for assessment of availability and requirement of water. They have put forward a multi pronged approach highlighting the need for completion of storage dams, interlinking of rivers, recycling of domestic water and of industrial used water, desalination of sea water and artificial recharge of ground water. A committee on Artificial Recharge of Groundwater has also been constituted. Availability of irrigation water is an important factor in increasing the food grain production but availability of water is decreasing each day due to overuse of underground water. The Central Ground Water Board has identified 1,065 assessment blocks as over exploited or critical. The strategy proposed for water recharge is to divert rain water into dug wells. Indian Council of Agricultural Research (ICAR) has set up one teaching cum demonstration model of water harvesting in each of 32 selected State Agricultural Universities and ICAR institutes to train 100 trainers and 1000 farmers every year.

Irrigation accounts for an enormous portion of the World's water use. Thus, more efficient utilisation of even a small amount of irrigation water would free large amounts of water for other uses. Experts recommend that we should ensure that our old water recharge systems are sustained and enhanced, new recharge systems are developed, harvest water where it falls, regulate groundwater use and massively promote conservation methods like drip irrigation and rice intensification. In Australia, water consumption in agriculture has been reduced by 30 per cent in the past 20 years by good agricultural practices. There is need to encourage modern methods of irrigation like drip irrigation. Micro management is also needed for efficient use of the irrigation water. Farmers should avoid one time flood irrigation in their fields and instead should give more number of irrigation at intervals for higher output. Weeds should be managed effectively in annual and perennial crops. Moisture conservation techniques like mulching should also be used in crops. Waste water of kitchen and toilet can also be recycled and channelled for farming or charging of the underground water resources. Irrigation efficiency can also be improved by computer controlled systems that supervise soil moisture and provide water only when necessary. Farmers can switch to more water efficient, drought resistant and salt tolerant crop varieties. Moreover, farmers can use organic farming techniques which result in higher crop yields per hectare and require only one fourth of the water and commercial fertiliser of conventional farming. Better farming techniques, such as leaving crop residue on fields and ground cover on drainage ways, intercropping, use of mulches, and low volume or drip irrigation, could reduce these water losses dramatically. Thus, there is need to efficiently conserve the available water resources and use them judiciously with modern technological tools.

Chapter - 8

Creating New Irrigation Potential to Boost Agriculture

Dr. Yashbir Singh Shivay & Dr. Anshu Rahal

Efficient use of water resource is basic to survival of the ever increasing population of a country, this is especially very crucial for India, where we are having less than 5% of the world's water resources and more than 18% world's population. Irrigation is one of the most important inputs required at different critical stages of plant growth of various crops for optimum production. The Government of India has taken up augmentation of irrigation potential through public funding and is assisting farmers to create potential on their own farms. Substantial irrigation potential has been created through major and medium irrigation schemes. In arid and semi-arid climatic conditions, the timing and amount of rainfall are not adequate to meet the moisture requirement of crops. Therefore, supplementary irrigation is essential to raise the crops, necessary to meet the needs of food and fiber for the growing population. Scientific irrigation water management provides the best insurance against weather induced fluctuations. This is the only way in which we can make our agriculture profitable and sustainable in the coming decades. The on farming irrigation management for different crops including efficient use of poor quality waters is an essential component of water management in irrigation command areas. It is felt necessary to include complete information on surface and sub surface drainage. The information pertaining to net sown area, gross sown area, net irrigated area and grass irrigated area over the period of time in India is given in The Table 1.

Water resources

Precipitation is the main source of renewable water supply. India receives a total annual rainfall of 400 km^3 with a variation of 100 mm per annum in western most regions to 1100 mm per year in eastern region. The water resources of the country, which occur as natural runoff in the rivers, are estimated to be 18,697 km^3 considering both ground water and surface water contributions. Due to various constraints of topography, distribution over time and space, 1,122 km³ water can be put to beneficial use, 690 km³ of which is due to surface water resources.

Irrigation potential

The total ultimate potential was earlier estimated at 113.8 million hectares, which has now been revised to 140 million hectares. The share of major and medium schemes that are surface water based is 58.5 million hectares, whereas that of minor schemes, based on surface water is 17.4 million hectares. The ground water based minor irrigation schemes are expected to contribute 64 million hectares compared with the earlier estimates of 40 million hectares.

Table 1. Cropped and irrigated areas from 1951 to 2008-09 in India

Year	Area (million hectare)			
	Net sown area	Gross sown area	Net irrigated area	Gross irrigated area
1950-51	118.75	131.89	20.85	22.56
1960-61	132.20	152.82	24.66	27.98
1970-71	140.27	165.79	31.10	38.19
1980-81	141.93	176.75	40.50	51.41
1990-91	141.98	182.24	49.87	65.68
2000-01	141.10	187.94	54.68	75.14
2005-06	141.46	193.32	60.79	84.26
2008-09	141.36	195.10	63.20	88.42

Command area development

There has been large scale irrigation development but there was short fall in utilization of the potential created. To focus attention on efficient utilization of the created resources, a multi disciplinary agency, the command area development authority was constituted in 1974-75. The command are development programme broadly covers :

(a) On farm development works;

(b) Introduction of rotational system of water distribution within the out let common (warabandi);

(c) Adoption of suitable cropping pattern roistering of irrigation system;

(d) Development of ground water for conjunctive use;

(e) Arrangement and supply of agricultural inputs and services including short term credit;

(f) Development of necessary infrastructure in the shape of roads, markets and ware housing/cold storages.

Command area development programme has been implemented in more than 100 irrigation projects with good results.

Basin-wise water resources

Monsoonal climate causes a highly skewed distribution of the resource availability and calls for its conservation in soil profile, aquifers, ponds, lakes, reservoirs and rivers for use during the lean period.

India is a very fortunate country to have many rivers whose total catchment's area is estimated to be 252.8 million hectares. Central water commission, Government of India, has divided the whole country in 20 river basins comprising 12 major basins, each having catchment's area exceeding 20,000 km^2 and 8 composite river basins combining suitably together ll the other remaining medium and small river systems.

Major basins are : i) Indus; ii) Ganga-Brahamputra Meghna; iii) Godavari; iv) Krishana; v) Cauvery; Mahanadi; vi) Pennar; vii) Brahmani-Baitarni; viii) Sabarmati; ix) Mahi; x) Narmada and xi) Tapi.

Likewise 8 composite river basins are : i) Subarnarekha-combining subarnarekha and other small rivers between Subarnarekha and Baitarni; ii) East flowing rivers between Mahanadi and Pennar; iii) East flowing rivers between Pennar and Kanyakumari; iv) Area of inland drainage in Rajasthan desert; v) West flowing rivers from Tapi to Tadri; vi) West flowing rivers from Tadri to Kanyakumari; Minor rivers draining into Myanmar (Burma) and Bangladesh. Catchments area and water resources potential of these basins is given in Table 2.

Table 2. Basin wise water potential in India

S.N.	River basin	Catchment area (km^2)	Water resources (km^2/year)	
			As per Central Water Commission 1993	As per NCIWRDP 1999
1.	Indus	321,289	73.31	73.31
	Ganga-Brahamputra-Meghna			
2.	Basin Ganga	862,769	525.02	525.02
3.	Brahamputra sub-basin	197,316	537.24	629.05
4.	Meghna (Barak) sub-basin	41,157	48.36	48.36
5.	Subarnarekha	29,196	12.37	12.37
6.	Brahmani-Baitarni	51,822	28.48	28.48
7.	Mahanadi	141,589	66.88	66.88

8.	Godavari	312,812	110.54	110.54
9.	Krishna	258,948	78.12	69.81
10.	Pennar	55,213	6.32	6.32
11.	Cauvery	87,900	21.36	21.36
12.	Tapi	65,145	14.88	14.88
13.	Narmada	98,796	45.64	45.64
14.	Mahi	34,842	11.02	11.02
15.	Sabarmati	21,674	3.81	3.81
16.	West flowing rivers of Kutch and Saurashtra including Luni	334,390	15.10	15.1
17.	West flowing rivers south of Tapi	113,057	200.94	200.94
18.	East flowing rivers between Mahanadi and Godavari	49,570	17.08	17.08
19.	East flowing rivers between Godavari and Krishna	12,289	1.81	1.81
20.	East flowing rivers between Krishna and Pennar	24,649	3.63	3.63
21.	East flowing rivers between Pennar and Cauvery	64,751	9.98	9.98
22.	East flowing rivers south of Cauvery	35,026	6.48	6.48
23.	Area of North Ladakh not draining into Indus	28,478	0	0.
24.	Rivers draining into Bangladesh	10,031	8.57	8.57
25.	Rivers draining into Myanmar	26,278	22.43	22.43
26.	Drainage areas of Andman, Nicobar & Lakshadweep Islands	8,280	0	0
27.	Total	3,287,260	1,869.37	1952.87
28.	Approximately Say	-	1870	1953

Total surface water resources of the country (yearly average streams flow) are about 1,869 km^2. Due to uneven distribution of rainfall, both spatial and temporal, only 37% (690 km^3) of the surface renewable water resources are estimated to be potentially utilizable. This low proportion is primarily due to low potentially utilizable water resources in the Meghna Brahamputra river basins. The Brahamputra River covers only 7.6% of the geographical area, accounting of 31% of the total renewable water resources.

According to Central Water Commission, potential utilizable for India are 690 km^3 of the surface water and 432 km^3 of the ground water (total 1,122 km^3 or BCM).

Water availability

India's total renewable per caput water resources in 1985 were 2,011 m^3 which stands at 1,820 m^3 on 31 March 2001 (Table 1). The total renewable water resources per person, excluding those people in the Brahamputra river basin, are only 1,345 m^3. About 103 million people living in the basins of Sabarmati, Kutch and Saurashtra and basin between Pennar and Kanyamkumari have per caput total renewable water resources below 500 m^3. More than 90 million people living in the 4 basins of Pennar, Tapi, Cauvery and east flowing rivers between Mahanadi and Pennar have per caput total renewable water resources below 1,000 m^3. The per caput renewable water resources in Indus, Ganga, Krishna and Subarnarekha basins are less than 1,700 m^3.

Table 3. Per capita water availability in India

Year	Population (million)	Per capita water availability (cubic metre)
1951	361	5,177
1955	395	4,732
1991	846	2,209
2001	1,027	1,820
2025	1,394 (projected)	1,341
2050	1,640 (projected)	1,140

Water demand

Country's water demand is dominated by irrigation needs. The total water demand for agriculture, domestic and industrial sectors of India in 1995 was estimated to be about 650 km^3; of this about 90% is withdrawn for agriculture sector. Different demand scenarios up to year 2050 AD have been estimated by the Ministry of Water Resources and the demands as projected for different years are given in Table 4. There will be a large gap even at the aggregate level between the water availability and requirement by 2050 AD, when the population of the country is expected to stabilize. The entire replinshipable ground water is also assumed to be utilized by 2025 AD. Therefore, to bridge the gap between water availability and requirement, inter basin water transfers by interlinking of rivers may be a viable alternative, which would also take care of the requirement of the water short areas, including drought prone areas.

Table 4. Estimates of future water demands (BCM) under different scenarios

Category	Year 2010			Year 2025			Year 2050		
	Low	**Medium**	**High**	**Low**	**Medium**	**High**	**Low**	**Medium**	**High**
Irrigation	489.0	536.0	556	619	688	734	830	1,008	1,191
Domestic	39.4	41.6	61	47	52	78	59	67	104
Industrial	37.0	37.0	37	61	67	79	69	81	116
Total	555.4	614.6	654	727	807	881	958	1,156	1,411

Irrigation demand

Irrigation demand of a region depends upon the areas irrigated with surface water and ground water, different crop water requirements and irrigation application efficiency. Irrigation demand may be worked out as given in Table 5.

Table 5. Irrigation with surface and ground water

		Surface irrigation area	Ground water irrigation area
Irrigation demands	Crop water requirements	+	
		Surface irrigation efficiency	Ground water irrigation efficiency

The irrigation demand ranges from 193 m^3 per person in Brahamputra basin to 1,617 m^3 per person in Indus basin. Irrigation efficiencies range from a low of 31% whereas most of the area is surface irrigated to a high of 62% (where most of the area is irrigated from ground water). Irrigation is the largest sector of water demand and irrigated agriculture shall further be called upon to produce a sizable portion of the food grains requirements for the growing population. Considering the rapid changes in the dietary habits and standard of living of the Indian population, it may be difficult to make correct estimates for future food grain requirements have been given in Table 6.

Table 6. Future food grains and water demands for irrigation under different scenarios

Year	Low demand scenario (Mt)	Water requirement (BCM)	Medium demand scenario (Mt)	Water requirement (BCM)	High demand scenario (Mt)	Water demand (BCM)
2010	249	489	265	536	271	556
2025	322	619	349	686	365	734
2050	469	830	539	1,008	605	1,191

Irrigation

Irrigation is the artificial application of water to partially meet the crop evapo-transpiration requirements. It is essential for sustaining crop productivity in many regions of the country mainly because the rainfall is inadequate and unevenly distributed to meet crop water demands. Irrigation water is a costly and scare input, and it is becoming more difficult to increase the area under irrigation to meet the demand for food, fodder and fiber for growing human and livestock population. The competing demands of water for other uses viz. urbanization and industrialization are also restricting the availability of water for crop production. Therefore, it is essential to optimize the use of water according to availability on sustainable basis in the decline water table areas, and to allow minimum loss of water by efficient water management techniques in areas where water table is continuously rising.

Irrigation requirements of some important crops

Irrigation requirement at the field level refers to the amount of water, exclusive or precipitation, required to mature the crops (Table 7). It is usually expressed in depth at the given time. It thus, includes the amount of water needed to meet the losses through evaporation and transpiration, both occurring simultaneously and hence termed evapo-transpiration (ET), application losses and the special needs. It does not include transit losses.

Table 7. Irrigation requirements of some important crops

S.No.	Crops (Kharif)	Crop duration (days)	Irrigation requirement (mm)
1.	Rice	130-140	700-800
2.	Sorghum	110	150
3.	Maize	110	150
4.	Sugarcane	330	700
5.	Summer pulses	70	210
6.	Pearl millet	90	150
7.	Sunflower	110	210
8.	Cotton	180	280
9.	Wheat	135	350
10.	Berseem	200	800
11.	Potato	110	450
12.	Lentil	135	150
13.	Mustard	150	150

14.	Barley	125	210
15.	Soybean	90	350
16.	Chickpea	150	150

Types of Irrigation

Various types of irrigation techniques differ in how the water obtained from the source is distributed within the field. In general, the goal is to supply the entire field uniformly with water, so that each plant has the amount of water it needs, neither too much nor too little.

1. Surface Irrigation

In surface irrigation systems water moves over and across the land by simple gravity flow in order to wet it and to infiltrate into the soil. Surface irrigation can be subdivided into furrow, border strip or basin irrigation. It is often called flood irrigation when the irrigation results in flooding or near flooding of the cultivated land. Historically, this has been the most common method of irrigating agricultural land.

Where water levels from the irrigation source permit, the levels are controlled by dikes, usually plugged by soil. This is often seen in terraced rice fields (rice paddies), where the method is used to flood or control the level of water in each distinct field. In some cases, the water is pumped, or lifted by human or animal power to the level of the land.

2. Localized Irrigation

Localized irrigation is a system where water is distributed under low pressure through a piped network, in a pre determined pattern, and applied as a small discharge to each plant or adjacent to it. Drip irrigation, spray or micro-sprinkler irrigation and bubbler irrigation belong to this category of irrigation methods.

3. Drip Irrigation

Drip irrigation, also known as trickle irrigation, functions as its name suggests. Water is delivered at or near the root zone of plants, drop by drop. This method can be the most water efficient method of irrigation, if managed properly, since evaporation and runoff are minimized. In modern agriculture, drip irrigation is often combined with plastic mulch, further reducing evaporation, and is also the means of delivery of fertilizer. The process is known as fertigation.

Deep percolation, where water moves below the root zone, can occur if a drip system is operated for too long of a duration or if the delivery rate is too high. Drip irrigation methods range from very high tech and computerized to low tech and labor intensive. Lower water pressures are usually needed than for most other types of systems, with the exception of low energy center pivot systems and surface irrigation systems, and the system can be designed for uniformity throughout a field or for precise water delivery to individual plants in a landscape containing a mix of plant species. Although it is difficult to regulate pressure on steep slopes, pressure compensating emitters are available, so the field does not have to be level. High tech solutions involve precisely calibrated emitters located along lines of tubing that extend from a computerized set of valves.

4. Sprinkler Irrigation

In sprinkler or overhead irrigation, water is piped to one or more central locations within the field and distributed by overhead high pressure sprinklers or guns. A system utilizing sprinklers, sprays, or guns mounted overhead on permanently installed risers is often referred to as a solid set irrigation system. Higher pressure sprinklers that rotate are called rotors and are driven by a ball drive, gear drive, or impact mechanism. Rotors can be designed to rotate in a full or partial circle.

5. Center Pivot Irrigation

Center pivot irrigation is a form of sprinkler irrigation consisting of several segments of pipe (usually galvanized steel or aluminum) joined together and supported by trusses, mounted on wheeled towers with sprinklers positioned along its length. The system moves in a circular pattern and is fed with water from the pivot point at the center of the arc. These systems are common in parts of the United States where terrain is flat.

6. Lateral move (side roll, wheel line) Irrigation

A series of pipes, each with a wheel of about 1.5 m diameter permanently affixed to its midpoint and sprinklers along its length, are coupled together at one edge of a field. Water is supplied at one end using a large hose. After sufficient water has been applied, the hose is removed and the remaining assembly rotated either by hand or with a purpose-built mechanism, so that the sprinklers move 10 m across the field. The hose is reconnected. The process is repeated until the opposite edge of the field is reached. This system is less expensive to install than a center pivot, but much more labor intensive to operate, and it is limited in the amount of water it can carry. Most systems utilize 4 or 5-inch (130 mm) diameter aluminum pipe.

Sub-Irrigation

Sub irrigation also sometimes called seepage irrigation has been used for many years in field crops in areas with high water tables. It is a method of artificially raising the water table to allow the soil to be moistened from below the plants root zone. Often those systems are located on permanent grasslands in lowlands or river valleys and combined with drainage infrastructure. A system of pumping stations, canals, weirs and gates allows it to increase or decrease the water level in a network of ditches and thereby control the water table.

Manual Irrigation using buckets or watering cans

These systems have low requirements for infrastructure and technical equipment but need high labor inputs. Irrigation using watering cans is to be found for example in peri-urban agriculture around large cities in some African countries.

Automatic, non-electric irrigation using bucklets and ropes

Besides the common manual watering by bucket, an automated, natural version of this also exist. Using plain polyester ropes combined with a prepared ground mixture can be used to water plants from a vessel filled with water. the ground mixture would need to be made depending on the plant itself, yet would mostly consist of black potting soil, vermiculite and perlite. This system would (with certain crops) allow you to save expenses as it does not consume any electricity and only little water (unlike sprinklers, water timers). However, it may only be used with certain crops (probably mostly larger crops that do not need a humid environment; perhaps e.g. paprika's).

Irrigation using stones to catch water from humid air

In countries where at night, humid air sweeps the countryside, stones are used to catch water from the humid air by condensation. This is for example practiced in the vineyards at Lanzarote.

Dry terraces for irrigation and water distribution

In subtropical countries as Mali and Senegal, a special type of terracing (without flood irrigation or intent to flatten farming ground) is used. Here, a stairs is made through the use of ground level differences which helps to decrease water evaporation and also distributes the water to all patches (sort of irrigation).

Sources of Irrigation Water

Sources of irrigation water can be groundwater extracted from springs or by using wells, surface water withdrawn from rivers, lakes or reservoirs or non conventional sources like treated wastewater, desalinated water or drainage water. a special form of irrigation using surface water is spate irrigation, also called floodwater harvesting. In case of a flood (spate) water is diverted to normally dry river beds (wadi's) using a network of dams, gates and channels and spread over large areas. The moisture stored in the soil will be used thereafter to grow crops. Spate irrigation areas are in particular located in semi-arid or arid, mountainous regions. While floodwater harvesting belongs to the accepted irrigation methods, rainwater harvesting is usually not considered as a form of irrigation. Rainwater harvesting is the collection of runoff water from roofs or unused land and the concentration of this water on cultivated land. Therefore, this method is considered as a water concentration method.

Problems in Irrigation

1. Competition for surface water rights.
2. Depletion of underground aquifers.
3. Ground subsidence
4. Under irrigation gives poor salinity control which leads to increased soil salinity with consequent builds up of toxic salts on soil surface in areas with high evaporation. This requires either leaching to remove these salts and a method of drainage to carry the salts away or use of mulch to minimize evaporation.
5. Over irrigation because of poor distribution uniformity or management wastes water, chemicals, and may lead to water pollution.
6. Deep drainage (from over irrigation) may result in rising water tables which in some instances will lead to problems of irrigation salinity.
7. Irrigation with saline or high sodium water may damage soil structure.

The central government initiated the Accelerated Irrigation Benefit Programme (AIBP) from 1996-97 for extending assistance for the completion of incomplete irrigation schemes. Under this programme, projects approved by the Planning Commission are eligible for assistance. Under the AIBP, Rs. 50,380.64 crore of central loan assistance (CLA)/ grant has been released up to 30 November 2011. As on 31 March 2011, 290 projects were covered under the AIBP and 134 completed. During 2010-11, an irrigation potential of 566.24 thousand ha is reported to have been created by states, from major/medium/minor irrigation projects under the AIBP, while the higher irrigation potential would help augment production and productivity, assured remuneration from such production is vital for development of agriculture. The government has also been creating irrigation potential through public funding and assisting farmers to create potential on their own farms. Substantial irrigation potential has been created through major, medium and minor irrigation schemes. The total irrigation potential in the country has increased from 81.1 million ha in 1991-92 to 102.8 million ha in 2006-07. The potential created so far is estimated to be 73.5 per cent of the ultimate irrigation potential (Table 8). Of the total potential created, however, only 87.2 million ha (84.9 per cent) is actually utilized.

Table 8. Irrigation potential created and utilized (million ha)

	1991-92	1992-97	1997-02	2002-07	Annual rate of growth (%)		
		Eighth Plan	Ninth Plan	Tenth Plan	Eighth Plan	Ninth Plan	Tenth Plan
Cumulative potential created (million ha)							
Major and medium	30.7	33.0	37.1	42.4	1.4	2.4	2.7
Minor	50.4	53.3	56.9	60.4	1.1	1.3	1.2
Total	81.1	86.3	94.0	102.8	1.2	1.7	1.8
Cumulative potential utilized (million ha)							
Major and Medium	26.3	28.4	31.0	34.4	1.6	1.7	2.1
Minor	46.5	48.8	50.0	52.8	0.9	0.5	1.1
Total	72.9	77.2	81.0	87.2	1.2	1.0	1.5
Per cent utilization							
Major and Medium	85.6	86.3	83.7	81.3	-	-	-
Minor	92.4	91.5	87.9	87.4	-	-	-
Total	89.8	89.5	86.2	84.9	-	-	-

The pace of creation of additional irrigation potential came down sharply from an average of about 3 per cent per annum during 1950-51 - 1980-90 to 1.2 per cent, 1.7 per cent and 1.8 per cent per annum, respectively, during the Eighth, Ninth and Tenth five year Plan periods. The rate of growth of utilization of the potential created declined to 1 per cent per annum during the Ninth Five Year Plan period and improved to 1.5 per cent per annum during the Tenth Five Year Plan Period. The average annual rate of utilization remained lower than the average annual addition to the irrigation potential resulting in the cumulative utilization witnessing continuous erosion. This not only amounts to an inefficient use of funds, but also a forgone income from irrigated lands.

Responding to the continuous decline in the rate of creation in irrigation potential, the Central Government initiated the Accelerated Irrigation Benefit Programme (AIBP) from 1996-97 for extending assistance in completion of irrigation schemes which had remained incomplete (Table 9). Under this programme, the projects approved by the Planning Commission were eligible for assistance. Further, the assistance, which was entirely a loan from the Centre in the beginning, was modified with inclusion of a grant component from 2004-05. The AIBP guidelines were further modified in December 2006 to provide for 90 per cent of the project cost as grant to special category States, DPAP/Tribal areas and KBK (Koraput, Bolangir and Kalahandi) districts of Orissa.

Table 9. Performance of AIBP Projects (Rs. Crore and area thousand ha)

Year	Central loan assistant/grant released	Total potential created under AIBP	Total potential created
1996-97	500.0	74.5	560.0
1997-98	952.2	182.0	645.2
1998-99	1,119.2	259.0	592.2
1999-2000	1,450.5	223.2	666.0
2000-2001	1,856.2	528.8	983.5
2001-02	2.602.0	442.8	1,214.6
2002-03	3,061.7	456.0	812.0
2003-04	3,128.5	447.0	1,004.0
2004-05	2,867.3	496.0	1,000.0[a]
2005-06	1,900.3	600.0	1,500.0[a]
2006-07[a]	2.302.0	932.0	-
Total	21,739.9	4,641.3	-

[a]Estimated

Under AIBP, the State Governments were provided Rs. 24,867.4 crore as CLA/grant for 229 major/medium Irrigation projects and 6,205 surface Minor Irrigation (MI) Schemes up to January 29, 2008. So far 91 major/medium and 4,605 surface MI schemes have been completed. In the current year, as on January 29, 2008, Rs. 3,127.5 crore has been released for AIBP.

To cover a larger area under irrigation, the Government sanctioned a National Project for Repair, Renovation and Restoration of Water Bodies directly linked to agriculture, in January 2005 with an estimated cost of Rs. 300 crore to be shared by the Centre and States in 3 :1 ratio. The water bodies having cultivated command area of more than one ha and up to 2,000 ha were included under the pilot scheme in one or two districts in each State. The Scheme was approved for 26 districts in 15 states. Central share of Rs. 179.3 crore has been released to the States till November 30,2007, covering 1,098 water bodies. The physical work for restoration has been completed for 733 water bodies and the work is in progress in the remaining 365 water bodies.

Following the pilot scheme, restoration of water bodies has also been taken up in States having considerable number of water bodies with the World Bank assistance. The World Bank Loan Agreement has been signed with Tamil Nadu for Rs. 2,182 crore to restore 5,763 water bodies having a cultivated command area (CCA) of 4 lakh ha. The Rs. 835 crore Andhra Pradesh Community Based Tank Management Project was signed with the World Bank in June 2007 for restoration of 3,000 water bodies with a CCA of 2.5 lakh ha. The project of Karnataka was signed for Rs. 259 crore with the World Bank in November 2007 for restoration of 1,225 water bodies involving a CCA of 0.52 lakh ha. The proposals from Orissa and West Bengal Governments have also been submitted to the World Bank.

Irrigation is one of the six components for the development of rural infrastructure under the Bharat Nirman and aims at creating the more irrigation potential in the coming time to mitigate the drought situations whenever it is being experienced and increased agricultural production to feed the burgeoning population in India.

Chapter - 9

Turning over Irrigation Management : Prospects and Constraints

Mamata Swain

In recent years proper utilisation of natural resources like land and water has become a matter of grave concern for policy makers and planners (Vaidyanathan, 1994). To meet the swelling demand for food due to growing population and in the face of scarcity of cultivable land, huge public investments are made on irrigation sector which is land augmenting in nature. It needs no emphasis that irrigation is a critical and crucial input required for agricultural production inasmuch as availability of irrigation water enables the use of other yield enhancing inputs like High Yielding Varieties of seeds, chemical fertiliser, organic manure and above all adoption of improved agronomic practices. Irrigation increases crop production by increasing crop yield, cropping intensity and making possible cultivation of high value and remunerative crops.

However, mere provision of irrigation facility to land does not ensure enhanced agricultural production. The productivity impact of irrigation is critically dependent on the way water is applied and utilised. The quality of irrigation service in terms of adequacy, timeliness, equity, dependability, predictability and convenience in its supply remarkably determines the yield from irrigation commands. For obtaining optimum yield water should be provided in time and in adequate quantity according to the crop water requirement of plants at its various growth stages, as water stress affects crop yield adversely. Irrespective of the location and size of the farm, water should be allocated equitably among the headenders and tailenders and also large farmers and small farmers. For planning the cropping pattern, farmers need to know in advance the timing and quantity of water supply. Water should be provided in a dependable, predictable and convenient manner without any uncertainty in its quantum of delivery and timing. Quality in irrigation service should be maintained over time and should be stable and sustainable in the long run.

In India in the post independence period, a major thrust has been given on creation of irrigation infrastructure by constructing big dams, reservoirs and canal networks. Though use of irrigation along with HYV seed and chemical fertiliser has considerably increased agricultural productivity and helped India in attaining self sufficiency in food production, overall performance of irrigation sector is considered sub optimal, inefficient and inequitable falling far short of expectations. Irrigation canals have remained exclusively managed by hierarchical government departments in a top down approach.

ACTIVE PARTICIPATION OF WATER USER

Irrigation has been considered as a technical enterprise aiming at construction of hardware like dams, reservoirs, weirs, barrages, canals etc. without scant regard to the software component or the socio-economic aspects and management part of the system. The role of the water user who is indeed the main protagonist in irrigation has not been pondered in their real dimensions and potential. The role of the farmer has been traditionally viewed as more passive than active in decision making in irrigation. His contribution to the overall success of irrigation in achieving substantial increase in food production has been traditionally underestimated. New trends advocate much more active participation of the water user in irrigation water management.

In a bureaucratically managed canal irrigation system it is extremely difficult on the part of irrigation technocrats to provide irrigation service catering to the need of numerous marginal and small farmers having fragmented land holdings scattered over extensive irrigation commands. Due, to rigid procedures followed in a bureaucratically managed system which lacks flexibility, the irrigation engineers are not in a position to take immediate decisions on the basis of observation of field conditions. Also, they lack knowledge and information on local resources and socio-economic conditions.

Due to highly subsidised water rates the irrigation sector is incurring use financial losses and is solely dependent on budgetary allocations. In the face of resource crunch and the inability of the irrigation sector to raise resources internally, it is becoming increasingly difficult to operate and maintain the extensive surface canal systems. In the absence of proper maintenance and repair, the conditions of irrigation structures have deteriorated. Irrigation departments do not have the resources to renovate, rehabilitate and modernise the existing physical structures. Also, a major chunk of operation and maintenance expenses is meant for meeting establishment costs which mostly include salary bills of the employees. A very own proportion is spent on works component. On the one hand the irrigation agency lacks the resources to repair the derelict irrigation systems, on the other the water users consider the can I structures as state property and its repair and upkeep is st te's responsibility. As a result, the operation and maintenance to the created physical structures have suffered causing system inefficiency. Water users also oppose any attempt to increase water rates as irrigation service is not satisfactory. Thus a improve the quality of irrigation service due to paucity of finance; and farmers are not ready to pay higher water rates as irrigation service is poor.

PARTICIPATORY MANAGEMENT

As an escape from this impasse the government is introducing far reaching irrigation sector reforms by bringing about institutional and organisational changes (Subramanian, 1997; Mitra, 1996). Of late, most of the state governments in India have taken policy decision to introduce Participatory Irrigation Management (PIM) and turning over the management of tertiary segment of the canals like minor/sub-minors/ distributaries to Water Users Associations (WUAs). It is contemplated that WUAs will be entrusted with the responsibilities of operation and maintenance of the tertiary units, distribution of water among water users and collection of water rates. The irrigation agency will make bulk sale of water volumetrically to WUA at minor/sub minor level and retailing of water to farmers will be the responsibility of WUA.

National Water Policy of India adopted in 1987 clearly envisages that Farmers should be involved progressively in various aspects of management of irrigation system, particularly in water distribution and collection of water rates. Recognising the need for systematic involvement and participation of farmers in irrigation management, the Government of Orissa in the state water policy, 1994 has incorporated the objective of handing over of operation and maintenance of irrigation systems to the users in due course. Under the World Bank assisted Water Resources Consolidation Project under implementation in Orissa attempts are under way to motivate farmers to form Water Users Association, so that the operation and maintenance of the downstream part of the system can be turned over to WUAs (Swain and Kar, 2000). The major functions of the WUAs will be to operate and maintain the distributory/minor canals, to ensure equitable water distribution among the WUA members, to collect irrigation fees and to advise the department on main system operation. To start with under the mandale of World Bank few pilot projects had been undertaken by Department of Water Resources in Rushikulya (Distributory, No. 11), Ghodahada, Derjang and Aunli irrigation commands to implement the System Improvement and Turnover (SIFT) programme. Since 1996 Government of Orissa is taking steps to form 723 WUAs or Pani Panchayats in 33 major and medium irrigation projects covering 3,32,000 ha. Of these 161 Pani Panchayats covering 74,000 ha. in 14 major and medium irrigation projects have already been registered and the operation and maintenance of the canals have been turned over to water users. The work of motivation and formation of Pani Panchayats have been entrusted to Water and Land Management Institute, Command Area Development Agency, Nabakrushna Choudhury Centre for Development Studies and some NGOs.

An attempt is made here to analyse the evolution of concept of Participatory Irrigation Management in context of India and its Ideological basis. The benefits of PIM to farmers as well as irrigation agency have been outlined. The possible difficulties that will be encountered in implementing irrigation system turnover to WUA in specific socio-economic context of Orissa are highlighted. Lastly a new approach for implementing PIM is indicated.

EVOLUTION OF CONCERN OF SYSTEM TURNOVER

In India the concept of PIM has evolved gradually through three distinct phases (Maloney and Raju; 1994 : 16). In 1980s the concept was in its nascent stage limiting to farmers participation through their representatives. It was felt then that in the decision making process of irrigation development, the views of farmers should be taken into account and they should be consulted in planning, design, construction, operation and maintenance of the system. However, mere farmers representation in scheme level committees could not yield much result. In the latter part of 1980s it was realised that farmers cannot have much stake in irrigation management without a formal structure/forum to express their views. Therefore, the catchword become Farmer's Organisation. It was recognised that for effective functioning the organisation needs to have the following four characteristics :

- Spontaneous support of the farmers under local leadership.
- Structured role to play.
- Authority to make and implement decisions at appropriate levels.
- Responsibility to raise its funds and manage its affairs.

In various states like Andhra Pradesh, Tamil Nadu and Maharashtra thousands of outlet associations/chak committees had been formed only in pen and paper but actually most of them became dysfunctional after a short period. By 1990s it became apparent that the concept of farmers organisation is not sufficient. Therefore, a radical concept of Farmer Management and System Turnover has evolved in which it is envisaged to entrust the WUAs with the responsibility of operation and maintenance of minor/distributory, allocation of water among farmers and collection of water charges from water users.

IDEOLOGICAL BASIS OF IRRIGATION SYSTEM TURNOVER

The concept of system turnover to water users is grounded in laudable ideologies like Democratisiation, Decentralisation. It is now widely recognised that the farmers who are the end users of irrigation water should participate in its management starting from planning, design, and construction to operation and maintenance of the system. Farmers have sufficient and accurate knowledge on their local resources like land and water. Existing social capital which includes local knowledge, skill, community network and kinship ties should be utilised in the management of irrigation systems. As the irrigation service is meant for the farmers and farm production, their views should be given due importance in the management of irrigation. The water users should have their say in the decision making process. It is difficult on the part of irrigation agency to look into the individual problems of numerous farmers and supply irrigation service catering to their specific needs. Therefore, it is proposed to delegate some work responsibility to Water Users' Association and management of the downstream part of the system should be turned over to water users. This attempt to decentralise irrigation management and empowerment of water users will undoubtdly bring about improvement in irrigation service by substantially reducing transaction costs of getting accurate information, negotiation and enforcement of contract cost (Baland and Platteau, 1996). The irrigation executives can devote their time for effective management of main canal system and other technical matters in which they have competence and comparative advantage.

In the era of liberalisation, delicensing and decontrol there is a growing realisation that the unnecessary bureaucratic control in management of irrigation system at tertiary levels should be reduced to improve irrigation efficiency and to check corruption and rent seeking behaviour. Usually the farmers believe that the canals belong to the government and they are the beneficiaries of the system. They do not have any role and responsibility in upkeep of the physical structures. In the changed institutional context irrigation will be considered as a common pool resource and will be managed by the farmers community (Sengupta, 1991; Singh, 1994) and its maintenance and sustainability will be the responsibility of the WUAs. The WUAs will be required to raise funds for the purpose. As public canals will be managed by WUAs, it is sometimes termed as privatisation as in Egypt. The WUA can raise funds by selling water, taking up commercial activities and from membership fees, share capital and subsidy.

BENEFITS OF FARMERS MANAGEMENT

Benefits that will accrue to farmers and irrigation agency due to farmers participation in irrigation management have been recorded by several authors (Singh, 1991; Maloney and Raju, 1994) which are mentioned below.

Benefits to Farmers

- Better water management at the tertiary level;
- Farmers flexibility in use of water, choice of crops and land use;
- More crop and more income;
- Optimal use of water in agriculture;
- Ensuring equity in water allocation;
- Helping resolve disputes on water distributions;
- Encouraging community responsibility for the management of assets;
- Forum for facilitating effective communication between farmers and government departments;
- Better collection of irrigation fees;
- More economic use of water, less wastage;
- Better maintenance reduction in the cost of irrigation;
- Encouraging collective management of agricultural input supply and marketing of agricultural produce;
- Less waterlogging because of careful use of water;
- Less opportunity for corruption; and
- Better mutual trust and understanding between farmers and officers.

Benefits to Irrigation Agency

- Improved relations with client farmers; less mistrust;
- Reduced complaints on inequitable distribution of water between head reach and tail end;
- Increase in irrigation efficiency, job satisfaction;
- More time to attend to technical matters and using their expertise;
- Less bothering about unauthorised outlets, obstructions and maintenance problems;
- Improvement of credibility of the irrigation agency and irrigation officers; and
- Better collection of water rates and saving on maintenance cost.

PROBLEMS IN ORGANISING FARMERS

Though the irrigation agency in Orissa has taken a policy decision to encourage farmers participation and attempts are under way to motivate farmers to form WUAs, the farmers response in this regard is not up to the satisfaction. The success of any organisation depends on seven fundamental factors as follows :

(i) Felt needs;

(ii) Common interest;

(iii) Collective effort;

(iv) Effective leadership;

(v) Bureaucratic commitment of the agency involved;

(vi) Political will of the party in power;

(vii) Financial viability; and

(viii) Legal support.

To remove the inequity and inefficiency in irrigation water use, the farmers should feel the necessity of forming an organisation or association as a remedial measure. It is also important that majority of water users should strive collectively for the formation of WUA. For motivating and organising the farmers there should be capable and popular leaders. The leader should be acceptable to the water users and trust worthy. The irrigation officials should have bureaucratic commitment to the cause of organising farmers to form WUA and to decentralise and delegate power to them. The political party in power should provide all out support for implementing the programme successfully. The WUA should be financially viable by raising its own resources from different sources. There should be necessary amendment of irrigation rules and acts to incorporate the role of WUAs in irrigation management.

Farmers will come forward to form WUAs and will be ready to take up the additional responsibility, if they are convinced that the benefits due to participatory management will exceed their costs of participation. As most of our farmers are not educated and lack vision to comprehend the future benefits due to participation, special care should be taken while motivating the farmers. They are to be convinced that the benefits due to participation will be substantial, tangible, quick yielding and sustainable. Otherwise the farmers will not evince interest in a programme introduced and implemented through a government agency. Though water is the most crucial input required for plant growth, the productivity impact of irrigation depends on use of other yield enhancing complementary inputs like HYV seeds, fertiliser, manure and modern agronomic practices. Therefore, other agricultural inputs should be made available to the farmers in time and as per requirement through WUAs.

It is also argued that in the specific socio-economic context of a caste and hierarchical society whether such collective effort will be forthcoming. Our society is a heterogeneous and hierarchical society having several caste and class cleavages. There are also political differences. The farmers having different political affiliations may have conflicts in interest and differences of opinion. It is a fact that it takes a long time to change peoples attitude and to create awareness about the importance of forming association for better water management. One must have the patience and perseverance to achieve the objective. A learning by doing approach should be followed to determine the model and modalities of forming WUA.

As a matter of fact in agriculture co-operative efforts are necessary, as there are several externality effects. If farmers at head reach use excess water, tailenders face water scarcity; if there is pest attack in one's field, the neighbouring field is affected. Therefore, a rational economic response is to internalise such externalities by making coalition or forming an association.

The WUAs need to have reliable sources of income, otherwise they can hardly be sustainable. Different categories of income to WUAs are :

- Sale of water;
- Sale/leading out common property resources;
- Fees and contributions;
- Commercial operations;
- Maintenance Contracts; and
- Subsidies.

THE RIGHT APPROACH

While introducing an institutional change in the management of an infrastructure, which is a vital ingredient for agricultural production and a common property resource, a careful and cautions approach should be followed. Therefore, few suggestions are made for propagating and implementing the idea of WUAs and system turnover.

(i) Select few projects in good operating conditions and preferably small projects having moderate water scarcity in the commands;

(ii) Undertake a feasibility study; Examine the castel, class conflict, groupism, political differences and history of confrontation and conflict if any;

(iii) Examine whether any indigenous community irrigation practices are prevalent in the area and exploit the possibility of using existing social capital. If it is seen that organisation of farmers is possible then a well designed strategy should be adopted;

(iv) A clear picture of government's intention and outline of the programme should be given to the farmers;

(v) As it is a new concept needing enough experimentation and experience before finalisation of its content and constituents in greater detail, the irrigation agency is not in a position to spent out the different components of the programme in concrete terms; the farmers should be informed accordingly. Otherwise frequent changes in the provisions will give a confusing picture to the farmers and they will lose confidence in the irrigation authority;

(vi) There should not be any haste and rigid time bound and target achievement objective. Such stipulations dilute the real content and intention of implementing the programme;

(vii) A detailed action plan should be prepared in consultation with the water users through Participatory Rural Appraisal method;

(ix) After getting confidence and sufficient experience a model by law, Memorandum of Understanding and necessary amendment in Irrigation Act should be made.

REFERENCES

Baland, Jean-Marie and J.P. Platteau (1996). Halting Degradation of Natural Resources : Is there a role for Rural Communities? Oxford : FAO and Clarendon.

Gooneratne, W. and S. Hirashima (1990). Irrigation and Water Management in Asia, New Delhi, Sterling.

Govt. of India, Ministry of Water Resources (1987). National Water Policy.

Govt. of Orissa (1994). State Water Policy of Orissa.

Maloney C. and K.V. Raju (1994). Managing Irrigation Together : Practice and Policy in India, Sage : New Delhi.

Mitra, A.K. (1992). Joint Management of Irrigation Systems in India - Relevance of Japanese Experience, Economic and Political Weekly, Vol. 27, No. 26, June 27, pp. A-75-A.82.

Mitra, A.K. (1996). Irrigation Sector Reforms - Issues and Approaches', Economic and Political Weekly, Vol. 31, No. 13, March 30, pp. A - 31-37.

Sengupta, N. (1991). Managing Common Property : Irrigation in India and the Philippines, Sage : New Delhi.

Singh, K. (1994). Managing Common Pool Resources : Principles and Case Studies, Delhi : OUP.

Singh, K.K. (1991) (ed.), Farmers in the Management of Irrigation Systems, New Delhi : Sterling.

Subramanian, A. *et a.,* (1997) (eds.), User Organizations for Sustainable Water Services, World Bank Technical Paper No. 354.

Swain Mamata and D.K. Das (1999). Emerging Trends and reforms in Irrigation in India : A Perspective of Orissa, New Delhi : MD.

Swain Mamata and G.C. Kar (2000) (eds.), Farmer and Local Participation in Irrigation Management, New Delhi : Commonwealth.

Vaidhyanathan, A. (1994). Better Resource Management for Poverty Alleviation, Economic and Political Weekly, Vol. 31, No. 23, pp. 1397-1404.

Chapter - 10

People's Participation in Irrigation Management

Jaya Chatterji

The past decade has witnessed a considerable amount of interest and action in the irrigation sector around the country. There have been widespread discussions, field level demonstrations have been undertaken, an institution has been set up specifically to further the cause and legislative action has been undertaken in a number of States.

There is all round agreement that efficiency, equity and sustainability will be best served if the irrigation system is managed by the users. A number of States have also passed Acts that promote management of irrigation systems by the water users.

The issue at present is the strategy to operationalise the Acts. How is management to be transferred? What will be the role of different agencies? Is there a common understanding of the changed roles and more importantly, do the different bodies have the requisite capacities to undertake these roles effectively?

Much of these strategies are based on what has been demonstrated or planned for in projects supported by India Canada Environment Facility. These projects are in Tamil Nadu and in Madhya Pradesh and this paper will therefore rely of the experiences from these two projects.

OVERVIEW OF THE IRRIGATION SECTOR

In India, about 80% of the accessible fresh water is used for agriculture. However, some of this is often used for household use, for fish production and for drinking water for livestock.

With rising population and increased industrial activity, there is a rising competition for scarce water and pressure to use water more efficiency and productivity.

The total cultivable area of the country is 142 mha. The irrigation potential, by the end of 2000 was about 90 mha. At the time of Independence, it was 22 mha. India has the second largest irrigated area in the world after China.

Under the Five Year Plans, huge investments were made in the construction of large irrigation projects. The thrust on the high water consuming high yielding varieties of the 1960s gave an impetus to this strategy. Since the Government constructed and owned the irrigation infrastructure, it inevitably took over the management responsibilities also. Farmers were completely excluded from the operations, maintenance and any other management decisions. For a very small cost, the farmers received irrigation water that raised the productivity of their land.

Over time, with the government failing to set aside sufficient funds for repairs and maintenance and the limited funds being managed by the centralized irrigation bureaucracy, the irrigation infrastructure went into a downward spiral of disuse and neglect. Year after year the agitation of farmers for timely supply of water and for repairs of the system become stronger. Larger farmers and those with nexus with the irrigation department managed to access more than their share of water. An assessment by the Government of India in 1995 indicated that the irrigation sector faced problems related to poor management, corruption, deteriorating physical infrastructure, lack of adequate financing, social inequity in access to water and lack of incentive to use water optimally. The extremely low cost of water encouraged wastage which in turn deprived those at the tail end of the system.

PROBLEMS WITH THE IRRIGATION SECTOR IN INDIA

The focus of the irrigation departments has been on construction and little or no maintenance work has been undertaken. This has led to ever increasing gaps in irrigation potential created and actual utilization.

Madhya Pradesh, for example, has a total irrigable area of 6.72 MHA and so far irrigated potential of 1.97 Mha has been created. However, actual utilization is only 0.98 Mha or about one half of the potential created. This includes irrigation schemes constructed after independence as well as tanks constructed during the Chandela reign in medieval period of history (about 900-1000 years ago).

Maintenance has been deferred for a long time and in some cases, as in the Chandela tanks, restoration has never been attempted.

Increasing levels of corruption has led to water being appropriated by the better off, more powerful farmers. The culture of might is right prevails. There is increasing level of conflict among farmers and between farmers and the irrigation department.

Tampering with the system is rampant with farmers breaking canals to access water in quantities beyond their legitimate share.

Lack of accountability of the irrigation department officials has led to poor water management, and to unreliable delivery of water both in quantum and in timeliness. The tailenders rarely get water and when they do, it is not timely.

Water rates for irrigation is extremely low and recoveries are poor, resulting in insufficient revenues to meet the operations and maintenance costs. In fact, currently, establishment cost of the Water Resources Department (WRD) in Madhya Pradesh exceeds the revenue from water charges by a large margin. Thus, whatever maintenance is attempted is met out of State Government's own budgetary resources. The precarious fiscal condition of the State Governments has resulted in the annual budgets of the departments being used more or less entirely for meeting administrative costs. The entire system is slowly collapsing.

There is no incentive for the farmers to use water judiciously. No extension advise is provided to the farmers.

The net result has been a decline in net irrigated area, low cropping intensities and low farm yields.

Changes Required

An organization of farmers with legal status must be transferred the rights to manage the irrigation system.

Such an organization should function in a democratic and economically efficient manner. The needs of the tailenders and small landholders should be taken into consideration. The requisite capacities will have to be built to take on the new responsibilities. A sense of ownership of the system and its management has to be instilled.

Such an organization should aim at optimum management of water.

The irrigation Department must reorient itself to deal on an equal footing, with a vibrant and dynamic farmers organization. At present, there is resistance from the staff to the change in role from an executing agency to that of a facilitator.

The irrigation infrastructure should be repaired and rehabilitated.

Potential for Bringing about the Required Changes

There are various positive elements that can facilitate the required change in Madhya Pradesh. At the same time, there are also sone negative elements. These are listed below :

Positive Elements

There is political will and support for participatory irrigation management and for handing over of the management rights to the WUAs. This had led to rapid progress in forming the WUAs, conducting elections and in imparting training to the office bearers.

There is a core group of senior officials in the Irrigation Department who are convinced about the need for this transfer. This group is responsive to the demands of the newly formed WUAs.

The officers who are to participate in the initiative to hand over management have been chosen with care. Thus, 120 officers have been chosen from about 5000, based on their performance track record.

A massive awareness raising programme has been initiated amonst the WUA members.

The Weaknesses

Initially it is the big, powerful and more articulate farmers who are seen to come forward in the WUAs. A concerted effort has to be made to ensure that the tailenders and small farmers find a voice in the decision making bodies. This will be a critical element in determining whether any real change takes place after the transfer to the farmers organization - the WUAs - or whether the net result is that the Irrigation Department officials are displaced by the bigger farmers.

After four years, the process of participation in Madhya Pradesh has not reached member farmers in an adequate way and is largely limited to the presidents of the WUAs. Decisions are not broad based and responsibilities are concentrated in the hands of very few individuals. There continues to be a lack of sense of ownership over the system, amongst the general farmers. If this continues, the efficiency of the system will fail to improve and the possibility of overcoming the weaknesses of the centralized management will be low.

No role whatsoever has been envisaged for women in the management.

The junior hierarchy of the irrigation department has not awakened to the new role for handover. Attitudinal changes and acquisition of participatory skills is required.

The take over of the management of the irrigation system is largely a government led initiative so far. It needs to be transformed into a people led movement.

How the Changes can be Brought About

Organization of Farmers

A number of States have passed legislation promoting transfer of management to farmers organizations. These States include Andhra Pradesh, Madhya Pradesh, Rajasthan, Karnataka, Tamil Nadu and Orissa. Most of them have three tiers - the Water Users Association [WUA] at the unit farmers level, Distributory Committee consisting of a federation of WUAs and Project Committee consisting of a federation of Distributory Committees. There are, in some instances, a State level Committee consisting of Distributory Committees.

Most of these States, notably Andhra Pradesh and Madhya Pradesh [MP], have conducted election to the Executive Committees of the various tiers. MP has proceeded to impart training to the Executive Committee members of the 1470 WUAs that have been formed across the State. The training has focused on the legislation and the rights and responsibilities of the WUAs. No training or reorientation has been planned or considered necessary for the functionaries of the Irrigation Department.

Capacities of the WUA Members

The WUAs are expected to have organizational, management, financial and social capacities to successfully manage the irrigation system. The WUAs have to function democratically and be sensitive to the needs of the weaker members.

The approach so far has been to entrust the function of building capacities to the Irrigation Department. However, the officials of the Department themselves lack the skills they are expected to transfer to the members of the WUAs. Hence in Madhya Pradesh, although the willingness was there, the ability was lacking and no progress could be made on this.

The strategy followed in the project supported by India Canada Environment Facility is to provide continuous capacity building support through capable and experienced non government organizations (NGOs). This is to be supplemented by training on specialized aspects, undertaken by experienced institutions or individuals.

The NGOs would train and the members of the WUAs and the Irrigation Department staff would learn, by doing. This would be followed by the NGOs supervising and providing corrective measures. Lastly, the NGOs would gradually withdraw and the WUAs and the Irrigation Department would take over.

With back - stopping services provided by the NGOs, the WUAs would be persuaded to meet regularly. All information especially those relating to financial matters and physical repairs progress, would be shared publicity and regularly. Once the system is established, it will be difficult to discontinue it.

Participation in the functioning of the WUAs will be broad based by assigning responsibilities to groups of people at the minor level and to individuals within that group. In exposure visits, in training, in representations, in short, in every activity, it will be ensured that sincere and capable persons are involved and not necessarily the office bearers.

All major decisions will be taken by general bodies and through open participation. For example, the question of where repairs would be undertaken and how would be decided through a ``walk through' of the system and an open discussion. The need to make an equitable distribution of water will be repeatedly emphasized.

It has been observed that farmers who can access the water appropriate as much as they can, beyond what is required by the crops, to insure against non availability at a later stage. To ensure that the farmers are ready to consider more judicious use is : (i) to announce the total availability of water for the season, at the beginning of the cropping season, (ii) ensure a cropping pattern in tune with the availability of water (iii) promote water conserving technologies that lead to higher production, and (iv) promote the use of low cost equipment that will assist the farmers to monitor soil moisture water.

The above interventions is a necessary condition that will release water that could be used to provide water to the tailenders and small farmers.

Another important requirement is to raise the cost of using water. Water charges need to be linked to the quantum used, rather than be levied on an acreage basis. If the WUAs are entrusted with the full responsibilities of maintaining and operation the system, they may find it viable to raise the water charges to at least the physical and managerial cost of providing it. Since passing the legislation in 1999, the government of Madhya Pradesh has raised the water rates by more than three times.

CONCLUSION

Till recently, the irrigation Department has strictly retained management in its hands. The mounting conflicts in distribution of water and the rapidly deteriorating condition of the system have been party responsible for the desire to hand over the management to the water users. It is partly governed by political expediency.

However, whatever the motivation, the trend towards transfer of management has started. To ensure that the transfer is real and not one of substituting the Irrigation Department officials by the bigger farmers, concerted efforts of the right kind has to e undertaken. There are challenges in this that will require a committed senior bureaucracy to overcome.

Chapter - 11

Irrigation System in the Context of PIM in India

Dr. Ramu & Dr. P. Jayashree

No living being, including plants can survive without water. Yet, many that have grown up in cities tend to take water for granted. Water is an elixir of life on earth. It is vital for the existence of all life forms and for all activities of human beings. Water resources have been the most exploited natural system since man strode the earth as colossus from the dark ages. It is supplied for purposes of drinking, irrigation and industrial uses. One critical problem confronting mankind today is how to manage the intensifying competition for water between expanding urban centres, traditional agricultural activities and in stream water uses dictated by environmental concerns.

In the agricultural sector, the prospects of increasing the gross cultivated area are limited by the dwindling number of economically attractive sites for large scale attractive sites for large scale irrigation and drainage projects. Over the last century, the global population has tripled, and water consumption has increased threefold (UNESCO, 2005). Water use in India is no exception to this general trend. The main cause of the increase are growing population and rising food demand. In an agrarian economy like India, the importance of water for agricultural productivity hardly needs any emphasis. India faces the dauting task of increasing its food production by over 50 per cent in the next two decades, and reaching towards the goal of sustainable agriculture requires a crucial role of water (Kumar 1998).

The concept of Participatory Irrigation Management (PIM) is closely linked to the concept of Irrigation Management Transfer (IMT). Indeed, IMT is a subset of PIM. IMT can be defined as the transfer of responsibility and authority for irrigation system management from government agencies to water users associations, or other private sector entities. This is a broad and rather vague definition. IMT may include transfer of decision-making authority (or governance). It may include transfer of ownership of scheme infrastructure (which is normally considered privatisation). It may include transfer of water rights from government to water users associations. Or it may only include turning over to water users a part of the management responsibilities, such as water delivery, canal maintenance and fixing the water fees, while final approval of operation and maintenance (O & M) plans and budgets are subject to government approval (FAO, 1999).

PIM may include the reordering of control over (claims to) water, the redefinition of boundaries and domains of governance and the construction of new entities (users, WUAs, etc.). There is however a large variety in the number of functions that can be transferred, the degree of transfer of the different functions, and the organisational setup aimed at after transfer. This is due to the fact that participation in irrigation management by water users can take a wide variety of forms. Farmers can be involved in various system management functions including planning, design, operation, maintenance, rehabilitation, resource mobilization, and conflict resolution. Moreover, they can be involved in these functions at various system levels; from the field channel to the entire system.

Almost all irrigation systems show some involvement of water users in the system management. When people speak of introducing PIM, they are usually referring to a change in the level, mode, or intensity of user participation that would increase farmer responsibility and authority in management processes (Svendsen *et al.,* 2000).

Since 1985 Ministry of Water Resources has been inspiring farmers participation in water distribution and management of tertiary system in the projects covered under the Centrally Sponsored Command Area Development Programme.

Recognising the need to provide legal backup to PIM in the country, Ministry of Water Resources commissioned an NGO, `Society for Peoples' Participation in Ecosystem Management (SOPPECOM); Pune to suggest suitable amendments in the existing irrigation acts which could be recommended to States for incorporation in their State Irrigation Acts.

SOPPECOM has been in the forefront of work relating to PIM and has successfully pioneered many action research programmes on formation of WUAs. The suggestions of SOPPECOM were circulated to States during June 1998. Conferences at National, State and Project levels have been organized for creating awareness on Participatory Irrigation Management amongst farmers and officials. Ministry of Water Resources has been organising National level training programmes on PIM in various parts of the country for CAD functionaries. In addition, matching grant is also being provided to States for organizing State and project level training programmes for farmers and field functionaries.

In this paper, the aim is primarily to explore the irrigation and drainage system in the context of participatory irrigation management (PIM) in India with trends in irrigation and drainage.

Irrigation Scenario in India

In the last forty years, the share of Indian agriculture in gross domestic product has decreased, but extensive use of HYV seeds, modern irrigation tech, and fertilizer have contributed in increasing the agricultural productivity and achieving self sufficiency in meeting food demand. Given increasing trend of population, policy makers find it imperative for India to achieve higher agriculture production and continue to meet the food security objective of the country; and indicated that irrigation will play a key role in future in achieving higher yield and sustaining the food security (Persaud *et al.,* 2003). During the last fifty years, gross irrigated area (GIA) of India has increased more than three folds from 22 to 76 million Hectares. Gross irrigated area is a straightforward multiplicative function of net irrigated area (NIA) and irrigation intensity (IRI), and thus the relevant equation which may arise is regarding the contribution of net irrigated area (NIA) relative to the irrigation intensity (IRI) in increasing the GIA.

There are state wise variations in irrigation; the level of irrigation is measured in terms of irrigation intensity and irrigation ratio, defined as NIA/NSA. Latest research shows high proportion of irrigated land of more than 70% in agricultural states like Punjab, Haryana and Uttar Pradesh where agriculture constitutes more than 30% of the state GDP. Among the southern states, proportion of irrigated land is below 30% in Karnataka and Kerala; while in Andhra Pradesh and Tamil Nadu, NIA/NSA is above 40%. Among the western states, Maharashtra has the lowest proportion of irrigated land where only 17% of the net cropped area is irrigated. Most of the eastern states are well endowed with irrigation where average NIA/NSA is 0.40. In the north-eastern state of Assam; however, less than 10% of net cropped area is irrigated. Many climatic factors like rainfall, drought affects irrigation.

Many states register a decrease in irrigation intensity (IRI), and much of the decrease are noticed in states like Himachal Pradesh, Orissa, Tamil Nadu and Maharashtra. In these states, the proportional irrigated area is not high except in Tamil Nadu. The opportunity cost of increasing the irrigation intensity is higher than increasing the net irrigated area. As a result, NIA has increased in these states with the development of minor irrigation. In West Bengal, however IRI has increased by 39% in the post 1997 period. One may argue that the higher opportunity cost of increasing the extensive margin leads to higher irrigation intensity. The alternative hypothesis is that high endowment of irrigation land increases the reliability of irrigation water and induces higher irrigation intensity. Hypothesis testing is done using regression. Increase in irrigated area could be the cause of higher irrigation intensity. We also hypothesize whether endowment of irrigated area is factor in the marginal effect on irrigation intensity.

Groundwater Irrigation

Groundwater irrigation in India developed during the period of green revolution and contributed much in increasing the gross irrigated area of the country. In the last five decades, groundwater irrigation has increased from 5 million hectares to 35 million hectares. The proportional area of groundwater to the net irrigated area has increased by 25% alone during the period 2000-2005. In the past, surface water irrigation played a significant role in increasing the net irrigated area. However, from mid sixties, the proportion of surface water to net irrigated area has decreased and in the last decade alone it has decreased largely by 23%. This is largely due to incompletion of planned irrigation projects and poor maintenance of the existing surface irrigation infrastructure. State wise there is wide variation in the source of net irrigated area.

In the northern and western states, the proportion of groundwater irrigation to net irrigated area is more than 70% in the post period 2000 while the share of groundwater and surface water is equal in the southern zone during the same period. Groundwater expansion growth is higher in the states like Assam and West Bengal. Assam has abundant and untapped ground water potential and the state Government has prioritised massive irrigation facility with the concept of Participatory irrigation Management.

Groundwater expansion growth rate is insignificant in Bihar while in Karnataka a decline in the proportion of groundwater irrigated area is observed NIA has increased by 10% the decline in groundwater proportion causing a declined in groundwater level. Groundwater accounts for 67% of the net irrigated area. Groundwater expansion has been growing at an exceptional rate in the recent decades. More reliable water delivery and declining extraction costs due to advances in technology and, in many instances, government subsidies for power and pump installation have encourages private investment in tube wells.

Groundwater irrigation, due to its lesser variation in its supply and higher the reliability in irrigated water supply, reduces the risk of investment in labour, seed, fertilizers, pesticides and other inputs and induces farmers to increase the irrigation intensity. Some states has experienced fast decline in groundwater level which leads to lower productivity of water and cause a decrease in irrigation intensity. Irrigation intensity is lower among southern states where the groundwater depletion problem is severe.

Mechanization

The mechanization involves judicious application of inputs by using agricultural machinery/equipment e.g. hand tools, bullock drawn equipment, power driven machines including the prime movers for performing various operations required for crop production activities. The mechanization ensures reduction of drudgery associated with various farm operations as also economize the utilization of inputs and thereby harmessing the potential of available resources. The adoption of mechanization is linked with endowment of irrigation.

The states with adequate irrigation facilities, the mechanization have progressed at faster rate in comparison to States that have scant irrigation facilities and dependent on monsoon. Irrigation reduces the farmers risk in investment on land and thus encourages adopting mechanization to increase production. Mechanization is a key factor, which can help in increasing the cropping intensity under irrigated conditions. The hypothesis is that higher proportional irrigated area encourages farmers to adopt mechanization, which facilitates in increasing the irrigation intensity.

Objectives and Necessity of PIM

1. To create a sense of ownership of water resources and the irrigation system among the users, so as to promote economy in water use and preservation of the system.
2. To improve service deliveries through better operation and maintenance.
3. To achieve optimum utilization of available resources through sophisticated deliveries, precisely as per crop needs.
4. To achieve equity in water distribution.
5. To increase production per unit of water, where water is scarce and to increase production per unit of land where water is adequate.
6. To make best use of natural precipitation and ground water in conjunction with flow irrigation for increasing irrigation and cropping intensity.
7. To facilitate the users to have a choice of crops, cropping sequence, timing of water supply, period of supply and also frequency of supply, depending on soils, climate and other infrastructure facilities available in the commands such as roads, markets cold storages, etc. so as to maximize the incomes and returns.
8. To encourage collective and community responsibility on the farmers to collect water charges and payment to Irrigation Agency.
9. To create healthy atmosphere between the Irrigation Agency personnel and the users.

Provision in National Water Policy (2002) & PIM Acts

Following modifications were made in the National Water Policy (2002) regarding the participatory approach to water resources management :

``Management of the water resources for diverse uses should incorporate a participatory approach by involving not only the various governmental agencies but also the users' and other stakeholders, in an effective and decisive manner, in various aspects of planning, design, development and management of the water resources schemes. Necessary legal and institutional changes should be made at various levels for the purpose, duly ensuring appropriate role for women. Water Users Association and local bodies such as municipalities and Gram Panchayats should particularly be involved in the operation, maintenance and management of water infrastructures/facilities at appropriate levels progressively, at appropriate levels progressively, with a view to eventually transfer the management of such facilities to the user groups/local bodies.

Recognising the need for sound legal framework for PIM in the country, the Ministry brought out a model act to be adopted by the State Legislatures for enacting new irrigation acts/amending the existing irrigation acts for facilitating PIM. In accordance with the model act eight State Governments, namely, Andhra Pradesh, Goa, Madhya Pradesh, Karnataka; Orissa, Rajasthan, Tamil Nadu and Kerala have enacted new acts. The legal framework provides for creation of farmers organisations at different levels of irrigation system as under :

(a) Water Users Association (WUA) : will have a delineated command area on a hydraulic basis, which shall be administratively viable. Generally a WUA would cover a group of outlets or a minor.

(b) Distributaries Committees : will comprise of 5 or more WUAs. All the presidents of WUAs will comprise general body of the distributaries committee.

(c) Project Committee : will be an apex committee of an irrigation system and presidents of the Distributaries committees in the project area shall constitute general body of this committee.

The Associations at different levels are expected to be actively involved in: 1. Maintenance of irrigation system in their area of operation; 2. Distribution of irrigation water to the beneficiary farmers as per the warabandi schedule; 3. Assisting the irrigation department in the preparation of water demand and collection of water charges; 4. Resolve disputes among the members and WUA; monitoring flow of water in the irrigation system etc.

Constraints In Implementation of PIM (Issues)

Lack of legal back up and policy changes : In many states, there is no or very little legal back up and clearcut policy decision at the Government level to take up PIM, which is a big impediment in implemention of PIM. For the actual irrigation management transfer and operation of PIM in an irrigation project, policy changes and legal back up are essential. This is important for distributing required quantity of water at minor/Distributory take off points, taking up correction of system deficiency, claim to get the maintenance funds proportionate to its portion transferred to associations, collection of water charges and retaining some portion of it for WUAs functioning, fixation of water rates, incentives to farmers, resolution of conflicts etc. Clarity on legislation is also required in certain States.

System deficiency : In older projects, there are many problems like deterioration of old control and measuring structures, leakages and seepage at various places, erosion of banks and beds, siltation and weed infestation. These are serious problems, hindering farmers to take over the system management on technical and financial considerations.

Uncertainty of water availability : This is another important aspect, as farmers will understandably be reluctant to take on the responsibility for managing the system unless deliveries of water are made reliable, flexible, practical and responsive to need. The engineers on their part may not be confident about ensuring supply of the requisite quantity of water to be WUAs, as would be obligatory in terms of the MOU signed between Irrigation Agency and WUA.

Fear of financial viability: Maintenance and operation of the system demands huge finances. Farmers have got the apprehension that in absence of surety of finance, it would be difficult for them to fulfill the requirement of funds for operation and maintenance. They feel that when Government is not able to handle the system with huge money available with them, how farmers would be able to do justice?

Lack of technical knowledge: Apart from the financial uncertainty, lack of technical input is one of the inhibiting factors to take over the system. When Government, having such qualified and senior Engineers, finds it difficult to manage the system, how untrained and uneducated farmers would be able to take up such a highly technical operation and maintenance work of big irrigation systems.

Lack of leadership : On account of limited exposure of the farmers to the rest of the world and PIM in particular, potent leadership is lacking, rather on account of limiting knowledge. At times so called local leaders give the negative or unclear version before other farmers which further create misunderstanding among the farmers bringing them sometimes into a fix.

Lack of publicity and training : Seeing believes; and knowledge brings confidence in people. This aspect is lacking and there is a constraint to adoption of PIM.

Demographic diversity : Due to variation in economic, ethnic, education levels etc. diversity of farmers, PIM is taking much time in this country. To handle this aspect deep study, analysis and solution need be found out.

Mega irrigation projects : World scenario gives an indication that there are smaller projects in the countries of the world, where irrigation project transfer has taken care for PIM. In India, there are huge projects having very large distribution system and cultivable common area sometimes more than 20 lakh hectares. Larger the project, complex would be its maintenance, operation and management aspects and so the formation and functioning of farmers associations for different necessary activities.

WUAs v/s Panchayats : In many of the areas, where WUAs have been formed, there is a clash of interest among Panchayats and WUAs on who is to own the system, particularly when watershed schemes are being handed over to the Panchayats. PIM in efficient systems : Some of the northern States have raised apprehensions that when their systems are runnin very efficiently, why not PIM should form an integral part of the system of distribution already in operation.

Future Prospects of Pim

It has now been realised that without active participation of beneficiaries, the irrigation systems cannot be managed efficiently. The experience shows that wherever farmers have been actively engaged, the overall management of irrigation system and the water use efficiency have significantly improved. The legal framework, which has been established in various States, will ensure systematic involvement of beneficiaries in the management of irrigation system at various levels. There has to be however, a provision for adequate financial support to these organizations to carry out their responsibilities. The PIM acts of various States do have provisions for the financial management of these associations.

For example acts of Andhra Pradesh and Madhya Pradesh Statesmention that the funds of the farmers organizations shall comprise of the following :

(i) Grants and commission received from the State Government as a share of the water tax collected in the area of operation of the farmers organization.

(ii) Such other funds as may be granted by the state government and central Government for the development of the area of operation.

(iii) Resources rose from any financing agency for undertaking any economic development activities in its area of operation.

(iv) Income from the properties and assets attached to the irrigation system.

(v) Fees collected by the farmers organization for the services rendered in better management of the irrigation system.

(vi) Amounts received from any other sources; and investment of private sector in distribution and ancillary/extension services.

REFERENCES

Basu, Kaushik (2000); ``Prelude to Political Economy : A study of the social and political foundations of Economics". UP, Oxford.

Biswas, A.K. (1996). Water for the developing world in the 21st century. Issues and implications. ICID Journal, Vol. 45, no. 2.

De Wrachien, D. (2001). Irrigation and drainage. Trends and challenges for the 21st century. Proceedings of the 19th European Regional Conference on Sustainable Use of Land and Water. June, Brno, Czech Republic.

Frederick, K.D., D.C. Major and E.Z. Stakhiv (1997). Water resources planning principles and evaluation criteria for climate change. In : Climatic Change. Frederick K.D. (ed.), Kluwer Academic Publishers, Wageningen. The Netherlands.

Johnson, S., Svendsen M. & Gonzalez F. (2002). ``Options for institutional Reform in the irrigation Sector. Discussion Paper prepared for the international Seminar on Participatory irrigation Management, 221-27 April 2002. Beijing.

Kanse, S. (2001). Land information System based on GIS in Adihalli-Myllanhalli Watershed. Project Report. Submitted to GDTA.

Kumar, P. (1998) : Food Demand and supply Projections for India. Agricultural Economics Policy Paper 98 - 01, IARI, New Delhi.

Persaud Suresh and Stacey Rosen (2003) : India's Consumer and Producer Price Policies : Implications for Food Security. Economics Research Service, Food Security Assessment.

Ragab, R., C. Prudhomme and N. Reynard (2000). Climate change and water management in the arid regions. The urgent need for non conventional water resources. Proceeding of the 2nd World Water Forum. March, The Hague, The Netherlands.

Ravindran, A. and Jaishankar, J. (2000). ``GISA Based Information System for Village Level Planning. GIS at development.

Sahoo, N.R. Jothimani P. and Tripathi, G.K. (2001). Multi criteria Analysis in GIS Environment, GIS at development pp. 38-40.

Smedema, L.K. (2000). Global drainage needs and challenges. Proceedings of the 8th International Workshop on Drainage. January, New Delhi, India.

Svendsen, M., Trava, J. and Johnson S.H. III. (2000). A synthesis of benefits and second generation problems. In : D. Groenfeldt, and M. Svendsen (eds), Case studies in Participatory Irrigation Management.

Vaidyanathan, A. (1999). ``Integrated Watershed Development : Some Major Issues, Society for the Promotion of Wastelands Development, Water Resources Management : Institutions and Irrigation Development", Oxford University Press, New Delhi.

Vuren, G.H. (1998) ``Farmers' Participation in Water Management", Proceedings in the 2nd Netherlands National ICID Day.

Chapter - 12

On-Farm Irrigation : The Science and Management

Dr. Manasmani Dev Goswami

International Water Management Institute's (IWM) Water Scarcity Study (1998), revealed that by 2025, 1.8 billion people would live in countries or regions with absolute water scarcity. Today most countries in the middle East and North Africa are identified as having serious water shortage. After two decades, these countries will be joined by Pakistan, South Africa and large parts of India and China. People will not have sufficient water resource to maintain their current level of per capita food production from irrigated agriculture, even at high degree of irrigation efficiency, and also to meet reasonable water needs for domestic. Industrial and environmental purposes. The (IWM) report clarifies that to meet the 2025 water demand primary water supply. The irrigation sector by far the largest water user today will still account for 69 percent of the total primary water supply. The primary water supply to irrigation must be increased by 17 per cent to meet the food needs.

New technologies have been developed to conserve and use water effectively. On farm development in a comprehensive manner is being taken up for the irrigation projects in developing countries like India. This includes providing separate irrigation and drainage channels, judicious distribution of water by constructing measurement, controlling and regulation structures, minimising the water losses by lining the channel or conveying the water through pipes etc. However, the inadequate design and operation of main system and the poor irrigation layout at the farm level are the technical ills of irrigation management. This leads to water loss to and unequal distribution. The adverse effects are water logging, salinity and low productivity. Researchers often suggest that to cope with demand for water, production should be increased through optimum utilisation and by studying water production functions; new technologies have to be developed for maximising water, use in areas of assured irrigation, coastal, bill and well irrigated area; need of human resource development; assessment of surface and groundwater resources to take up development activities.

On farm Irrigation

Irrigation is the artificial application of water to soil for the purpose of crop production. The application is meant for supplementing the water available from rainfall and the contribution to soil moisture from ground water. The On farm Irrigation deals with the science of water application inside the farm. The amount and the timing of rainfall in earth are not adequate to meet the moisture requirement of crops and irrigation is essential to raise crops necessary to meet needs of food and fiber. Scientific management of irrigation water provides the best insurance again weather induced fluctuations. Today, the world caters for rapid expansion of irrigation to cope with demands of growing population. In the present day profitable agriculture, the integrated policy for water resources management should also underline the additional requirements of the growing industry, human and livestock consumption, provide for flood control etc.

Whatever be the source of water in irrigation fields, the efficiency of the system depends on the application of sound principles in design and construction of the utilisation structure, usually the well, and the characteristics of the water lifting device in relation to the source of water. Devices for irrigation range from age old indigenous water lifts to highly efficient pumps. Pumps operated by electric motors or engines show prominence in large scale farm irrigation. Shallow tube wells with low to medium centrifugal pumps are used in small holdings and shallow iso bath region. Deep tube wells with submersible pumps are used in large holdings irrigation where ground water table is very deep. Canal irrigation can command huge network of farms.

Water Requirement and irrigation Efficiency

For on farm irrigation management with reference to crop planning, the estimation of water requirement of crops is of prime importance. The crop water requirement is defined as the quantity of water, regardless of its source, required by a crop or diversified pattern of crops in a given period of time for its normal growth under field conditions at a place. This includes the losses due to evapo-transpiration or consumptive use plus the losses during the application of irrigation water (unavoidable losses) and the quantity of water required for special operations such as land preparation, transplanting, leaching etc. The irrigation demand, however, refers to the crop water requirement, exclusive of effective rainfall and contribution from soil profile. The irrigation requirement of an outlet command area includes the irrigation requirement of individual farm holdings and the losses in the conveyance and distribution system.

The design of the irrigation system, the degree of land preparation and the skill of the irrigation are the principal factors influencing effectiveness of irrigation. The irrigation efficiency indicates how efficiently the available water supply is being used based on different methods of evaluation. The ratio of crop yield to the amount of water depleted by the crop in the process of evapo-transpiration is termed as crop water use efficiency. On the other hand, the ratio of crop yield to the total amount of water used in the field is termed as field water use efficiency. Of the two indices, the concept of crop water use efficiency is of fundamental interest while field water use efficiency in of greater practical importance.

Status

It is necessary to have a thorough knowledge about physical and chemical characteristics of the soil meteorological behaviour and economical conditions of the inhabitants before putting a land under irrigation in order to avoid hazards of developing waterlogging. Salinity and alkali problems. The physical status is viewed from the point of texture, permeability, non capillary porosity, water table depth, soil depth, topography, water holding capacity etc. The chemical status of land are characterised by soluble salt content, exchangeable sodium and pH.

Loamy soil texture is ideal for irrigation. Clayey texture hinders leaching and results in poor aeration, while sandy texture increases water application losses. High values of water holding capacity, depth of soil and non capillary porosity are desirable characteristics from both irrigation and drainage points of view. Higher content of soluble salts in soil results in salinity problems and those of exchangeable sodium in alkali problems, pH value beyond 9 is undesirable on account of alkali hazards. Neither is the pH below 4.5 good as availability of nutrients to plants becomes a limiting factor.

Irrigation Methods

The type of soil, the topography of the land, the water supply and the crop to be irrigated determine the right method of irrigation to be used. Whatever be the method of irrigation, the design of the system is necessary for the most efficient use of water by the crop. In the surface method of irrigation, water is applied directly to the soil surface from a channel at the upper reach of the field.

Water may be distributed to the crops in border strips, check basins or furrows. Two general requirements of prime importance to obtain high efficiency in surface method of irrigation are properly constructed. Water distribution system will provide adequate control of water to the fields and proper land preparation to permit uniform distribution of water over the field. In the sprinkler method of irrigation, water is sprayed into the air and allowed to fall on the ground surface somewhat resembling rainfall. WIth careful selection of nozzle sizes, operating pressures and sprinkler spacing, the amount of irrigation water required to refill the crop root zone can be applied nearly at a rate to suit the infiltration rate of the soil, thereby obtaining efficient irrigation. The drip irrigation method is becoming increasingly popular in areas with water scarcity and salt problem. The plants are watered with the help of emitters or drippers frequently and with a volume of water approaching the consumptive use of the plants where conventional losses such deep percolation, runoff and soil water evaporation are minimised.

Irrigation Scheduling and Management

The farm irrigation requirement depends on the irrigation needs of individual crops which are characterised by critical stages of growth, and the requirement regarding the number of irrigations and their timings vary widely for different crops. Excess water in the root zone or inversely shortage of water during the most sensitive time of crop growth is harmful and affects the production. Therefore, the right amount of water at right time in right technical direction is essential for best farm output. The irrigation management to evaluate scheduling has therefore great importance.

Rice occupies more than 40 per cent of the irrigation area in India. It is a semi aquatic plant and hence its water requirements are many times more than other food crops. It is a major consumer of the water resources of the country and thus needs careful water management in order to increase its efficiency of water use. The irrigation scheduling for some common crops are wheat which has the time of sowing from 5th to 15th November and water requirement approximately 45 cm will have number of irrigation as 25th, 45th, 70th, 90th, 105th, 125th and 140th day reckoned from the date of sowing its critical stages of growth are crown root, initiation, tillering, jointing, booting, flowering, milk and dough stages. Potato which has sowing time normally 3rd week of October and 45 cm crop water requirement will have the irrigation on 1st, 13th, 25th, 37th, 49th, 65th, 80th, 97th, and 109th day after sowing. The scheduling of irrigation and the management in the field should be done by the farmers themselves. They should acquire the technical know-how about on-farm irrigation system, water use and distribution and particulars of crops. A group of farmers having representatives from different sections/ categories may form water users association at field level to monitor judicious use of water and functioning of the farm irrigation system.

Chapter - 13

Dynamics of Lift Irrigation

G. Sheshnagiri & S. Subhash

One of the implicit assumptions for the advocacy of irrigated agriculture is that water is perse beneficial to the farmers as it increases the productivity of land and incomes. So, no farmer would abstain from the use of water, when it is available in adequate quantities. Of late it is suggested, if the management of water allocation and distribution is done by the farmers themselves at the tertiary and quaternary levels of the canal system, benefits to farmers would be greater than they are at present. Today most public irrigation systems are managed by the bureaucracy right from head works to the farm gate. Experience shows that bureaucratic management at the tertiary level is inefficient and mismanagement due to inadequate and untrained manpower, inadequate physical system, and paucity of knowledge of modern irrigation methods on the part of irrigators. In order to improve the efficiency and better management sixth and seventh five year plan emphasized establishment of water users association like water user co-operatives societies.

The purpose of this study is to examine peoples participation in the management of Lift Irrigation Co-operative Society in Vyara Taluka, in Gujarat State, a tribal dominated area, to draw out lessons for community management of small lift irrigation project. The study is important because it deals with :

(1) The poorest among the poor who are not in a position to invest on a lift irrigation.

(2) People whose idiom and world view are not in tune with the dominant civic culture, its technology, its people, its pace or institutions but who need to be helped if co operative management is to succeed.

BACKGROUND OF THE STUDY

The study was carried out to examine the impact of lift irrigation on tribals of Panihari village in Vyara Taluka. The study being an explorative and qualitative as a methodological tool used, was carried out in this tribal, village with formal, as well as informal unstructured and focused interview with adopters and non-adopters of the lift irrigation scheme and Participatory Research Appraisal techniques.

OBJECTIVES

The main objectives of the study are:

- To identify the institutional interventions.
- Empowerment of tribals.
- Social and Economic Impact of the lift irrigation scheme and that on tribals livelihoods.
- To find the scope and limitations of replication of this successful Lift Irrigation (LI) scheme in other parts of the country.

PROFILE OF THE VILLAGE

Panihari village forms a part of Taluka Vyara, District Surat. The total population according to census of 1991 was, 1,957 of which male consists of 1,039 and 918 females. There are 450 families residing in the village, the area of the village is 369.22 hectares and connected by road. The village has one primary school and health centre each. The farming area is 316.39 hectares and has a minor canal passes through the village. The area under irrigation with private water facilities was 5 hectare. The cultivated idle land is 105 hectare (1991 census). The land, which cannot be used for cultivation is 59.12 hectares.

Description of the Lift Irrigation Scheme

Gujarat Water Resources Development Corporation Limited (GWRDC) under the control and guidance of irrigation department, Government of Gujarat, undertake the projects of medium and major Lift Irrigation schemes. In the beginning, the farmers were not having facility of irrigation from canal. So therefore it was decided to provide irrigation facility by implementing Lift Irrigation schemes. Government of Gujarat selected 78 Lift Irrigation Schemes, one such major project was Ukai Dam on Tapti river in Southern Gujarat which had an irrigation potential of 14965 hectares on Ukai Left Bank Main Canal, the scheme which had taken up by the GWRDC. The Pilot Irrigation Scheme selected to initiate the project was the scheme at the village Panihari. The scheme benefited the farmers of Panihari and Bharatpur villages having gross command area of Lift Irrigation Scheme is 478 acres.

The command area was divided into number of blocks, varying from 8-20 hectares of land. RCC pipelines were laid to carry the water from the main lift to the field. The scheme started operation in the year 1980 under the management of GWRDC. Initially only few farmers came forward to avail the scheme. Realising the benefits accruing to the other farmers, the number of farmers using the scheme has increased over the years.

The Lift Irrigation Scheme operated by GWRDC, started running into problems because of inefficient service, meanwhile the farmers of the area demanded to run the scheme on their own through a co-operative society in the place of GWRDC. The LI Scheme of Panihari was transferred to the management of Co operative Society in the year January 1998. At present the Co-operative Society is able to run the scheme successfully, which consists of 100 adopters. The Co-operative Society also undertakes the distribution of seeds, fertilizers and agricultural implements to its members. This society charges Rs. 2.50 for 10,000 liters of water.

INTERVENTIONS IN THE AREA

To execute Lift Irrigation System in the area there arose a necessity of intervention by different agencies to achieve efficient and speedy services to adopters of this scheme. In the Panihari LI Scheme the agencies that had intervened were GWRDC and Vyara Taluka Western Command Area Water Distribution Co operative Society Ltd, Panihar. These two agencies play important and distinct role. GWRDC, being the sole authority of water and canal, maintains the LI scheme. Co-operative society has taken up the responsibility of distribution of water and agricultural kit. The core function of the Co-operative Society being water distribution was the focus of the study. It was found that Co-operative which is supposed to act as multi/purpose Co-operative society, and multi functions such as providing credit to farmers and educating and motivating farmers to take up new technologies in the farming was involved only in water distribution. But the society was unable to play these kind of roles due to agreement with GWRDC it has entered, accepting that it will not take up any other activity other than water distribution and distribution of agricultural kit. Another important note to be taken into consideration is that the rates of water distribution between the GWRDC and Co-operative society are totally different, the Co-operative Society charging higher rate than the GWRDC.

ADOPTERS AND NON-ADOPTERS

All the adopters of LI Scheme are members of Co-operative society. The adopters are mainly small, medium, and marginal farmers. Non adopters are those who are financially weak and low motivated farmers who either lease out their land to the large farmers and they get the one third of the produce from the cultivator. There are 10 women farmers on the LI scheme. Women farmers are facing difficult in cultivating land due to many reasons like new agricultural practices and financial difficulties.

IMPORTANCE OF LIFT IRRIGATION ON LIVELIHOODS OF ADOPTERS AND NON-ADOPTERS

The impact can be summarized as given below :

- *Social Sustainability* in which the marginal farmers start playing important role in the day to activities in decision making process.
- *Financial Sustainability,* which the tribal farmers got through the LI Scheme, by increasing the incomes.
- Under the Technical Sustainability, promotion of indigenous and new technologies with respect for and sensitivity to existing practices, strong and targeted capacity building programme.
- Asset generation and human capital development, level of participation and expansion of choice.
- Stability of the farming system increases the freedom of farmers to take decision regarding the crop pattern (Cash Vs. Subsistence Crop or Market or Subsistence Crop).
- The farmer also has the facility of Agricultural Produce Marketing Committee to sell their produce in a regulated market.
- Families and households have been enable to fulfill social and family obligations such as marriage of children, providing hospitability to household/village and community guests.

There have also been general improvements related to heath, sanitation, housing, common facilities as also an improvement in the level of food security especially for the marginalized groups in the village. Feeling of empowerment, general, improvements in self esteem, confidence, and ability to innovate. Reduction in the social distance between groups of different social status, feelings of social isolation both within the community, as well as with reference to the wider world has decreased. The community has become more socially inclusive, with greater interaction between members of different social categories. It was also observed in order to educate their children, some women farmers who leased out their land, sending the children to school to educate them, which reveal's the level of awareness about the importance of education.

The Economic impact could be traced out to the extent that their incomes have increased considerably, with supplementary income by Animal Husbandry which has made them to afford many basic things to their life such as an vehicle, RCC House, Telephone, etc. To conclude, the LI Scheme has revolutionized the livelihoods of tribal farmers by giving them social and economic status in the society for further progress.

PROBLEMS FACED

Rotation System

This is the system evolved by the GWRDC to meet the demand of water to farmers by fixing the time and allotting that time to a particular time so that that farmer can make use of the water when their turn comes, however, several farmer are facing the water shortage due to long gap during which the crop in his field may worsen resulting in low yield, and hence some of the rich farmers have their own tube-wells, mean while small and marginal farmers continue to face this problem. Some of the suggestions to the problem evolved during our course of study is that, some of the farmers can be made in a group and ask them to grow same crop, the success of this observation depends on land size, since many small farmers are growing the rice and groundnut. To add, Government of Gujarat has recently directly GWRDC not to release water for paddy and groundnut in hot season, which has further exacerbated the situation.

Conflict among Farmers

The first come first serve basis in the distribution of water has created minor conflicts among the farmers. Many farmers believe that they are discriminated because of the rotation system. Conflict arises due to the blockage of main pipes through which water passes to the kundi or outlet. Young farmers do this usually, and the co operative is not taking any action by making the pipes fool proof.

Politicisation of the Co-operative Society

Many farmers complained abut the politicisation of the society due to which their grievances are not met in time, due to which several farmers are suffering from water shortage. The farmers are divided on political lines in the society leading to minor conflict at the personal level and at societal level.

Non-availability of Credit

Since the co-operative has entered into an agreement with GWRDC that it will not venture into any other activities other than water distribution and distribution of agricultural kit, the small farmers are badly hi due to the non availability of credit and hence the society can change the contract with the government and can provide the credit to the farmers.

Lack of Extension Facilities

The farmers especially small farmers in the village are not using the HYV seeds which show the laxity on the part of the any of the agricultural university or agricultural department to promote extension programmes in this village. If rejuvenated, many farmers lives can be improved considerably by adopting new technological practices in agriculture.

REPLICATION AND RECOMMENDATIONS

Replication of this scheme which is successful at Panihari Village, if implemented after a brief survey and profile of farmers and their land, so that benefits can be quantified. The LI should be implemented in the area where the farmers are open to new methods of farming and technology. To make water distribution system effective it should be taken up by a neutral body to avoid inefficiency and corruption on one side and politicisation on the other side.

CONCLUSION

Water Co-operative societies have promising future in the command area of irrigation projects of all kinds. It offers ideal solution of the rather complex problem as distribution of irrigation water on the basis of equity. It also makes easy introduction as the discipline of rotational water distribution and sale of water in bulk on volumetric basic. Water co operatives should be encouraged to grow very fast and in great number with a view to optimize the benefits of irrigation projects and there by increase the rate of overall status of social life and economic prospects from good to better and best in days to come.

Chapter - 14

Irrigation : A Necessary Input to Boost Agriculture

Dr. Anita Modi & Sanjay Kumar

Agriculture is the backbone of our country. Development prosperity and stability of agriculture depend on the efficient management of soil and water systems. In the context of water systems, irrigation is a pre condition for the success of agriculture. The importance of irrigation in agriculture is also amply clear from the fact that a major chunk of available water in the world is used in the agriculture sector. At the global level, 65 percent of the total available water is used in agriculture sector, 23 percent in industries and the remaining 8 percent is used for domestic and other purposes. Again, the need of excessive water for the production of agricultural output is quite obvious from the fact that about 1000 tones of water is required for the production of one ton of wheat. Thus, assured irrigation system is essential for the generation output, employment, and income and capital formation in agriculture.

Water More Valuable Than Land

In our country, about 83 percent of the available water is used for irrigation in agriculture. It is estimated that the demand of water for the purpose of irrigation will further increase. The significance of irrigation in our country also enhances because of the fact that 58 percent of our population depends on agriculture and its related activities for their livelihood. Highlighting the importance of irrigation Sir Charles Treve-lyan has aptly remarked, ``irrigation is everything in India; water is even more valuable than land." Irrigation is not only an insurance against the vagaries of nature but also helps in raising the productivity of land.

Since independence, emphasis has been put on the expansion and development of irrigation infrastructure to boost agriculture production. The actual irrigation potential created has rises from 22.6 million hectares in 1950-51 to around 102.8 million hectares in 2006-07. About 42.4 million hectares of irrigation potential has been created through major and medium projects whereas over 60.4 million hectares through minor irrigation projects. It is noted that about 74 percent of the total exploitable irrigation potential has potential from all available sources is estimated about 139.9 million hectares, in which the estimated share of major and medium projects is about 58.3 million hectares and 81.6 million hectares through minor irrigation projects.

Our government has implemented many policies to assure water for irrigation in agriculture. National Water Policy was adopted in 1987 to control the excessive exploitation of water and conserve the water sources and resources so that the development of agriculture can be ensured. In the same way, Bharat Nirman Yojana was initiated for the period 2005-06 to 2008-09 for the infrastructural development of rural areas. In this program, irrigation is considered as the main component of agriculture. The target was laid down to create irrigation potential of 10 million hectares.

Irrigation Potential

Despite the achievements in the irrigation system, there are some critical issues which should be analyzed. First of all, the under utilization of existing irrigation system is a cause of concern. It is estimated that about 85 per cent of the created irrigation potential is actually been utilized and the rest is lying idle. According to official estimation, the actual utilization of created irrigation potential is about 87.2 million hectares only. The main factors responsible for this dismal picture of under utilization are poor maintenance and operation of irrigation projects, lack of complementary facilities like field distribution system, water control structures, leveling of land, roads, credit, and marketing and proper water drainage system. It is also observed that inadequate water conveyance system is also causing the problems of wastage of water, water logging, floods and salinity.

It is urgently required to implement the multi cropping system and alternative cropping pattern for deriving the maximum benefits from large scale investment in irrigation system. It is found that most of the irrigated area is under the single crop system. In Ninth Five Year, to lessen the gap between created irrigation potential and its actual utilization, Commands Area Development Programme, Institutional reforms and participation of farmers in management of irrigation projects etc. have been adopted. It is heartening to note that Water Resources Ministry has started farmers participatory research programme with the cooperation of agriculture universities and research sections to disseminate the knowledge of latest techniques to the farmers. In Eleventh Five Year Plan, the target of increasing the irrigation facilities by 25 lakh hectare annually is laid down so that the dependence of agriculture on monsoon can be reduced. For the optimum utilization of irrigation potential, a provision of Rs. 5000 crore is also made by Thirteen Finance Commission for the period of 2011-12 to 2014-15.

Another main concern about irrigation is that since 1970s, the dependence on ground water for irrigation has increased leaps and bounds. The excessive extraction of ground water for irrigation has endangered the bank of ground water. The subsidized rate of electricity, easy availability of loans for boring purposes, pump sets, lack of knowledge to farmers about scientific agriculture system and the absence of rain water conservation methods are the factors responsible for the depletion of ground water. About 60% of irrigation in the country is done with ground water.

It is estimated that a 10 percent increase in the efficiency of water use can irrigate additional 140 lakh hectare land. To achieve this objective, it is essential that necessary information about efficient water use must be provided to the farmers and complementary irrigation facilities must be ensured. The participation of farmers in the management and maintenance of irrigation system should be raised by conferring on them some type of co ownership in the irrigation system. The private sector should also be encouraged to participate in the development of irrigation sector. Lastly, a comprehensive watershed management plan should be formulated and effectively implemented for the successful operation of irrigation sector. To make efficient use of irrigation system and to minimize water wastage; it is desirable to promote the sprinkler method of irrigation. All these measures will go to a long way to make the country ``water rich" and ``agriculture rich".

The history of development of irrigation in India can be traced back to prehistoric times. In an agrarian economy like India, irrigation has played a major role. Vedas and ancient Indian scriptures made references to wells, canals, tanks and dams which were beneficial to the community and for their efficient operation and maintenance the responsibility was of the State. Civilization flourished on the banks of rivers and the water was harnessed for sustenance of life. According to the ancient Indian writers, the digging of a tank or well was amongst the greatest of the meritorious acts of a man. Vishnu Purana enjoins merit to a person who effected repairs to wells, gardens and dams. The irrigation technologies during the Indus Valley Civilization were in the form of small and minor works, which were operated by households to irrigate small patches of land and did not require a collective effort. Nearly all the irrigation technologies prevalent then still exist in India with little technological change and are continued to be used by households in rural areas.

The spread of agricultural settlements to less fertile area led to emergence of large irrigation works in the form of reservoirs and small canals. While the construction of small schemes was well within the capability of village communities, large irrigation works emerged only with the growth of states and empires.

In south, perennial irrigation began with construction of the Grand Anicut by the Cholas as early as second century to provide water for irrigation from the Cauvery river. Wherever the topography and terrain permitted, it was an old practice in the region to impound the surface drainage water in tanks or reservoirs by throwing across an earthen dam with a weir, where necessary, to take off excess water, and a sluice at a suitable level to irrigate the land below. Some of the tanks got supplemental supply from stream and river channels. The entire land scape in the central and southern India is studded with numerous irrigation tanks which have been traced back to many centuries before the beginning of the Christian era. In northern India too there are a number of small canals in the upper valleys of rivers which are very old.

Irrigation during Medieval India

In medieval India, rapid advances took place in the construction of inundation canals. Water was blocked by constructing bunds across steams. This raised the water level and canals were constructed to take the water to the fields. Ghiyasuddin Tughluq (1220-1250) is credited to be the first ruler who encouraged digging canals. However, it is Firoz Tughluq (1351-86) who, inspired from central Asian experience, is considered to be the greatest canal builder before the nineteenth century. As agricultural development was the pillar of the economy, irrigation systems were paid special attention.

Irrigation under British Rule

Irrigation development during British rule began with the renovation, improvement and extension of existing works. The Government also ventured into new projects, like the Upper Ganga Canal, the Upper Bari Doab Canal and Krishna and Godavari Delta Systems, which were all river diversion works of considerable size. The period from 1836 to 1866 marked the development and completion of these four major works. In 1867, the Government adopted the practice of taking up works, which promised a minimum net return. Thereafter, a number of projects were taken up. These included major canal works like the Sirhind, the Lower Ganga, the Agra and the Mutha Canals, and the Periyar Dam and canals.

The recurrence of drought and famines during the second half of the nineteenth century necessitated the development of irrigation to give protection against the failure of crops and to reduce large scale expenditure on famine relief. Significant protective works constructed during the period were the Betwa Canal, the Nira Left Bank Canal, the Gokak Canal, the Khaswad Tank and the Rushikulya Canal. Between the two types of works, namely productive and protective. the former received greater attention. The gross area irrigated in India under British rule by public works at the close of the nineteenth century was about 7.5 m.ha. Of this, 4.5 m.ha. came from minor works, like tanks, inundation canals etc. The area irrigated by protective works was only a little more than 0.12 m.ha.

At the Time of Independence

The net irrigated area in the Indian sub continent, comprising the British Provinces and Princely States, at the time of independence was about 28.2 m.ha. The partition of the country resulted in the apportionment of the irrigated area between the two countries; net irrigated area in India and Pakistan being 19.4 m.ha and 8.8 m.ha respectively. Major canal systems, including the Sutlej and Indus systems went to Pakistan. East Bengal, now Bangladesh, which comprises the fertile Ganga Brahmaputra delta region also went to Pakistan. The irrigation works which remained with India, barring some of the old works in Uttar Pradesh and in the deltas of the South, were mostly of protective nature, meant more to ward off famine than to produce significant yields.

Irrigation Development Now

At the central level the union Ministry of Water Resources is responsible for development, conservation and management of water as a national resource, i.e. for policy on water resources development and for technical assistance to the states on irrigation, multipurpose projects ground water exploration, and exploitation, command area development, drainage, flood control, water logging, sea erosion problems, dam safety and hydraulic structures for navigation and hydropower. It also oversees the regulation and development of inter state rivers. These functions are carried out through various Central Organizations. Urban water supply and sewage disposal is handled by the Ministry of Urban Development whereas Rural Water Supply comes under the purview of Department of Drinking Water under the Ministry of Rural Development. Hydroelectric power and thermal power is the responsibility of the Ministry of power and pollution and environment control is that of the ministry of Environment and Forests.

Water being a State subject, the State Governments have primary responsibility for use and control of this resource. The administrative control of responsibility for development of water, rests with the various State Departments and Corporations. Major and medium irrigation is handled by the irrigation/water resources departments. Minor irrigation is looked after partly by water resources departments, minor irrigation corporations. Zilla Parishads/Panchayats and by other departments such as Agriculture. Urban water supply is generally the responsibility of public health departments and panchayats take care of rural water supply. Government tubewells are constructed and managed by the irrigation/water resources department or by tube well corporations set up for the purpose. Hydro-power is the responsibility of the State Electricity Boards.

Chapter - 15

Development of Irrigation Potential No more Myth

K.G. Suresh

There have been a large number of projects which have been languishing for want of funds. The effort under Bharat Nirman is to identify all such projects and target their completion to create 10 million hectares of additional irrigation capacity. This was started by prime minister Manmohan Singh while launching the government's ambitions Bharat Nirman programme in New Delhi which seeks to change the face of rural India in the coming years economically. Socially and infrastructurally.

The Government, he said, has identified major and medium irrigation projects amounting to four million hectares, which could be completed, as well as 2.8 million hectares that can come from irrigation.

In fact, under the irrigation component of Bharat Nirman, the target of creation of additional irrigation potential of one crore hectare in four years (2005-08) is planned to be met largely through expeditious completion of identified ongoing major and medium irrigation projects. Irrigation potential of 42 lakh hectares is planned to be created by expeditiously completing such ongoing major and medium projects.

The irrigation component of Bharat Nirman was discussed at length at a conference of state irrigation Secretaries. Based on the discussions there, the state Governments were requested to furnish details of projects proposed to be included under Bharat Nirman. Since, then, several state Governments have furnished to the Water Resources Ministry preliminary assessments of project wise irrigation potential and requirements thereof for inclusion under the programme.

A whopping Rs. 944.18 crore has been released so far as grant under the Accelerated Irrigation Benefit Programme (AIBP) and the target of 600,000 hectares of irrigation potential is expected to be created this year alone.

As per a 1997 survey, the per capita water availability in the country was 1967 cubic kilometers. The annual precipitation volume including snowfall stood at 4000 Cu. Km while the average annual potential flow in rivers in pegged at 1869 cu.km.

The estimated utilizable water resources in 1122 cu.km with surface water resources accounting for 690 cu.km and ground water resources comprising 432 cu.km.

The ultimate irrigation potential (UIP) for the country has been estimated as 139.88 Mha (Million hectares), which include potential through Major and Medium Irrigation (MMI) projects (58.46 Mha), surface water based minor irrigation (MI) schemes (17.42 Mha) and ground water development (64.00 Mha). A potential of 101.3 Mha has been created in the country so far which consists of 37.1 Mha under medium and major projects and 64.2 under minor projects.

However, the created potential has not been fully utilized and the gap between the created and utilized potential has been estimated to be of the order of 4.99 Mha.

Realising the gap between the irrigation potential created and the potential utilized, the Government, under Bharat Nirman, plans to restore and utilize irrigation potential of ten lakh hectares through implementation of extension, renovation and modernization of schemes alongwith Command Area Development (CAD) and water management practices.

The Centrally Sponsored CAD programme was launched in 1974 on the recommendations of the National Irrigation Commission set up in 1972 and a Committee of Ministers constituted in 1973 with the objective to bridge the gap between the irrigation potential created and that utilized through micro level infrastructure development and efficient farm management besides to enhance agricultural production and productivity and to improve socio-economic conditions of the farmers.

In April, 2004, the Programme was restructured and renamed as `Command Area Development and Water Management (CADWM) programme.

Among the 12 components covered under the CADWM programme are soil and topographic survey, planning and designing of On Form Development (OFD) works; correction system deficiencies above the outlet upto distributaries of 4.25 Cumec (150 cusec capacity); construction of field drains, intermediate and link drains for letting out surplus water; reclamation of waterlogged areas of irrigated commands using conventional techniques and including bio drainage wherever applicable and renovation and desalting of existing irrigation tanks including the irrigation system and control structures with the designated irrigation demands.

The Programme also provides for institutional support to Water Users Association as also Research and Development activities, including training of senior level officers, conferences, workshops and seminars to be directly arranged by the Ministry.

The CADW unit of the Union Water Resources Ministry has been monitoring the CADWM Programme through periodical progress reports, detailed accounts (submitted by the states with the release proposals), field visits, meetings, consultants visit etc.

With a view to further strengthen the monitoring of the Programme, the States have been advised to constitute a Committee at state level having representation from all the concerned State and Central government Department in it. Many of the States have already formed such committees and many others are going through the process.

For the country as a whole 66% of the ultimate irrigation potential of major and medium projects has been created. 388 major and medium projects, which were taken up prior to or during IXth Plan, are still ongoing which would result in creation of 12.1 Mha of additional irrigation potential. In addition, the States have proposed 204 major and medium projects during Xth Plan and potential likely to be created is of the order of 4.99 Mha.

So far, 173 major and medium 4.169 minor and 21 Extension, Renovation and Modernisation (ERM) projects have been provided Central loan assistance under the Accelerated Irrigation Benefit Programme. The potential creation through projects supported under AIBP has been found to be 0.35 Mha per year with about 0.47 Mha per year in the last two years. Going by the present Plan, the projected creation of irrigation potential through AIBP is 0.50 Mha per year in the remaining period of Xth Five Year Plan.

The projects for ERM of major and medium irrigation schemes are also being implemented with arrangements similar to that for completion of ongoing major and medium schemes. The implementations of ERM projects alongwith CADWM help in sustaining the created facilities and in improving utilization.

Another major component of the Bharat Nirman Programme is Ground Water (GW) development. At present, urban areas utilize 50 per cent ground water while in the rural areas, it is 85 per cent. Irrigation accounts for 55 per cent whereas industries utilize 50 per cent.

The ultimate irrigation potential to be created is based on the assessed replenishable ground water after fully accounting for the domestic and industrial uses. There are considerable areas in the country with unutilized ground water resources. It has been assessed that ground water is still available for utilization in many parts of the country, particularly in Eastern India, Madhya Pradesh, Chhattisgarh and specific pockets of Andhra Pradesh, Karnataka, Maharashtra and J&K.

Under Bharat Nirman, irrigation potential of 28 lakh hectares is planned to be created through GW development. The remaining target for creation of irrigation potential of 10 lakh hectares is planned to be created by way of minor irrigation schemes using surface flow.

However, the Government is also aware that the increasing pace of ground water development to meet the growing demands of water in agriculture, industrial and domestic sectors, has brought problems of over exploitation of the resource, continuously declining water levels, sea water ingress in coastal areas and ground water pollution in different parts of the country.

In Punjab, Haryana, Rajasthan, Gujarat and Tamil Nadu, the rechargeable quantum of ground water has been exceeded and mining of static reserves has commenced.

The failing ground water levels in various parts of the country have threatened the sustainability of ground water resource, as water levels have gone deep beyond the economic lifts of pumping. The speedy and uncontrolled development of ground water resources has resulted into increase in over exploited and critical dark blocks from 253 to 1065 between 1985 and 2004.

	Targets Under Bharat Nirman	**Unit in Million Hectares**
SN.	**Component**	**Target**
1	Completion of ongoing major and Medium irrigation projects	4.2 Mha
2	Minor Irrigation Scheme	2.8 Mha
	a) Surface Water	1.0 Mha
	b) Ground Water	1.8 Mha
3	Enhancing utilization of completed Projects	2.0 Mha
	a) Extension, renovation and modernisation of major and medium projects	1.0 Mha
	b) Repair, renovation and restoration of water bodies/ERM of minor irrigation schemes	1.0 Mha
4	Ground water development in area with Unutilized ground water potential (for benefit of small and marginal farmers and tribals and Dalits)	1.0 Mha

Achievements - Upto March 2006
(As reported by State Governments)

Unit in Thousand Hectare

S No.	**State Name**	**Achievement upto March 2006**
1	Andhra Pradesh	70.79
2	Arunachal Pradesh*	0.266
3	Assam*	1.503

4	Bihar	Not reported
5	Chhattisgarh	53.261
6	Goa*	1.224
7	Gujarat	Not reported
8	Haryana	Not reported
9	Himachal Pradesh*	2.223
10	J & K	Not reported
11	Jharkhand	14.847
12	Karanataka	74.563
13	Kerala	Not reported
14	Madhya Pradesh	Not reported
15	Maharashtra	Not reported
16	Manipur	Not reported
17	Meghalaya*	0.716
18	Mizoram	Not reported
19	Nagaland*	0.875
20	Orissa	24.59
21	Punjab	49.665
22	Rajasthan	164.58
23	Sikkim*	0.8
24	Tamil Nadu*	5.917
25	Tripura	4.788
26	Uttar Pradesh*	351.806
27	Uttaranchal*	4,684
28	West Bengal*	3,455
	Total	**830.553**

Note :- States with (*) mark reported progress up to December 2005

As natural replenishment of ground water reservoir is slow and is unable to keep pace with the excessive continued exploitation of ground water resources in various parts of the country, there is an urgent need to augment the natural supply of ground water through artificial recharge. The artificial recharge techniques aim at increasing the recharge period in the post monsoon season for about three months by providing additional recharge. This provides sustainability to ground water development during the lean season.

Amongst the prominent technique of artificial recharge to ground water is rain water harvesting. Rain Water Harvesting is the technique of collection and storage of rain water at surface or in sub surface aqufer, before it is lost as surface runoff. The augmented resource can be harvested in the time of need.

The Government recently constituted the Artificial Recharge of Ground Water Advisory Council to specifically study the issue. The first meeting of the Council, inauguarated by Prime Minister Manmohan Singh was attended among others by Union Water Resources Minister Saifuddin Soz and leading experts including M.S. Swaminanthan, Sunita Narain and Rajendra Singh.

The creation of irrigation potential through minor irrigation (both surface and ground water) schemes considerably varies from state to state. While full potential has been tapped in some of the states, it is comparatively very low in other states.

Minor irrigation through surface water covers water sources (tans and reservoirs) with a cultivable command area of less than 2000 hectares. About 70 per cent of the ultimate potential through surface water based minor irrigation schemes has since been created.

The Report of the National Commission for Integrated Water Resources Development had pointed out that the carrying capacity of the tanks had decreased over time for a variety of reasons and that their restoration as also of other sources was a priority task.

Hence, a key strategy of the irrigation component of the Bharat Nirman Programme is development of 10 lakh hectares of irrigation potential by way of repair, renovation and restoration of water bodies and extension, renovation and modernization of minor irrigation schemes.

The National Project for the Repair, Renovation and Restoration of Water Bodies used by farmers, especially in dry land, remote and tribal areas, was launched to augment the storage capacity of water bodies, including lakes, tanks and reservoirs, and restore their lost or wasted irrigation potentials beginning with a pilot project covering 16 districts in nine states at an estimated cost of Rs. 300 crores.

Today, the programme is being implemented in 23 districts in 13 states through pilot projects. Twenty thousand water bodies with a command area of 1,47 million hectares involving an estimated cost of Rs. 4,481 crore has been identified in the first phase. The funding pattern involving Centre, states and multilateral agencies have been finalized and the State Governments have been asked to sign Memoranda of Understanding for the same.

The project period would be between seven and 10 years. The Detailed Project Reports prepared by the states would spell out the graded monitoring mechanism. The state monitoring committees may co-opt Central Water Commission (CWC) representatives from the regional offices. Besides, the states would specify in the reports periodic evaluation to take the scheme forward.

The Union Water Resources Ministry has also established, through the CWC and the Central Ground Water Board, a superimposed monitoring with appropriate frequency of the projects taken up in the states in a structured manner.

At the Central level, the monitoring structure would include a combination of on site examination of works and off site analysis of the state's monitoring reports.

In a developing country like India, it is natural that the demand for water is consistently on the rise. Geography and vagaries of weather have also contributed to the variation in availability of water. Burgeoning population besides rapid industrialization and urbanization too are having their telling effect on the availability and quality of water.

Notwithstanding the water sharing disputes among various states and prophecies of a third World War on account of Water, the scenario in India remains optimistic due to the far sightedness and vision of its planners and policy makers. The ambitious schemes envisaged under the ``Bharat Nirman programme and its sincere implementation gives us a ray of hope that an additional irrigation capacity of 100 lakhs (10 million) hectares is neither a myth nor a mirage but a realistic goal.

Chapter - 16

Drip Irrigation and Farm Productivity

K. Srinivasulu & M. Adinarayana

The average farm productivity is very low in rainfed and limited irrigated regions and only 30% of the arable land is having irrigation facility in the country. Generally, farmers adopt conventional methods of irrigation like flooding, furrow irrigation, border irrigation etc where water losses are more and water use efficiency is low. More than 50 per cent of water is going waste in different forms such as conveyance loss, leaching loss, percolation loss and evaporation loss. There are incidences where the water losses were up to 70 per cent when applied through flooding.

In rainfed areas, where underground water is available for irrigation through bore wells or runoff water harvested from the farm and stored in the farm pond and irrigation methods are conventional, the water losses are more. More and more water tapped from the underground or farm pond leading to the scarcity of water at critical stages of the crop and terminal moisture stress to crop, results in crop failure. This phenomenon is more apparent in low rainfall or drought years.

Under these situations, micro irrigation like sprinkler or drip irrigation will suit well to reduce the losses and increase in the area under irrigation. The sprinkler irrigation is efficient in reducing all losses except evaporation losses. But it needs high pressure to operate. This is more suitable for close growing field crops. Whereas, for the wide spaced field crops or orchards, drip irrigation is the best method in which evaporation losses also get reduced to some extent.

What is Drip Irrigation

Drip irrigation is localized application of water slowly drop by drop near the plant surface through small openings called drippers or emitters with variable discharge rates as per the requirement of the crop plants.

Drip system contains pumping unit, mains, sub mains, laterals, drippers or emitters etc. The water used for drop irrigation should be free from all foreign materials. Use of impure water causes clogging of drippers. Using filters will avoid the clogging problem. Number and type of filters depends on the impurities present in the irrigation water used.

Drip irrigation is becoming popular due to the decreasing water resources. It appears to be a viable technology in the years to come. The merits and demerits of the system are as follows :

Merits of Drip Irrigation

Saving in irrigation water

In the drop irrigation, all the losses of irrigation water, conveyance, percolation, seepage and run off losses are eliminated. Evaporation losses are reduced to some extent and water saving is the highest of all the irrigation methods. With this saved irrigation water, the area under irrigation can be extended.

Increased Crop Yield

The crop yields in the drip irrigation system are higher than the crop yields under irrigation with conventional methods. The water content of the soil varies between field capacity and wilting point in conventional methods of irrigation. Whenever the soil water content is less, the plant has to spend more energy to obtain water. In the drop irrigation system, the soil water content is kept at higher level throughout the crop growth period hence the plant needs to spend less energy compared to the conventional methods of irrigation. Thus the saved energy is useful for productive purposes.

Water with higher Concentration of Salts can be used for Irrigation

In drip irrigation method, the soil water content is kept higher throughout the crop growth period, the irrigation water having more salts can also be used for irrigation through drip system without marked adverse effect on crop growth. In other words, the water that is not suitable through conventional methods of irrigation due to relatively high salinity, can be used for irrigation through drip irrigation without any adverse effect on the crop. However, continuous use of such water will cause salinity hazard.

Adoption to undulated Topographies

On irregular topographic areas i.e. hilly terrains, conventional methods of irrigation can not be possible due to variation in elevation. Under these situations, drip system can be adopted without any problem, because in this system, water is carried through small pipes called drippers/emitters.

Less Weed Infestation

Weeds, weed propagules and weed seeds are carried away from one place to another along with flooding irrigation water (e.g. Cuscuta) in conventional methods of irrigation, whereas in the drip irrigation system, no such transfer of weed propagules will be observed.

Only part of the ground area will be wetted in the fields with drip irrigation system leading to reduced germination of weed seeds and growth compared to the conventional systems of irrigation where the total ground area is wetted.

Hence the weed intensity, weed index and weed dry matter are less in the fields where drip irrigation system is adopted, compared to the conventional irrigation systems.

Application of herbicides can also be done through drip system along with irrigation water, but it needs proper calibration of quantity of herbicides. At present, no herbicides are specifically labeled for application through irrigation water by drip system.

Low Incidence of Diseases

Conventional irrigation systems, which wet the entire soil surface and increase relative humidity of micro climate favour, the spread of diseases like root rots and damping off cause great loss to the farmers.

Under high soil moisture and relative humidity conditions, following diseases spread faster due to rapid formation and dispersal of zoospores.

- Root rot of cotton — *Phymatotrichum omnivorum*
- Black root rot of tobacco — *Thielaviopsis basicola*
- Collar root rot of citrus — *Phytophthora* spp.
- Damping off of papaya — *Pythium* spp.

Root knot and other nematode disease

- Sclerotium root rots — *Sclerotium rolfsii*
- Armillaria root rot — *Armillariella mellea*

Reduced Fertilizer Requirement

As there are no water losses through percolation and leaching, losses of nutrients along with the leaching water will be minimized in drip irrigation system. Thus the fertilizer use efficiency will increase and fertilizer requirement will be reduced.

Uniform Distribution of Water

In the conventional irrigation methods like flooding, border, furrow methods of irrigation, the opportune time for infiltration of water is more at one end where irrigation started and less at another end, leading to more water intake at starting end and less water intake at another end.

In drip irrigation system, distribution of water in all directions of the field is uniform as the opportune time is same for the entire field.

Utilizing Water sources with Slower Flow Velocity

Water sources with slower flow rate can be better utilized in the drip irrigation system, which is not possible in other methods of irrigation.

Demerits

Though, there are several advantages with the drip irrigation, which is efficient in saving irrigation water, there are some possible draw backs/ difficulties in adoption of this system on long term basis.

Accumulation of salts in the Root Zone

In conventional methods of irrigation, the percolating or leaching water, which is going waste, takes away the accumulated salts in the top layer and reduces the salt content of the top layer of the soil.

In drip irrigation system, water is used to wet the shallow root zone only. All the losses of water through percolation and seepage will be arrested. The salts that are present in the irrigation water will be left in the top layer of the soil as the water is lost through evapotranspiration only. As a result, all the salts that are present in the irrigation water will be accumulated in top layer of the soil and create salinity hazard in the long run.

However, in drip irrigation system, the effect of salinity does not visualize early, as the soil is kept at higher level of available moisture (field capacity) status throughout the crop growth period due to which the salt concentration is reduced. Whenever dry spell prevails or watering through drip system is stopped, the crop plants may suffer from salt injury in the fields where drip system is followed for longer period of time.

Drip irrigation provides both water and essential plant nutrients (through fertigation) at the shallow depth of the soil, the root zone of the plant confined to the shallow depth of the soil. In contrast, in the conventional irrigation methods, the plant extracts water and nutrients from deeper layers of the soil and the root zone of the plants gets extended to the deeper layers of the soil in search of water and nutrients. Whenever cyclone or Gale/passes through the crop fields, there will be a possibility of lodging of crop which is fed through drip irrigation system. This envisages the susceptibility of the crop plant fed through drip irrigation system to the lodging due to shallow root system.

Establishment of the drip system needs all the components to life the water, filter the water and distribute the water, irrespective of the extent of the area. Majority of the dry land farmers are having small holdings with poor resource base. The high initial investment hinders the farmer to opt for drip irrigation system.

Mechanical damage to the plastic laterals/drippers by rodents is a problem and needs attention to control the rats.

The small openings of drippers will be clogged by soil particles, organic matter, algal particles or chemical precipitants etc. Hence frequent cleaning of drippers is necessary.

The clogged particles can be removed by flushing of filters, laterals, mains and drippers periodically. Injection of acids (HCI & H_2SO_4) into the drip system controls clogging due to chemical precipitation and biological particles. Acid treatments are required when Ca and Mg content in the irrigation water exceeds 50 ppm.

The laterals spread in the field for delivering water near plant surface obstructing the field operations like inter cultivation operations aimed to control weeds and conserve moisture.

In light or sandy soils, where the lateral movement of the water is very less and downward movement or vertical movement is very high, too many number of drippers are required to cover the field which is not desirable.

Despite the demerits of drip irrigation, it is gaining momentum year after year. Nearly, 4.5 lakh ha was under drip irrigation system in India during 2002-2003. Among the states, Maharashtra is first in the rate with 1.2 lakh ha.

Though there are several advantages with the drip irrigation system, in majority of states, the percentage of farmers adopting drop irrigation has decreased soon after removal of subsidy. This confirms the high initial investment and inability of farmers in establishing drip irrigation system due to their poor resource base. Hence, the governmental and non governmental organizations should encourage the farmers in adopting the highly efficient system of irrigation by providing subsidy.

In drip irrigation system, water will be used more efficiently, and there will be saving in irrigation water to the extent of 40-60 per cent. The saved irrigation water due to the adoption of drip irrigation system can be used to extend the area under irrigation. The productivity of the farm will grow and will be sustained due to increase in area under irrigation, reduced weed growth, reduced pest and disease incidence. Similarly, the profitability of the farming enterprise will also be increased due to increased productivity as well as reduced input costs like fertilizer.

Though the establishment of the drip system is expensive, the increase in productivity will pay back the cost of establishment of the drip irrigation system in a period of 4 to 5 years. Thus there is tremendous scope for drip irrigation system in enhancing and sustaining the productivity of the farm as a whole in rainfed and limited irrigated regions.

Chapter - 17

Irrigation and Agriculture

Surinder Sud

Irrigation has been on the priority list ever since the planned development began in India. Huge investment, both public and private, has gone into creation of irrigation infrastructure trough major, medium and minor irrigation projects. As a result, the gross irrigation potential has expanded from the pre-plan level of 22.6 million hectares in 1950-51 to a little cover 89 million hectares now. But the country's total irrigation potential through all sources, estimated at 113.5 million hectares earlier, has now been reassessed at whopping 139.9 million hectares. So far, only about 64 per cent of this potential has been gainfully harnessed.

However, due largely to poor utilisation of the created potential, the net irrigated area is only about 55 million hectares (2000-01) which is only abound 40 per cent of the net cultivated area. Thus, crops over about 60 per cent of the area still depend on rains, causing instability in overall agricultural production.

This apart, the efficiency of the use of irrigation water, too, is not up to the mark. In the case of flooding of the field, the most common method of applying irrigation water to the crop; the water use efficiency, on a country wide average, is only around 50 per cent. Though the efficiency of new water application systems like sprinkler irrigation and drip irrigation is fairly high, being 80 per cent and 95 per cent, respectively, their use is still confined to a very small area due largely to the high initial cost of installing such systems.

Indeed, the importance of irrigation for raising as well as stabilising agricultural production was realised way back in the 19th century itself. As a result, some large scale irrigation works were taken up, chiefly as commercial ventures. And these did prove lucrative enterprises. Till independence, the irrigation sector was a net source of revenue for the government. This was because the then British government had adopted the policy of taking up only those works that promised some returns.

However, the recurrence of droughts and famines during the second half of the 19th century forced the government to undertake some irrigation projects to provide protection to crops against monsoon failures.

The real objective behind this was to save expenditure on famine relief. As a result, projects like Betwa Canal, the Nira Left Bank canal, the Gokak canal, the Khaswad Tank and the Rushikulya canal wee taken up as crop protection irrigation projects. But, the priority continued to be for the irrigation projects that could generate revenue for the government. The irrigation sector, therefore, continued to be a source of revenue for the official exchequer.

In the planned development era after the Independence, the focus shifted from purely revenue generation to rapid harnessing of water resources for multiple benefits. Consequently, the state governments were encouraged to expeditiously formulate and implement water resources projects for specific purposes like irrigation, flood control, hydro-power generation, drinking water supply, industrial and various other miscellaneous uses.

This policy resulted in creation of a huge infrastructure of various kinds of dams, barrages, hydro power structures, canal networks, etc. The most significant impact of all this is the creation of a total water storage capacity of whopping 200 billion cubic metres, most of which is available for irrigation, besides serving other purposes.

Apart from environmental and rehabilitation issues that have been impeding implementation of major and medium irrigation projects, paucity of resources has been the most formidable bane of this sector. The failures to meet the targets, resulting in inordinate delays in the completion of irrigation projects, have resulted in whopping escalation in costs, making further progress in this sector all the more difficult. Many of even the operational projects not catering to the potential command areas for want of resources for operation and maintenance. Lack of resources of the state governments and flawed water pricing policies are largely to blame for the current dismal state of this vital sector.

In fact, a large number of river valley projects, including multi-purpose and irrigation projects, have spilled over from one plan to another chiefly for want to adequate funds allocation by the state governments responsible for implementing the projects. Though no new irrigation and multi purpose projects of any significance have been launched since the 7th plan still there were as many as 171 major, 259 medium and 72 extension. renovation and modernisation schemes that spilled over from the 8th plan to the 9th plan. Of these, five projects were pending since the 1st plan, seven projects each from the 2nd and 7th plant. 12 projects from the 3rd plan, 18th from the 4th plan and 20 from the 8th plan.

The spillover cost of the pending projects was estimated at whopping Rs. 75,690 crore (Economic Survey 2005-06). The capital cost of creation of irrigation potential through medium and major irrigation projects is estimated to have risen several folds from around Rs. 40,000 a hectare in the early 1970s to over Rs. 1,90,000 a hectare in the 1990s at the constant 1995-96 prices.

In fact, in real terms, the allocation of government resources for canal irrigation has almost been stagnating since the mid 1980s. The sare of irrigation in the total plan expenditure has been below 10 per cent in the 1980s as well as 1990s. This had led to a perceptible fall in the rate of creation of canal irrigation potential. It dropped from about a million hectares a year between 1974 and 1980 to merely 0.4 million tonnes a year in the 8th plan (1992-97).

It was basically for these reasons that the Accelerated Irrigation Benefit Programme (AIBP) was launched in 1996-97. The main aim of this programme was to accelerate the work on those irrigation projects where large investments had already been made to that these could be completed without further delays.

Subsequently, the Central funding of the AIBP was linked to the introduction of irrigation sector reforms by the states. While most states, barring those falling in the special category due to their backwardness, got the Central plan assistance in the ratio (Central : State) of 2 : 1, the reforming states (which agreed to raise water charges expenses) got the assistance in the ratio of 4 : 1.

The other source of surface irrigation is the traditional tranks and other surface water bodies. The country has about 1.5 million tanks dotting the countryside in all the states. The density of these tanks is relatively higher in states like Andhra Pradesh, Karnataka, Madhya Pradesh, Maharashtra, Tamil Nadu and West Bengal. But many of these tanks suffer from lack of proper maintenance. Besides, the growing siltation has, over the years, reduced their irrigation capacity. While in the 1960s, an area of about 4.8 million hectares was estimated to be irrigated through tanks, this acreage is now reckoned to have dropped by 1.7 million hectares due to drying up of the tanks or other reasons.

A massive programme for the repair, renovation and restoration of tanks and other water bodies was envisaged in the 2004-05 budget. For this purpose, a pilot scheme to be implemented in the 10th plan was approved in January 2005 with an estimated cost of Rs. 300 crore. The funding of this state sector scheme is shared by the Centre and the state governments in the ratio of 75 : 25. The Union water resources ministry had approved projects in 23 districts of 13 states involving a total cost of Rs. 262.91 crore till April 2006. These projects are in the states of Andhra Pradesh, Chhattisgarh, Gujarat, Himachal Pradesh, Jammu and Kashmir, Jharkhand, Karnataka, Kerala, Madhya Pradesh, Orissa, Rajasthan, Tamil Nadu and West Bengal. More projects are likely to be brought under this scheme subsequently to step up the overall irrigation command area from surface water bodies.

Indeed, minor irrigation is gradually becoming more important because of the several advantages that it offers. These include small investment, simpler components, quicker rewards and easier management. It is also farmer friendly as it can respond to individual needs. As a result, the bulk of the minor irrigation potential has been created through private investment and is based largely on groundwater exploitation. But several government departments, such as those of agriculture, rural development, irrigation, social welfare, etc., have also been promoting minor irrigation programmes.

An ambititious scheme to expand the network of wells was launched in 1988-89 by the Ministry of Rural Areas and Employment. Called a Million Wells Scheme, it has helped construct over 12.6 lakh wells with an expenditure of over Rs. 4.728 crore.

Besides, the Agriculture Ministry has been prompting the financial institutions like commercial banks, regional rural banks, cooperatives and the National Bank for Agriculture and Rural Development (NABARD) to provide credit to the farmers for the construction of wells and tubewells. The Water Resources Ministry has been assisting the state governments in the preparation of minor irrigation schemes that could quality for external financial funding.

A worrisome aspect of the irrigation sector is the large gap between the irrigation potential created and its utilisation. A Centrally sponsored Command Area Development (CAD) scheme was launched in 1974-75 to narrow down this gap. This was subsequently expanded to include water management aspects as well to ensure efficient utilisation of water with a view to optimise farm production in irrigation command areas. Till November, 2005, about 310 projects having a total command area of around 28.85 million hectares had been included under this programme.

Irrigation has now been included as one of the six components for the development of rural infrastructure under the massive Bharat Nirman programme. Under this, the target has been fixed to create an additional irrigation potential of 10 million hectares in four years (2005-06 to 2008-09). This is sought to be achieved largely through expeditious completion of identified on going major and medium irrigation projects, besides other measures. On completion, these major and medium projects alone would add about 4.2 million hectares to the total irrigation potential in a short period.

The irrigation component of the Bharat Nirman also seeks to reduce the gap in the creation and utilisation of irrigation potential. For this, it purpose to restore and utilise irrigation potential of one million hectares through extension, renovation and modernisation of the existing schemes and by introducing more efficient water management practices in the irrigation command areas.

This apart, the Bharat Nirman also envisages to tap the available ground water for irrigation in the areas where this source is not being adequately exploited. As such, it has set a goal of creating irrigation potential of about 2.8 million hectares through ground water development.

In addition to this, about one million hectares of irrigation potential is planned to be created by way of minor irrigation schemes by using surface flows. Another one million hectares irrigation potential is proposed to be made available through repair, renovation and restoration of water bodes and extension, renovation and modernisation of minor irrigation schemes. The focus of all the minor irrigation programmes would be to benefit the small and marginal farmers.

However, while the need for further expansion of irrigation to boost agricultural growth is paramount, equally significant is the necessity of addressing the problems that irrigation often brings in its wake. These are manifested in growing menace of water logging and soil salinity in the canal irrigated areas and fast depletion of ground water in the tubewell irrigation regions.

The reasons for these problems, *prima-facie,* appear to be inadequate drainage, excessive use of water in the canal command areas and indiscriminate exploitation of ground water in well irrigated areas. But, behind all these factors is the unrealistic pricing of water in the canal irrigated areas and cheap or free supply of electricity for running tubewells. Water charges have not been revised in several states are giving free or subsidised power to the agriculture sector for running the tubewells. These may be populist measures but these are now proving counter productive.

Also related to this is the question of an individuals right to unrestricted use of water under his piece of land. The legal provisions of granting such rights haves, obviously, become outdated, especially considering that the underground water is a common aquifer and over exploitation by one can deprive the other of his share, especially in areas where the water table is receding very fast.

Such policies prompt indiscriminate and inefficient exploitation of water, giving rise to ecological problems, besides telling upon soil health and consequential productivity constraint. Unless these issues can be taken care of through water sector reforms, the sustainability of the irrigation sector, from both economic and ecological viewpoints, may be in jeopardy.

Chapter - 18

Drip Irrigation System - An Overview

Dr. A.M. Michael

"Drip or trickle irrigation is an efficient irrigation method of recent origin, which is becoming increasingly popular in areas of water scarcity and poor quality irrigation water. In drip irrigation, water is applied frequently at low rates from a low pressure delivery system comprising of small diameter plastic pipes fitted with outlets, called emitters or drippers, directly to the land surface close to the plant where the roots grow. Drip irrigation is best suited to water scarcity areas and regions where irrigation water quality is marginal or low (saline water). Due to the frequent application of irrigation water, salts are pushed to the periphery of the moist zone, away from the root spread area. It is suitable to almost all types of soils. In clay soils with low infiltration rates, water is to be applied slowly to prevent surface water ponding and runoff. In sandy soils with high infiltration rates, higher dripper discharge rates will be required to ensure adequate lateral wetting of the soil. Drip irrigation is advantageous on lands with undulating topography without undertaking major land leveling operations and on slopes where the

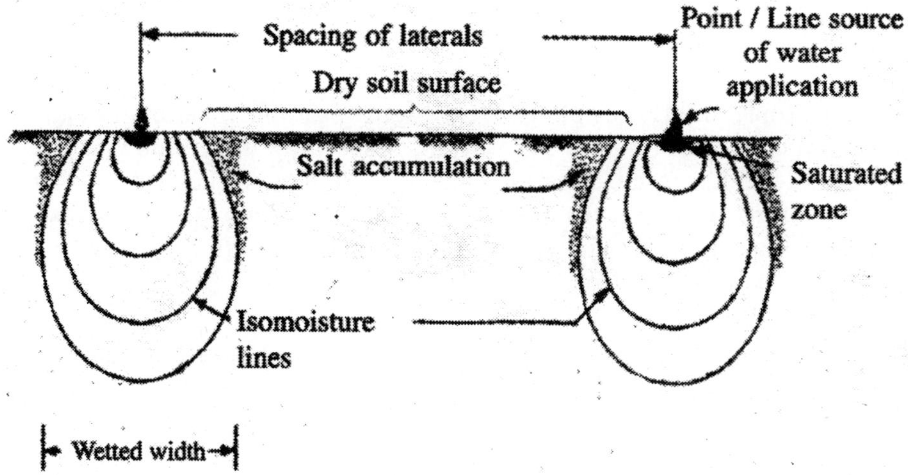

Fig. 1. Profiles of moisture front advance in drip irrigation in medium textured soils.

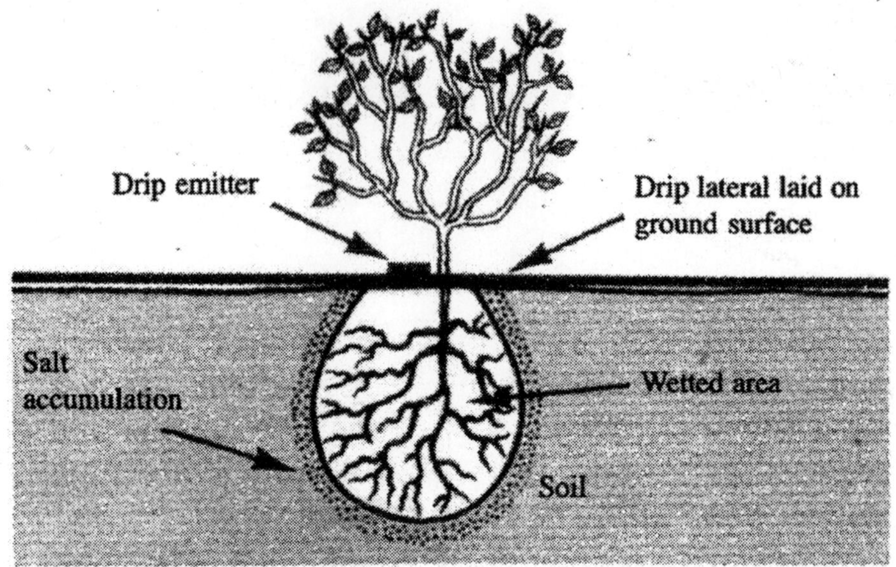

Figure – 2. Movement of salts to the periphery of the wetted bulb in point source of water application (drip irrigation).
(Source: Adapted from Lake Irrigation Co. Bakerfield, California, USA)

Soil depth is limited and the crop value is high. The labour requirement is low in drip irrigation. Drip irrigation is suitable for almost all orchard crops, plantation crops and most of the row crops. The method has been observed to be highly economical and beneficial in water scarcity areas to grow orchard and plantation like coconut, tea, coffee, cardamom, citrus, grapes, banana, papaya, mango, guava, pineapple and pomegranate; row crops like sugarcane, cotton, groundnut, sapota, strawberry and vegetable crops including tomato, potato and other widely spaced vegetable crops, and flower plants. It is highly suitable for nursery raising and establishment of forestry plantations, especially under the wasteland development programme. It is also suitable to irrigate sugarcane cotton and groundnut. However, from economic considerations, drip systems are not adapted to close growing field crops like cereals and pulses, as the number of laterals required are high which result in high costs of the system.

The terms *drip irrigation, trickle irrigation* and *micro-irrigation are* synonymous. Drip irrigation is a micro-irrigation method in which the rate of water application is very low and without any pressure, i.e., drop by drop. While drip irrigation and trickle irrigation have identical meaning, the term micro-irrigation encompasses other low pressure water application methods like microsprinkler, bubbler and spray systems in which water is emitted in the form of small sprinkling, tiny streams or miniature sprays through fixed diffusers or applicators, localizing the water on the soil in separate spots as well as indigenous methods of micro-level application of water to plants. The term *localized irrigation* is sometimes used in place of drip/trickle/micro-irrigation, as the systems cause wetting which is localized to the plant root zone. The devices for water application are in the form of emitters strip tubings, jets or mini-sprinklers. The service pressure of the system is of the order of 1 bar or less. Fertilizers and other chemicals can be applied efficiently to individual plants or a group of plants, using the drip system. Although the terms drip irrigation and trickle irrigation are used interchangeably.

Drip irrigation is based on the basic concept of irrigating only the root zone of the crop, rather than the entire land surface on which the crop is grown. Water flows from the emission points through the soil by capillarity and gravity. The soil moisture content of the crop root zone is maintained at near optimum levels to facilitate optimum crop growth and production. This will involve frequent applications of small quantities of water. Irrigating only a part of the land surface minimizes evaporation losses, reduces weed growth - especially in widely spaced crops and minimizes interruption of cultural operations like inter-cultivation and harvesting.

Major advantages of drip irrigation systems: The major advantages of drip irrigation, as compared to other methods of irrigation include the following:

(i) ***Higher crop yields:*** Studies on crop performance with drip irrigation, in comparison to other irrigation methods have consistently shown yield increases ranging from 20 to 50% or more in drip irrigated crops, namely, fruit crops (banana, grapes, citrus, pomegranate, papaya, pineapple, watermelon and most other fruits), vegetable crops (cabbage, cauliflower, okra, tomato, potato, onion, chilli, radish, brinjal, bottle gourd, French bean and most other vegetable crops grown in the tropical, sub-tropical and temperate regions) and commercial crops like cotton and sugarcane. The increase in crop yield is mainly due to the maintenance of optimum soil moisture in the crop root zone throughout the growing period of the crop with the better control of irrigation water application in the drip system,

(ii) ***Improved quality of the harvested produce of the crop:*** Drip irrigation results in improved "quality of crop produce due to the maintenance of optimum soil moisture conditions in the root zone throughout the growing season, resulting in uniform crop growth. In addition, damage and loss due to water contact with foliage are eliminated, resulting in better quality of produce.

(iii) ***Sayings in irrigation water:*** With the elimination of water application losses such as deep percolation, runoff and evaporation and reduction in the volume of the soil which is wetted (limited to the crop root zone), there is substantial saving in the amount of water required for irrigation, as compared to other methods of irrigation. Loss of water from the land surface which is not occupied by the crop is eliminated. As compared to sprinkler irrigation, the loss of water due to evaporation from the plant foliage is eliminated. Further, the losses in water conveyance such as seepage is eliminated with the adoption of drip method of water application, as compared to the unlined open channel water conveyance system commonly used in surface irrigation methods. The savings in the quantity of irrigation water resulting from the adoption of drip irrigation usually range from 20 to 50% or more, when compared to surface methods of water application and to a lesser extent when compared to sprinkler irrigation. Further, the irrigation water requirement of crops under drip irrigation is substantially low during the initial stages of crop growth. The saving in water by introducing large scale application of drip irrigation in a project area increases the irrigation command area of the project and provides opportunity for diversification of crops and introduction of high value' crops in the cropping system,

(iv) ***Increased efficiency in fertilizer use:*** A drip irrigation system is an effective means of applying fertilizers and other plant nutrients into the crop root zone where they are needed. The efficiency in fertilizer application increases greatly when fertilizers and nutrients are applied with the drip system, as compared to other methods of water application.

(v) ***Reduced energy consumption:*** Drip irrigation results in reduced energy requirement in pumping, compared to sprinkler irrigation, due to the low operating pressure required by the drip system. Further, the quantity of water required for irrigation is low under the drip system, which further reduces the energy requirement.

(vi) ***Tolerance to windy atmospheric conditions:*** Wind has no adverse influence on drip irrigation, as compared to the sprinkler system. In the drip system, the laterals with the emitters are laid at the land surface and only the point of water application is wetted which is hot influenced by the wind. The evaporation loss from the land surface is also negligible, as the wetted area is under the plant foliage.

(vii) ***Reduced labour costs:*** In irrigated agriculture using the drip system, the labour requirements are low due to reduced cost of field preparation, elimination of fertilizer application as a separate operation, low requirement of weeding and fewer harvesting rounds due to more uniform ripening of the crop.

(viii) ***Improved disease and pest control:*** Plant diseases and pests are lower in drip irrigated crops, as compared to other methods of water application. Bacteria, fungi and other organisms are reduced as the 'above ground' plant parts remain dry. As the application of water is limited only to the crop root zone, the areas surrounding the plant remain dry and hence the weed growth in the crop field is greatly reduced. With little or no weed growth, the intensity of pest and disease incidence is reduced. With the elimination of surface runoff resulting from drip irrigation, the spread of disease causing organisms by water movement is reduced,

(ix) ***Feasibility of irrigating undulating terrain and sloppy land:*** Hilly and undulating terrain cannot be irrigated with surface irrigation methods without extensive land levelling operations. Drip and sprinkler methods can be designed to operate efficiently on almost any topography without incurring the expenditure in land levelling. Drip irrigation can be designed to suit changing field gradients.

(x) ***Suitability for problem soils:*** In heavy clay soils, with low infiltration rates, low rates of water application can be selected with the drip system to match the soil characteristics. On the other hand, in sandy soils with high infiltration rates, water can be applied in small quantities frequently, avoiding deep percolation losses,

(xi) ***Improved tolerance to salinity:*** *With* the adoption of drip irrigation systems, it is possible to use more saline water for irrigation, as the root zone soil column is kept at higher level of moisture content, resulting in reduced soil moisture stress. The contact of water with the plant is minimized, compared to other methods of irrigation. In general, drip irrigation will reduce the sensitivity of most crops to saline soils and waters, due to its ability to maintain low water tension in the root zone. The frequent application of water with drip irrigation reduces the concentration of salts in the root zone by moving salts away from the root zone to the edges (perimeter) of the wetted bulb formed by the point source of water application in the drip system (Fig. 2). This process, called *micro-leaching,* prevents the harmful combination of high soil salinity and low moisture availability.

(*xii*) ***Promotes congenial soil physical conditions in the root Zone :*** The maintenance of soil moisture at nearly constant and optimum levels by renewing the water supply to the root zone nearly at the same rate as it is used by the plant results in low soil suction and facilitates water and nutrient uptake by the plant and high soil hydraulic conductivity. The soil, on the other hand, is never saturated in a properly managed drip irrigation system, and adequate aeration is maintained throughout the growing period of the crop. The low rate of water application makes drip irrigation suitable even in soils with low infiltration characteristics. In comparison to sprinkler irrigation, where the losses due to over-irrigation and irrigating areas not occupied by crops, including field edge losses, (may be as high as 20% or more), these types of losses are almost eliminated in drip irrigation. Further, no water is lost due to the transpiration of weeds, as irrigation is limited to the crop. The drip system facilitates better control of the irrigation water supply, as compared to the sprinkler irrigation system. There is no formation of soil crust in drip irrigation,

(*xiii*) ***Promotes better uniformity in irrigation water application:*** The large number of irrigation points (emitters) in a drip system provide better uniformity of application of water. Low discharges and low pressure heads in the water distribution network enable the use of smaller pipes of lower pressure ratings at reduced costs. Fertilizers, pesticides and other chemicals may be injected into the system and applied in small quantities, as needed, with irrigation water.

(*xiv*) ***Easy in operation:*** Drip irrigation system is comparatively easy to operate. Irrigation can be continued throughout the day and the night, regardless of wind, daytime, temperature or cultural practices,

(*xv*) ***Facilitates automation:*** Automation could be easily incorporated into the drip irrigation system

(*xvi*) ***Adapted to irrigate crops in green houses:*** Drip irrigation system is well adapted to irrigate crops grown in covered green houses with no wetting of the walls or the cover.

Problems associated with drip irrigation systems

In spite of the major advantages described earlier. There are several problems associated with drip irrigation systems:

(*i*) ***Emitter clogging:*** A major problem associated with drip irrigation systems is the clogging of emitters, unless preventive measures are taken. Emitter outlets are very small and can become clogged easily by suspended materials (sand and silt), precipitated dissolved salts (mainly carbonates), rust and other iron oxides and organic matter (including plant roots, algae and other minute animals). Clogging can reduce emission rates and cause poor uniformity of water application. Prolonged clogging can cause severe damage to the crop. Clogging also increases the maintenance costs, as it becomes necessary to check, repair or replace clogged emitters and other components. Chemical treatment and proper filtration of water can prevent or correct emitter clogging. The first requirement is to select emitter devices which may require less maintenance. Well-designed filtering systems, when properly maintained, can solve the problem caused by suspended material (sand and silt) and in some cases the organic matter. An effective means of dissolving carbonates is by means of dilute acids (mostly HNO3 or HC1). However, it will require an interruption in irrigation water application. Flushing by water or air under pressure is feasible if the pipes and joints can withstand the pressure. *Pulse irrigation,* namely, irrigating at a higher rate for a very short time, is effective.

(ii) ***Restricted root development of crops:*** Drip irrigation normally wets only a part of the root zone. Hence, the root distribution is almost limited to the moist zone. Many factors, involving soil and plant characteristics, management practices (amount and frequency of water application) and the design of the irrigation system (number of emitters per plant, placement and discharge rate of emitters) influence the root development of crops. The concentrated distribution of roots may reduce the plants ability to withstand strong winds. Further, the ability of the plant to withstand drought, resulting from any breakdown in the irrigation system, is greatly reduced as water of the wetted zone gets depleted soon and the surrounding region is dry. Careful planning and operation of the irrigation system and prevention of breakdowns will reduce the magnitude of the problem.

(iii) ***Salt accumulation at the root zone periphery:*** Where high salinity water is used for irrigation in arid regions, salts tend to accumulate towards the periphery of the wetted zone and the interface between the irrigated and non-irrigated zones in the soil. As the root zone is kept constantly at high moisture level, there may not be any major impact on the crop, but in the next growing season these salts, if not leached away, may damage the succeeding crop if planted on the interface having the salt patch. In regions where drip irrigation is practiced in the dry season with rainfall exceeding about 300 mm in the following

rainy season, the salts will generally be leached. Like in other methods of irrigation, good natural or artificial drainage is needed in drip irrigated areas as well. Otherwise, artificial leaching is to be provided once every one or two years.

(iv) ***Damage from rodents and other animals:*** There are several kinds of burrowing animals which can cause damage to surface or buried polyethylene laterals. They include rats, mice, squirrels, dogs and rabbits. When present in large numbers, these animals can cause heavy damage to drip irrigation systems by chewing holes in the lateral lines. Use of repellents to keep the animals away and baiting or trapping them are possible control measures. Rats and squirrels can be controlled by adopting subsurface drip system. Rats generally never cross a wet surface. Keeping the system moist is a possible remedy to control it. A chemical repellent which tastes or smells bad to the animal can be injected through the system or laid down with the laterals during installation. There are a number of chemicals available which are obnoxious to animals, including anhydrous or aqua ammonia and a number of insecticides which can be injected into the drip system. Rodent damage can also be prevented by the use of polyvinyl chloride (PVC) laterals. Ants and beetles can damage thin-walled polyethylene tubing. Ant damage typically takes the form of holes chewed through the sides of strip tubing and enlargement of orifices in strip tubing. Some types of insects will build a cocoon inside emitter outlets in water distribution tubing. Cocoons are built when the system is shut down for a period of time. These insects are to be controlled only during prolonged shut down periods. Ant damage can be successfully controlled with chlorinated hydrocarbon insecticides. However, these chemicals are usually toxic and persist in the environment and should be used with great care. A common recommendation is to use lateral tubes with sufficient wall thickness (not less than 15 mils (0.38 mm) to reduce damage by ants and insects.

(v) ***High cost of drip irrigation systems:*** The cost of drip irrigation system is high when compared to surface or portable sprinkler irrigation systems. Drip irrigation systems are expensive because of their requirements of large quantities of piping and filtration equipment to clean and water. However, the cost of drip irrigation systems varies considerably depending on the crop and terrain. Steep terrain may require several pressure regulators in the system. In many situations the benefits of the drip system will usually overweight the cost of the system, when compared to other methods of irrigation. Selecting widely spaced crops of high market value is necessary

to increase the benefit cost ratio in drip irrigation. Under average field conditions, the major components of the total cost of the system lie in the number of laterals/and emitters of the system. Evidently both these items will be low in hards and other widely spaced row crops.

APPLICATION OF FERTILIZERS AND CHEMICALS THROUGH DRIP IRRIGATION SYSTEMS

Water soluble fertilizers can be effectively and efficiently applied through drip irrigation systems (Fertigation). Reduced labour, equipment and energy costs and higher fertilizer use efficiency are the major benefits of fertigation, compared to the conventional methods of water application. The success of drip irrigation, to a good degree, is due to the improved supply of nutrients to plants, which is a unique capability of fertigation. Proper control of the time of application, the concentration of fertilizers/ chemicals, proportion of the nutrients and the location of the point of application are possible with the fertilization systems used in pressurized irrigation systems. Plant protection chemicals can also be applied effectively using the same facility. However, materials such as inorganic forms of phosphorous that form chemical precipitates and cause clogging should not be injected into drip systems.

A detailed treatment on the subject of fertigation is presented in Figure 3 illustrates the layout of a drip irrigation system with fertilization unit.

The requirement of fertilizer application, including the type of fertilizers to be injected should be considered in designing a drip irrigation system. Some types of fertilizers are not suitable for application through drip systems, because of volatilisation of gaseous ammonia, low water solubility and problems with the chemical quality of irrigation water. Therefore, fertilizer injection equipment should be designed with an understanding of the chemical composition of the fertilizer to be used. Nitrogen is relatively problem free.

Nitrogen. Anhydrous ammonia and aqua-ammonia can be injected into the irrigation water, but volatilisation is a major problem. Another problem with ammonia injection is the rise of hydroxide ion concentration in water. Ammonia increases the pH, which causes soluble calcium and magnesium to precipitate in irrigation water and coat the inside of the pipes and plug emitters. This problem, however, can be solved by injecting a water-softener ahead of the ammonia gas. However, the process may increase the cost of fertilization.

Ammonium sulphate and ammonium nitrate are very common fertilizers. In the former, all the nitrogen is in the ammonium form, and in the latter about 28% by weight of the fertilizer is ammonium nitrogen and 8% is nitrate nitrogen. Urea is a soluble nitrogen fertilizer. It does not react with water to form ions. Urea and ammonium nitrate are mixed in water to give a fairly concentrated liquid fertilizer mixture. All of these nitrogen materials can be injected with no side effects in the water or irrigation system.

Both urea and nitrate nitrogen stay in solution and move with the soil water. However, ammonium nitrogen behaves differently. Because it is a positively charged ion, it enters into cation exchange reactions in the soil. A small change in either soluble constituents alters the relative amount of the ions in exchangeable form. In the exchangeable form, ammonium is immobile. Because, cation exchange reactions are very rapid, ammonia applied in irrigation water is immobilized almost instantly on contact with soil and remains on or near the soil surface (US Soil Cons. Service, 1984).

Ammonium applied in water readily converts into exchangeable ammonium and generates an equivalent amount of cations in solution. In semi-arid and arid 'regions, soils are usually neutral to alkali (pH 7 to 8.20) depending on how much free lime or calcium carbonate is present. In such soils, any exchangeable ammonium that exists at the soil surface is likely to volatize. Ammonium is sensitive to temperature and moisture. Water evaporates rapidly from soil after irrigation, and ammonium is susceptible to gaseous loss at this time (US Soil Cons. Service, 1984).

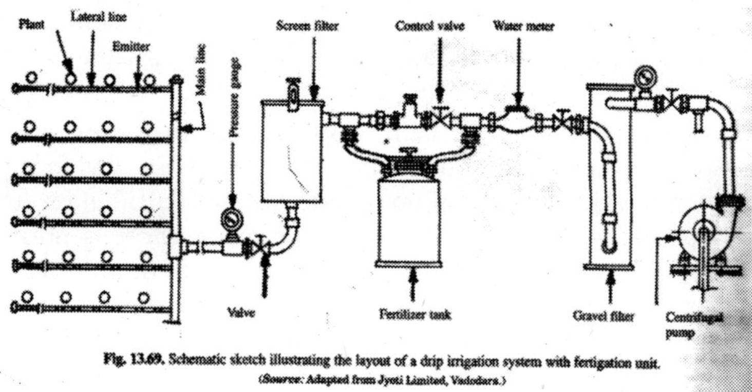

Fig. 3. Schematic sketch illustrating the layout of a drip irrigation system with fertigation unit. (Source: Adapted from Jyoti Limited, Vadodara)

Phosphorous: Phosphorous is difficult to apply by injection into the drip system. Treble-superphosphate, which is commonly used is usually classified as water soluble, but is only moderately soluble. Hence, it is usually not recommended for injection in irrigation water, unless special precautions are taken. Several kinds of ammonium phosphates are soluble in water. Ammonium phosphate sulphate, ammonium phosphate and di-ammonium phosphate are suitable for injection in drip systems when the soil requires nitrogen and phosphorous. Phosphoric acid is another form of soluble phosphorous. The quality of irrigation water is to be determined before injecting phosphorous into the drip system. If the irrigation water has a pH above 7.5 and a high calcium content, the injected phosphorous will precipitate as calcium phosphate and may clog the emitters and restrict the flow in the pipeline. Under such situations, phosphoric acid could be used to meet phosphorous needs. Flushing the drip irrigation system with a solution of either sulphuric acid or hydrochloric acid immediately after applying phosphoric acid prevents clogging. Organic phosphate compounds like glycerol-phosphoric acid can also be injected into the drip irrigation system without any precipitation problem.

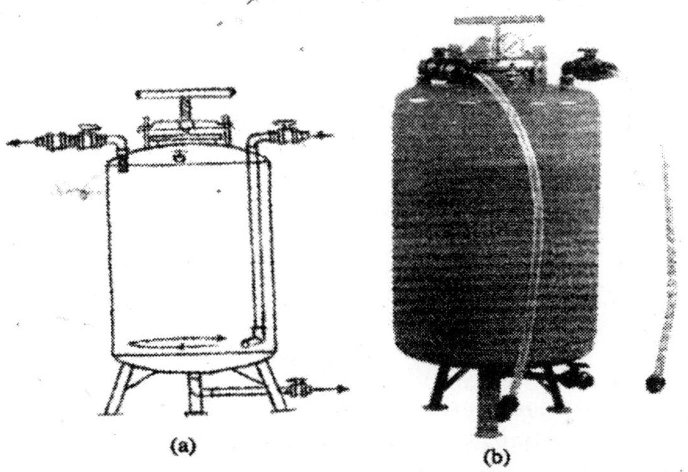

Fig. 4. Drip Irrigation Systems (a) Sketch of the tank assembly, (b) The tank to which are fitted inlet and outlet tubes. (Source: Jain Irrigation System

Potassium: Potassium is easy for injection through drip irrigation systems. The commonly used potassium oxide is very soluble. The fertilizer moves readily into the soil and is not easily leached away.

Trace Elements: The common trace elements like magnesium, zinc, boron, iron, and copper can be applied through drip irrigation systems. The application rates of trace elements should be based on proper analysis of soil and water, as trace elements applied in excessive quantities can react with salts in the water and can be toxic to plants.

Pumping Unit

Drip irrigation systems have pumping units to lift water from wells, streams, canals, ponds and reservoirs and provide pressure for the operation of the system. The pumps can be powered either by electric motors or internal combustion engines. The types of pumps used in drip irrigation include volute centrifugal pumps, vertical turbine pumps and submersible pumps.

DRIP IRRIGATION SYSTEMS

Many problems with drip irrigation systems arise out of poor installation. The main items in the installation of drip irrigation systems are the installation of the head assembly (Control head), comprising the pumping set, non-return valve, water meter, filters, fertilization equipment, and flow control, air release and pressure release valves. The network of water distribution pipes including the main line, sub-mains, and laterals require careful attention in joining and installation. General requirements in installation, operation and maintenance were included in the description of the various components of drip irrigation systems.

Installation of filters and fertilization unit

The major points to be considered in the installation of the filtering system are the use of minimum number of accessories like elbows and reducers. Normally, the pump delivery pipe should be connected directly to the hydrocyclone or the media filter, followed by the fertilization equipment, and the screen filter, all of which are installed in the main pipe. Arrangements for backwashing of the filter and the disposal of the contaminated water are essential requirements. Suitable facility to collect or to dispose off the bypass water is necessary. Sufficient space is to be provided for the operation of filter valves and other valves. Hard surface or masonry/concrete foundation is to be provided for the sand filter and the hydrocyclone, so that they will not collapse due to vibration and load. For a screen filter, a strong support by using G.I. fittings is necessary to avoid its vibration due to load. Use of 'hold-tight' over the threads of G.I. fittings and application of proper mixture of 'M-seal' over the joints uniformly are necessary to prevent leakages.

Installation of mains and sub-mains. It is a desirable practice to lay the mains and sub-main pipes underground, at a depth not less than 45 cm to avoid damage due to vehicles and equipment. The United States Soil Conservation Service has recommended the following minimum cover of earth over for various pipe sizes (Fred Hamisch, 1977):

The PVC pipes are usually joined by solvent welding. Pipe sizes of 4 cm diameter and above are usually connected by gasketed joints. Thermal properties of PVC pipes require that an allowance for expansion and contraction be made when the pipeline is assembled. This is normally accomplished by 'snaking' of the PVC pipe in the trench. PVC pipe is affected by sunlight and heat. The pipe must be buried, or otherwise protected from these elements. Otherwise, it will become brittle and lose many of its valuable properties. The pipe should be laid in the trench and allowed to come to within a few degrees of the lowest possible temperature before backfilling. When the trench contains rock, hard pan, or boulders, which might damage the pipe, it should be surrounded with good clean fill material to a depth of 15 cm over-the pipe with no stones larger than 2.5 cm (Fred Hamisch, 1977). The remaining portion of the. Trench may be filled with the available excavated material, avoiding large stones.

The assembly procedures for PVC solvent weld pipes are available from the manufacturers of PVC pipes. An important point to be kept in mind in making a good joint is to avoid leaving any excess solvent on the joint. No excess solvent should be allowed on the outside or the inside of the joint. The solvents used in making the joints are flammable and toxic. Hence, care must be taken when handling them. Solvent jointing requires a certain amount of skill which can be acquired.

The installation procedures for gasket jointed pipes are the same as for solvent-weld PVC pipes. However, the assembly procedures are simplified, as gasket joints do not require solvent cement application. In both the cases, thrust blocks should be used at all changes of direction, tees and bends, changes of pipe size, deadens and valves. The size and shape of the thrust block will depend on the operating pressure, pipe size and soil type. Normally, a thrust block is a stiff concrete poured between the pipe and the undisturbed bearing wall of the trench.

The main points to be kept in mind while connecting mains and sub-mains are the removal of mud and other impurities in the pipe before fitting, fitting a gunmetal gate valve at the start of the sub-main, and providing flush valves at the end of sub-mains. The main and sub-mains are made firm by putting soil near the flush valve and at different places so that the pipe will not move while making a drill hole into it. The rubber grommets in the holes made in the sub-main are fitted in such a way that the groove in it goes inside the pipe. The take-off is fixed tightly in the grommet. The loose fitting of the takeoff indicates the breakage of grommet. PVC pieces which fall into the pipe should be flushed out before commissioning the irrigation system.

Punching of lateral and fixing of drippers: The following procedures may be followed in the punching of laterals:

(i) Pass water through the polytube so that it gets bulged and easy for punching,

(ii) Punch the lateral sideways.

(iii) The dripper position is fixed according to design,

(iv) All the drippers should come on the same straight line,

(v) Do not fix the drippers until a complete lateral line is punched,

(vi) Punching should start from the sub-main.

Fixing of drippers: *(i)* In case of drippers having arrow marks, the arrow head should be towards the sub-main, *(ii)* While fixing the dripper, push it into the lateral and pull it slightly. *(iii)* Close the end of the lateral by fitting end cap. (ICAR, 2003 - contribution by Dr. Ashwani

Kumar). However, in medium and large scale installations the drippers should preferably be installed in the factory.

Testing of the drip irrigation system: After the irrigation system is installed, it is tested by adopting the following procedure: *(i)* Backwash the filter till clean water comes out through its flush valve, *(ii)* Ensure that all the valves are open before the testing. *(iii)* Close the flush valve after the sub-main is completely flushed. *(iv)* When the laterals are completely Flushed, close them with end caps, *(v)* Check the pressure on the gauge installed on the filter, *(vi)* If excess pressure is observed, open the by-pass valve slowly till the desired pressure is obtained, *(vii)* Measure the discharge at a minimum of 3 different places. For this the volumetric method may be used, *(viii)* Check the working of the air release valve on the sub-main, *(ix)* After fitting the entire system, install the venturi on the filter (ICAR. 2003 - contribution from Dr. Ashwani Kumar).

Operation of the Drip Irrigation System

The proper operation of a drip irrigation system involves the following steps for the owner/operator:

- *(i)* Acquiring complete information and instructions from the designer and dealer,
- *(ii)* Determining when and how long to irrigate.
- *(iii)* Checking the water meter readings and recording the figures,
- *(iv)* Accurately setting the hydraulic metering valve,
- *(v)* Operating the head valve to begin irrigation,
- *(vi)* Checking the system along all components for proper operation, beginning with pressure readings at the header,
- *(vii)* Checking the emitters, at least on a random basis,
- *(viii)* Setting the chemical and fertilizer injection equipment. (US Soil Cons. Service, 1984).

Maintenance of Drip Irrigation Systems

Reliable performance of a drip irrigation system depends on preventive maintenance that includes proper filtration, pipe flushing, field check of leakage in any component of the system, and monitoring the effective functioning of mechanical devices. General maintenance includes periodic checking of drippers, including their discharge, wetting zone, and leakage of pipes and filter gaskets in the lids, flushing valves and fittings. The placement of drippers should be periodically inspected. In case the placement is disturbed, they should be put at the proper location. Filter is the heart of the drip irrigation system. Filter failure will lead to clogging of the entire system. Various.

Flushing of sub-mains and laterals: After installation/repairs, the system should be flushed systematically, beginning with the main line and proceeding to the sub-mains, manifolds, and laterals. The main lines and then the sub-mains should be flushed one at a time with the manifold or riser valves turned off. Closing the valves on all lines, except the one being flushed, allows a large flow of water. The manifolds should be flushed with all the lateral riser valves turned off. Finally, the lateral hoses should be connected and flushed for about an hour at each operating station.

Sometimes silt escapes through the filters and settles in sub-mains and laterals. Further, some algae and bacteria lead to the formation of slimes/pastes in the pipe and laterals. To remove these silts and slimes, opening the flush valve should flush the sub-mains.

Removing the end stops flushes the lateral lines. By flushing, even the traces of accumulated salts will also be removed. Stop the flushing once the water going out is clean.

Fine sand, silt and clay tend to settle in the low-velocity section of the system, at the ends of manifolds and laterals. Emitters' receiving high concentrations of fine contaminants are susceptible to clogging; therefore, periodic flushing is a recommended part of a good maintenance program. Annual flushing is enough for many systems, but some water and emitter combinations require more frequent flushing to control clogging.

(o) ***Chemical treatment.*** Clogging or plugging of drippers will be due to precipitation and accumulation of certain dissolved salts like carbonate, bicarbonate, iron, calcium and manganese salts. Clogging is also due to the presence of microorganisms and the related iron and sulphur slimes due to algae and bacteria. Clogging or plugging is usually avoided/cleared by chemical treatment of water. Chemical treatments commonly used in drip irrigation systems include addition of chloride and/or acid to the water supply.

Acid treatment: Hydrochloric acid (HC1) is injected into drip systems, at the rate suggested. The acid treatment is performed till a pH of 4 is observed and the system is shut for 24 hours. Next day the system is flushed by opening the flush valve and lateral ends.

Chlorine treatment. Chlorine treatment in the form of bleaching powder is performed to inhibit the growth of organisms like algae and bacteria. The bleaching powder is dissolved in water and this solution is injected into the system for about 30 minutes. Then the system is shut off for 24 hours. After 24 hours, the lateral ends and flush valves are opened to flush out the water with impurities. Bleaching powder is directly added into the source at a rate of 2 mg/ litre.

Standardization of Drip Irrigation Systems

Indian standards for different facets of drip irrigation systems are formulated by Irrigation and Farm Drainage Equipment and System Sectional Committee of the Bureau of Indian Standards. Most of the Indian Standards on irrigation equipment are based on corresponding international standards, which have been adopted after suitable modifications to meet indigenous requirements.

Design of Drip Irrigation System

Design of a drip irrigation system broadly involves the estimation of water requirement of the crops to be irrigated, number of drippers and laterals, diameter of the main, sub-main and lateral pipes, and the size of the pumping unit. The major difference between drip irrigation and other methods of irrigation is that not all the area is irrigated. Under normal conditions in widely spaced crops, a minimum of 30% of the area is irrigated when the drip system is adopted. For mature trees, about 75% of the area may have to be irrigated, while nearly the entire area is irrigated in case of closely-spaced crops. The necessary data to be collected includes soil type, infiltration characteristics of the soil, type of crops, consumptive use of water by crops, water quality, and climate, Class A pan evaporation data, and the availability of funds. A topographical map of the area to be irrigated, with contour lines at a spacing of about one metre is necessary in planning and layout of the system.

A check list of procedures in designing a drip irrigation system is given below:

(i) Prepare on inventory of available resources and operating conditions,

(ii) Determine the water requirement to be met by the drip system.

(iii) Determine the appropriate type of the system.

(iv) Determine the type and design of emitters (on-line/ in-line and the specific type required),

(v) Determine the capacity of the pumping system,

(vi) Decide on the economic sizes of the piping system,

(vii) Determine the maximum and minimum operating pressures and the minimum efficiency requirement,

(viii) Determine the appropriate filtering system to suit the site conditions.

(ix) Determine the requirement of the fertilization system,

(x) Plan field evaluation.

(xi) Prepare drawings, specifications, cost estimates, schedules and instructions for proper layout, installation, operation and maintenance.

References

EPC Irrigation Ltd. Leaflet, EPC Drip Irrigation, B-20 MIDC, Ambad, Nasik, Maharashtra, P. 18

Jyoti Limited, Vodadara, Leaflet, Jyoti Drip Irrigation System, Jyoti Ltd. Industrial Area, Vadodara, Gujarat, P. 4

Karmeli; D and Keller, J., (1974). Evaluation of Drip Irrigation System, Proc. Second Drip Irrigation Congress, San Diego, California, USA.

National Bank for Agriculture and Rural Development (1992), Model Scheme for Financing Drip Irrigation For Availing Bank Finance, NABARD, Mumbai, p.30

Netafim Irrigation Equipment and Drip System, Leaflet, Netafim Drip Irrigation Systems, Kibbutz Hatzerim Israel, P. 4.

Patel, V.C. and Angadi, S.S. (1997), Drip Irrigation, University of Agricultural Sciences, Dharwad, Karnataka, P. 40

Rajput, T.B.S. and Neelam Patel (2002), Drip Irrigation Manual, Water Technology Centre, IARI, New Delhi, P. 53.

Chapter - 19

Water Lifts and Pumps : An Overview

Dr. S.B. Verma, Dr. Ajay Kumar
& Dr. Arvind Kumar Shrivastawa

Lift irrigation requires that water be raised from its source to the field surface. Whatever the source of water in a lift irrigation project, the efficiency of the system depends on the application of sound principles in the design and construction of the utilization structure, usually the well and the characteristics of the water lifting device in relation to the source of water. Devices for irrigation water lifting range from age-old indigenous water lifts to highly efficient pumps. Pumps operated by electric motors or engines have come into prominence in all large scale lift irrigation schemes. This is because high output and efficiency levels can be attained and controlled, using mechanically-powered water lifts.

Basically, there are four principles involved in pumping water. These are atmospheric pressure, positive displacement, centrifugal force, and movement of columns of water caused by the difference in specific gravity. Numerous kinds of pumps are available, all of which embody one or more of these principles. The classification of commonly used irrigation water lifts is shown in the schematic chart on the following page.

Selection of a suitable water lifting device for a particular situation depends on the characteristics of the source of water and the lifting device, the amount of water to be lifted, the depth to the pumping water level, type of power available and the economic status of the farmer.

Man and animal power are the major energy sources in small-scale irrigation pumping and domestic water supply in many regions of developing countries. From a purely economic standpoint, the cost of raising water with man or animal power is substantially higher than those which are driven by electric motors and engines. Studies have revealed that the cost of raising a unit quantity of water is the lowest in case of pumps operated by electric motors. The relative cost for engine-operated pumps is about 15 to 25 per cent more than that of electric motor-driven pumps. The cost of lifting water by animal power is about 10 to 20 times more than with electric motors. Manually-operated water lifts are the costliest, being 3 to 4 times more than animal-operated devices. In spite of the economic disadvantage, man and animal power continue as important energy sources in domestic water supply and small-scale irrigation in developing countries. Low economic status of the rural masses often necessitate the use of man and animal power in water lifting in developing countries.

Application of Non-conventional Energy Source in Pumping

Fossil fuels (petrol/diesel) and electricity generated at centralized grid power systems have been the conventional energy sources in pumping for irrigation and water supply. In spite of their outstanding merits, fossil fuels have some major limitations. They are nonrenewable and their supply is bound to reduce and approach exhaustion with continued exploitation. The cost of diesel/petrol and their distribution to remote areas may sometimes be prohibitive in many regions. Their indiscriminate user result in environmental hazards like atmospheric pollution. Likewise, the cost of generating electricity at hydro-electric and thermal stations, transmitting it over long distances and distribution to individual small landholdings for operating pumps is becoming increasingly high. Hence, there is an ever increasing effort to adopt non-conventional energy sources in pumping, wherever feasible.

Fig. 1. A typical windmill installation for irrigation.

The four major non-conventional energy sources which have established their feasibility in water pumping are wind, micro rrydro-power, sunlight and biomass. All these energy sources are renewable, are available to the user at the pumping point and are environmentally safe. The pumping units suitable for non-conventional energy sources are the windmill operated reciprocating pumps, micro-water power operated hydraulic rams, photo-voltaic power operated centrifugal pumps and biogas-engine operated centrifugal pumps. All of them are currently limited to small scale use.

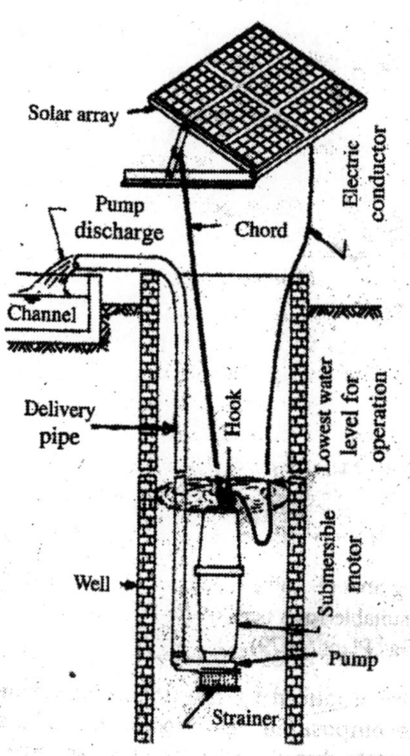

Fig. 2 : Photo-voltaic pumping set using submersible motor and pump installed in a shallow open well

Solar Photo-voltaic Pumping System

Solar energy, with its virtually infinite potential and free availability, represents a non-polluting and inexhaustible energy source which can be developed to meet the energy needs of mankind in a major way. The solar-pump unit consists essentially of a solar array, a direct-current electric motor and a pumping unit (Fig. 2). The other components are the electrical control and some mechanism for tracking the array against the sun. Two types of pumping sets are used with photo-voltaic systems, a vertical centrifugal pump coupled to a submersible DC electric motor (Figs. 2 and 3) or an ordinary volute centrifugal pump close-coupled to a horizontal DC electric motor. However, the submersible pump unit is more suitable for the photovoltaic system. This arrangement eliminates the suction pipe and foot valve and results in higher efficiency of the pumping unit. The submersible pump is made leak-proof by a mechanical seal. In case a volute centrifugal pump is used, care is taken to limit the pump suction within about 4.5m to maintain a high lever of pump efficiency.

Fig. 3. A solar photo-voltaic pump in operation.

Photo voltaic power generation: Photo voltaic cells, frequently referred to as solar cells, convert the light part of the solar spectrum (sunlight) into electricity. They are the most rapidly expanding energy sources in the world. The solar cell operates on the principal of the photo-voltaic effect - the creation of charge carrier within a material by the absorption of energy from the incident solar radiation. The solar cell is made of two thin layers of a semi-conductor material appropriately doped with impurity atoms so as to give one layer a negative electrical bias (n-bias) and the other a positive bias (p-bias). Sunlight falling on the cell reaches the junction between the two layers in the form of photons of energy and knocks electrons across the junction. This results in the development of a potential difference across the two layers. Direct current (DC) electricity can be drawn across the two layers through an external circuit.

The efficiency of solar cells in converting incident solar energy into electrical energy depends on the illumination spectrum intensity, material of construction and design of the cell, atmospheric temperature and dustiness of the sky around the solar panel. Silicon is the most commonly used material in manufacturing solar cells. A solar cell behaves like a low voltage (0.5 volt) battery whose charge is continuously replenished at a rate proportional to the incident solar radiation. Connecting multiples of cells into series parallel configuration results in photo-voltage modules or solar arrays with high currents and voltages. The power developed by a solar array ranges from 80 to 120 watts/sq m area of the panel. The output of the solar array varies with the incoming radiation, orientation of the panel and other factors. The solar panel is mounted on a suitable frame which can adjust the array to the direction of the sun rays. The angle of the array has to be changed 3 or 4 times a day.

Biogas Engine Powered Pumping Plants

The potential of biomass as an energy source is being increasingly realized. Biomass constitutes a significant, clean and renewable energy source. The biomass in the biological system may be classified in two broad categories: terrestrial biomass {organic residues and higher plants) and aquatic biomass (fresh-water aquatic plants, seaweeds, micro-algae and floating marine plants). Biogas is a mixture of gases containing methane, carbon dioxide, hydrogen and traces of a new other gases produced by the anaerobic fermentation of easily decomposable cellulosic materials. Animal manure (cattle dung) and municipal sewage have been the main materials used for producing biogas. The process has the advantage that animal and human waste can be used to generate energy while, at the same time, retaining their nutrient value for use as organic fertilizer. The production of methane gas from crop residues and aquatic plants like water hyacinth have also been attempted with considerable success. Water hyacinth is one of the major weeds leading to the blocking of many drainage canals and even irrigation canals and tanks. Their removal and use as a source of energy provide a substantial energy source and aid in preventing blocking of water bodies.

Biogas plants: Biogas plants may be classified into two, namely drum type and drumless type. The conventional biogas plant, originally developed at the Indian Agricultural Research Institute, New Delhi in 1935, is the drum type. It consist of a masonry digester (fermentation tank) with an inlet pipe on one side for feeding cattle dung mixed with water into the plant and an outlet pipe on the other side for discharging the spent slurry. The gas collects in a gas holder or drum made of mild steel. The gas holder is inverted over the slurry and moves up and down with the accumulation and discharge of gas.

Operation of biogas plants: To operate the biogas plant, a mixture of cattle dung or other animal excreta and water, in the ratio 1:1 is added as slurry to fill the digester. In a new plant, the production of gas may start in 5 to 10 days in summer, and 15 to 20 days in winter. When fresh dung is added into the digester, the digested slurry overflows into a collection pit.

Biogas engine pumping set: Ordinary (petrol/gasoline) engines can be adapted to run on biogas. In case of compression-ignition engines (diesel) adapted to run on biogas, a small part of the fuel continues to be diesel. A popular commercial make of biogas-run internal combustion engine working on the diesel cycle is designed to use about 80 per cent biogas and 20 per cent diesel oil. Minor modifications are made in the combustion chamber of the conventional diesel engine to adapt it for use with biogas. The engine is started on diesel oil and after warming, is made to run on the biogas-diesel mixture for continued operation. The gas consumption of biogas engines is about 450 litres (0.45 m^3) per brake horse power per hour of operation.

Hand Pumps for Domestic Water Supply

Properly installed hand pumps serve as a means of providing clean drinking water in rural areas. Handpumps are important in villages which are not provided with piped water supplies. The practice of drawing water from exposed open wells with a rope and bucket, coupled with bathing and washing near wells and in ponds used for drinking water supplies, are health hazards in rural areas of developing countries. Most governments have launched large programmes to provide potable water through hand pumps and piped water supplies.

The type of handpump to be selected will depend on the depth of ground-water, the yield of the well, constructional features and efficiency of the pump, and availability of facilities for maintenance and repair. The per capita provision of drinking water under government-sponsored domestic water supply projects in most states in India, vary from 25 to 70l/day/person. The discharge of handpumps varies from 500 to 1500 litres/ hour, depending mainly on the depth to ground water. India Mark II hand pumps [Fig. 4] to suit various pumping situations are widely used in developing countries for domestic water supply. They are comparatively trouble, free.

Handpump installation : Handpumps may be installed in bore wells (Fig. 4), filter points and open wells. If the pumping water level is within 6 m during the dry season, it is advantageous to install a shallow well lift pump, with the pump chamber located at the ground surface on top of the well. A deep-well lift pump with its cylinder submerged in water, is selected when the ground-water level is below 6 m. While installing a hand-pump in an open well, it is essential to close the top of the well with a removable slab made of reinforced concrete. The cover should slope towards the periphery of the well in order to drain away the waste water. The well cover will prevent the entry of pollutants from the ground surface into the well.

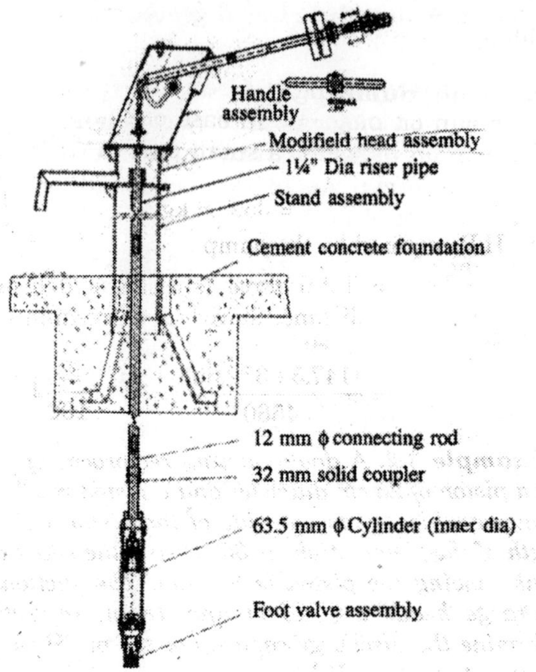

Fig. 4. Deep well hand pump assembly and installation.

When ever a hand pump is installed in a bore well or filter point, it is essential to provide a permanent platform, about lm square, around the well. The platform, may be of reinforced concrete or masonry. Brick or stone masonry may be used, depending on their relative costs. Provision of suitable drainage through a pipe or lined channel of semi-circular or rectangular cross-section is essential in a hand-pump installation. The drain should lead the waste water to a natural channel or ditch.

Maintenance and repair of hand-pumps: A large number of handpumps remain out of order due to inadequate arrangements for repair and maintenance. There is a need for training village youth, including artisans, in the repair of hand-pumps. An adequate stock of parts exposed to wear and tear should be maintained within easy reach of a group of hand pumps. The parts of hand-pumps most exposed to wear and tear are the cup leather washers of the plunger, valve seats, gaskets, bolts and nuts, cotter pins and sockets. A stock of a few section of the pump rod and arrangements for threading the rods are also essential.

CENTRIFUGAL PUMPS Amongs modern pumps, centrifugal pumps are most widely used in irrigation practice. They are simple in construction, easy to operate, low in initial cost and produce a constant steady discharge. The wearing parts are few. They are adapted to direct motor or engine drives without the use of expensive gears. This type of pump is well adapted to usual pumping service such as irrigation, water supply and sewage service. Having no valves, the pump can handle liquids having solids in suspension, provided it is constructed to suit such conditions.

Principles of Operation of Centrifugal Pumps

A centrifugal pump is a rotary machine consisting of two basic parts-the rotary element or impeller and the stationary element or easing (Fig. 5 to 7). The impeller is a wheel or disc mounted on a shaft and provided with a number of vanes or blades usually curved in form. The vanes are arranged in a circular array around an inlet opening at the centre. In some pumps, a diffuser consisting of a series of guide vanes or blades, surrounds the impeller (Fig. 6). The impeller is secured on a shaft mounted on suitable bearings. The shaft usually has stuffing box or seal where it passes through the casing wall (Fig. 7). Stuffing box packings are generally made of materials such as asbestors or organic fibre. The casing surrounds the impeller and is usually in the form of a spiral or volute curve with a cross-sectional area-increasing towards the discharge opening.

A centrifugal pump may be defined as one in which an impeller rotating inside a close-fitting case draws in the liquid at the centre and by virtue of centrifugal force throws out the liquid through an opening at the side of the casing. The underlying hydraulic principle in the design of an impeller is the production of high velocity and the partial transformation of this velocity into pressure head. In operation, the pump is filled with water and the impeller is rotated. The blades cause the liquid to rotate with the impeller and in turn impart a high velocity to the water. Centrifugal force causes it to be thrown outward from the impeller into the casing. The outward flow through the impeller reduces the pressure at the inlet, allowing more water to be drawn in through the suction pipe by atmospheric pressure or an external pressure. The liquid passes into the casing where the high velocity is reduced and converted into pressure and water is pumped out through the discharge pipe. This conversion of velocity energy into pressure energy is accomplished either in a volute casing or in a diffuser casing.

By changing the form of the vanes, different characteristic are obtained. By enlarging the diameter of the inlet eye and the width of the impeller, the quantity of water that the pump delivers against a given head is increased.

The number of vanes used in an impeller varies with the operating conditions. In general, the higher the head, the more the vanes used; and greater the rate of pumping, the fewer the vanes. Too few vanes provide poor guidance for the water; too many cause excessive frictional resistance. The minimum number of vanes is usually three and the maximum about twelve.

When the flow enters at one side of an impeller, thrust is developed which must be overcome hydraulically or by mechanical means. The common methods of balancing the end thrust hydraulically are by providing for suction at both ends of the impeller and, in case of multi-stage pumps by using back to back mounting of impellers in pairs. In most centrifugal pumps, including vertical turbines, the thrust is taken up by means of thrust bearings.

VERTICAL TURNINE PUMPS

A vertical turbine pump, also called a deep well turbine pump, is a vertical axis centrifugal or mixed flow type pump comprising of stages which accommodate rotating impellers and stationary bowls possessing guide vanes. The bowl assemblies are nearly always located beneath the water surface and hence the deep-well turbine pumps are adapted to seasonal fluctuations in water level in the well. They are specially adapted to tube wells where the pumping water level is below the practical limits of a volute centrifugal pump. The comparatively small diameter of turbine pumps suit their installation in tubewells.

Vertical turbine pumps are adapted to high lifts and have high efficiencies under optimum operating conditions. They have, however, higher initial cost and are more difficult to install and repair, as compared to volute pumps.

As with all centrifugal pumps, the pressure head developed depends on the diameter of the impeller and the speed at which it is rotated. In the deep-well turbine pump, the diameter of the bowl and that of the impeller inside it are restricted by the relatively small diameter of the tubewell. Hence, the pressure head developed by a single impeller is not large. Additional head is obtained by adding more bowl assemblies or stages.

SBMIERSIBLE PUMPS

A Vertical turbine pump close-coupled to small diameter submersible electric motor is termed a submersible pump. The motor is fixed directly below the intake of the pump. The pump element and the motor operate entirely submerged. Such an installation eliminates the long vertical shaft in the column pipe. The performance characteristics of the submersible pump are similar to the vertical turbine pump. Efficiency is increased by the direct coupling of the motor and its effective cooling by submergence in water.

The principal advantage of the submersible pump is that it can be used in a very deep tube-well where a long shaft would not be practical. These pumps are also less affected by deviations in vertical alignment of wells. As the submersible pump has no above-ground working parts, it can be used where flooding may be a hazard. It is also adaptable to locations like public grounds where an above-ground pump house would be inconvenient.

The submersible pump consists of a pump and motor assembly, a discharge column, a head assembly and a waterproof cable to conduct the electric current to the submerged motor.

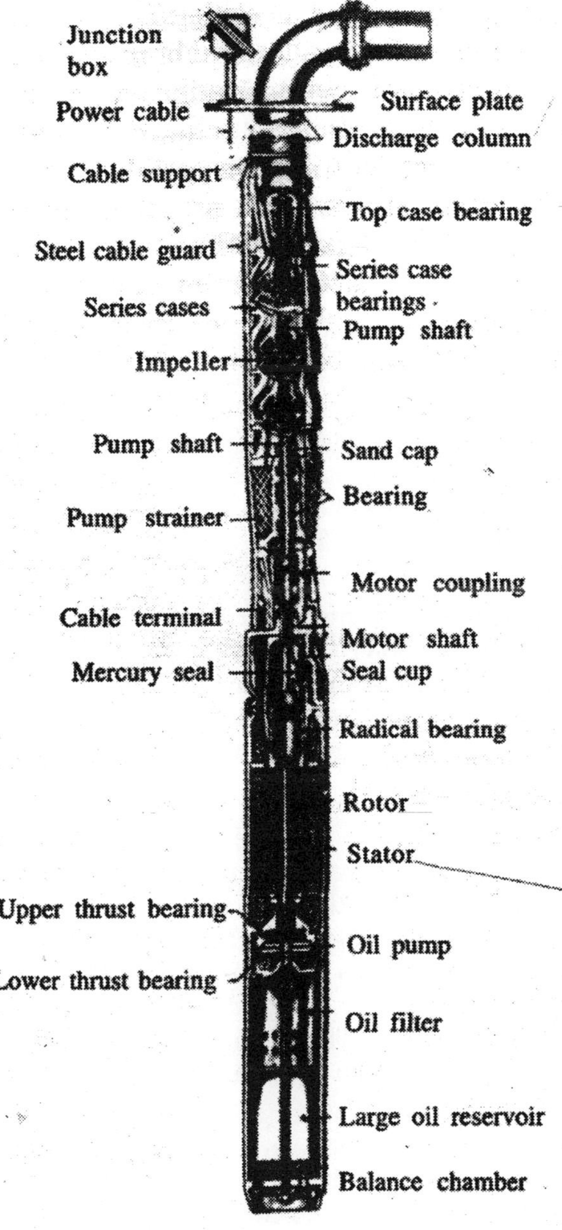

Fig. 5. Submersible pump showing parts.

Pump element: The construction of the pump element is similar to the vertical turbine pump. The propelling shaft, usually made of stainless steel., is very short and on it are mounted the bronze impellers. Impellers are usually made of bronze and may be closed or semi-open. The closed-impeller is generally used where it is necessary for the pump to develop high pressure. Water enters the pump through a screen located between the motor and the pump.

Electric motor: The submersible electric motor has the same diameter as the pump bowl but it is much longer than an ordinary motor. It is a squirrel cage induction motor which may be of the dry or wet type. The dry motor is enclosed in a steel case filled with a light oil of high dielectric strength. A mercury seal placed directly above the armature prevents oil leakage or water entrance at the point where the drive shaft passes through the case to the impellers.

Submersible Pump Construction

Wet motors are those in which the well water has access to the inside of the motor, with the rotor and bearings actually operating in water. In this type of motor, windings of the stators are completely sealed off from the rotor by means of a thin stainless steel inner liner. A filter around the shaft is required to prevent the entrance of abrasive material into the motor. The wet type motor should be filled with water during installation so that the bearings will have sufficient lubrication when the motor is first started.

The stator windings are continuous for the whole length of the motor. The rotors are made in sections on a continuous shaft, with bearings between them to guide the shaft and maintain correct alignment. The electric cables, leading from the motor to the starting box on the ground surface, are water proofed and are placed outside the discharge pipe.

Propeller Pumps

Propeller pumps are often referred to as axial flow pumps. The principle of operation is similar to that of a boat propeller, except that the impeller is enclosed in a housing. This type of pump is particularly adapted to handling large volumes of water at comparatively low heads. Propeller pumps are not easily clogged by foreign materials and suspended sediments in water. They are usually limited to pumping heads of around 1 to 2.5 metres and are available in sizes ranging from 20 to 120 cm in diameter.

The impeller, also known as the propeller, operates in a cylindrical casing which is an extension of the pump discharge column. A flared entrance below the propeller is used to cut down the entrance losses, and guide vanes above smooth out the disturbances caused by the propeller.

Water is moved up by the lift of the propeller blades. Each blade of the propeller helps to impart a velocity in the direction of the shaft. The number of blades is usually 3 to 5. They are set on the shaft at angles determined according to the head and speed. The propellers are usually made of bronze and are cast in one piece. Sometimes, however, the blades are cast separately and may be threaded on their hub ends to receive a nut for the purpose of attaching them to the hub. With this arrangement the blade angle may be adjusted to suit the operating conditions. The propeller is balanced hydraulically and statically. The blades are carefully cast and scraped to reduce skin friction. They are keyed to the drive shaft and are accurately positioned by a locking collar and nut. A cone-shaped cover is usually installed over the locking nut to eliminate eddies and to prevent the entrance of sand or grit into the lower pump bearings. Diffusion casings are usually provided on propeller and mixed flow pumps to convert into pressure the tangential component of the velocity of discharge from the impeller. The discharge column, discharge head and pump drive of propeller pumps are, in general, similar to the vertical turbine pump. Propeller pumps are commonly of the oil-lubricated type.

Operating Characteristics

A slight increase in the pumping head causes a large decrease in the quantity of water delivered by a propeller pump. From the standpoint of efficiency, it is necessary that the propeller pump is operated at the rated head as far as possible. The power required is increased in a propeller pump as the head is increased, and the capacity is reduced. Overload is likely to occur when the discharge valve is nearly closed instead of open, in contrast to be centrifugal pump. Because of the steep horse power curve at shutoff, propeller and mixed flow pumps are started against an open discharge.

Propeller Pump Installation: The propeller pump should be installed on a firm, strong foundation. The foundation should be able to support the weight of the pump evenly on all sides of base plate and allow the unit to hang vertically.

For drainage installations, the pump is usually installed in a sump with an automatic float control. The automatic operation of the electric motor is by means of start and stop collars on a float rod. Automatic switches are provided for the start and stop of the pump at the required flood stages. For large drainage installations, two pumps with different capacities may be necessary, one for handling surface runoff during the rainy seasons and the other for seepage or flow from tile drains.

When the source of water contains appreciable amounts of floating material, some type of strainer or screen should be installed to include the floating material that may damage the propeller if drawn into the pump. Provision is made to attach a small strainer that can be attached to the suction bowl. These strainers are satisfactory when the water pumped is comparatively free of floating vegetation and small debris. When the source of water supply contains this type of foreign material, the small strainer is likely to get clogged. Under such situations, it is desirable to construct some type of screen around the inlet, so as to increase the areas for straining out the small debris. Sometimes, it may be necessary to provide trash racks at some distance away in the drainage channel.

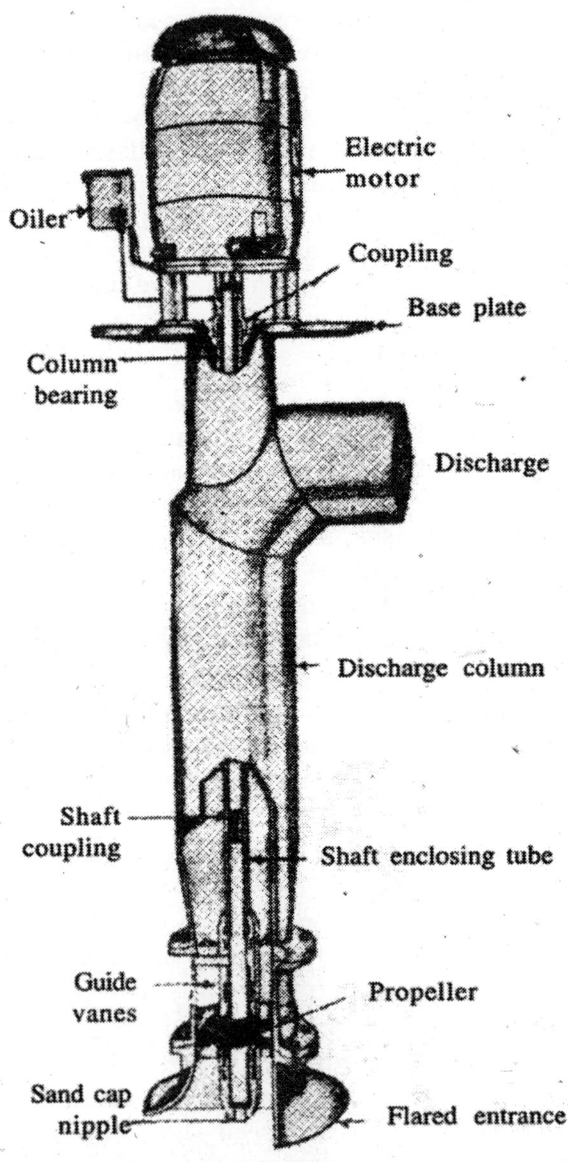

Fig. 6. Propeller pump with the propeller and lower shaft assembly exposed

Propeller pumps are not suitable for suction lift. The impeller bowl must be submerged with the pump operating at the proper submergence depth, the submergence being the distance from the pumping water level to the lowest part of the suction pipe. The minimum submerged required varies from 65 to 90 cm for pumps of size 20 to 35 cm. Failure to observe the required submergence depth may result in cavitations. It is also important that proper clearances are maintained between the end of the suction pipe and the side walls and the bottom of the pit or pump intake bay. The minimum clearance between the pump and the side wall of the intake structure varies from 30 to 50 cm, and from the bottom of the sump floor to pump inlet end between 20 to 30 cm, for pump size of 20 to 30 cm. Inadequate clearances can result in the lowering of pump efficiency.

Mixed Flow Pumps

A mixed-flow pump combines some of the features of both the vertical turbine pump and the propeller pump. The head is developed partly by the centrifugal force as in a conventional centrifugal pump and partly by the lift of the impeller vanes on the liquid as in a propeller pump. It is applicable for high discharge medium head conditions. The head usually varies from 5 to 10 metres.

The main difference between mixed-flow and propeller pumps is in the 'construction of the impeller. The impeller blades of the mixed-flow pumps are designed to give an outward thrust to the water, in addition to imparting to it a velocity upward. Above the impeller, curved vanes guide the water to the column and straighten the flow. The mixed flow pump can work satisfactorily with two or three stages at the bottom of the shaft. The construction of the column assembly and pump drive is similar to the propeller pump. When clear water is being pumped, a water lubricated column assembly is suitable. Oil-lubricated pumps are needed when the water contains suspended sediments and impurities.

JET PUMPS

Jet pumps which are used in irrigation practice consist of a combination of a centrifugal pump and a Jet Mechanism, or an enjector. The centrifugal pump, coupled with the electric motor or engine, is placed at the ground surface and furnishes the driving head and capacity for the jet pump placed in the well bellow the water surface. The main parts of a jet pump are a nozzle and a venture. At the pump delivery, a portion of the high pressure water returns through the pressure pipe to activate the nozzle in the ejector.

The nozzle is shaped so that it smoothly but abruptly reduces the area through which the flow must pass, thus increasing the velocity of flow. This creates a low pressure area around the venture which draws more water from the well. The gradual enlargement in the venture tube to the full diameter of suction pipe reduces the velocity of water with a minimum of turbulence. The vacuum created by the impeller of the pump placed at the ground surface draws the flow through the delivery pipe at the desired pressure. The additional supply of water which is obtained from the well is discharged past the control valve while the volume required for producing the flow is recirculated through the pressure pipe. The control valve is set to maintain the necessary' pressure to produce flow at the existing pumping head.

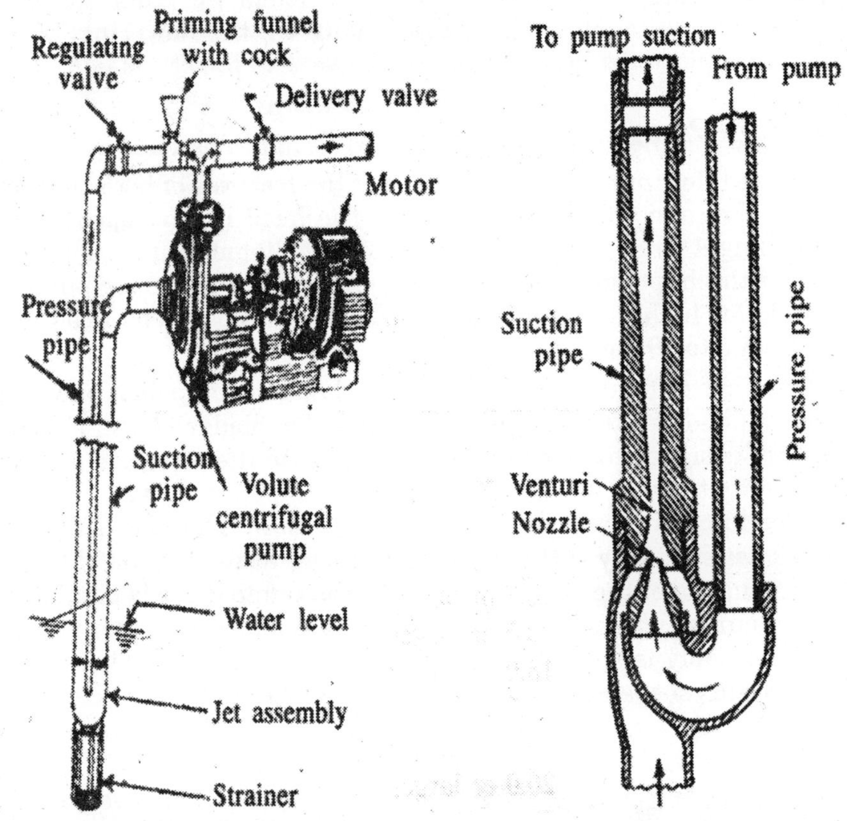

Fig. 7. Jet pump assembly with electric jet assembly motor driven centrifugal pump

The most important factor governing the capacity and efficiency of a jet pump is the selection of the jet to correspond to both the type of the pump and depth of low-water level at the source. Jet pumps are by nature inefficient. The highest efficiency usually obtained is about 35%. The efficiency of the pump is influenced mainly by the nozzle throat ratio. For small jet pumps used in domestic water supply, the spacing between the nozzle and the throat is equal to one nozzle diameter. The length of the throat is about six throat diameters.

Adaptability: Jet pumps are generally used for low capacity deep well pumping where it is necessary to locate the moving parts of the pump and the prime mover at the ground surface. The following are the main advantages of a jet pump:

(1) Its adaptability to be installed in wells as small as 5 cm inside diameter. (When the well diameter is too small to admit two pipes, **a** single pipe ejector may be used to connect the piping to the pump head. When this is done, the jet should always be submerged and the well casing extended to the low water level).

(2) Its high lift which cannot be reached by ordinary centrifugal pumps. The suction lifts usually obtained with low to medium size jet pumps vary from 12 to 18 metres.

(3) Accessibility of all moving parts which are located at the ground surface.

(4) Simplicity combined with relatively low cost and easy maintenance.

(5) Adaptability to being installed with the moving parts offset from the well.

In some locations, where water levels are subject to large seasonal variations, or where severe corrosion or plugging at the nozzle occurs, jet pumps are not very satisfactory.

AIR-LIFT PUMPS

Principle of Working of Air-lift Pumps

An air-lift pump operates by the injection of compressed air directly into the water inside a discharge or eductor pipe at a point below the water level in the well. The operating principle of the pump is shown in Fig. 8. The injection of the air results in a mixture of air bubbles and water. This composite fluid is lighter in weight than water so that the heavier column of water around the pipe displaces the lighter mixture forcing it upward and out of the discharge pipe. The piping assembly used for air-lift pumping from a well consists of a vertical discharge pipe called educator pipe and a smaller air pipe.

An air-lift pump for pumping or developing a well is arranged commonly with the air pipe inside the eductor pipe as shown in Fig. 8. It is also possible to locate the air pipe outside the eductor pipe if there is space in the bore hole, as otherwise there is considerable friction loss when the air pipe is located inside a small diameter eductor pipe. Both the eductor pipe and the air pipe must be submerged in water in the well with 40% or more of their lengths extending below the pumping level. The energy that is available to operate the air-lift pump is that which is contained in the compressed air.

The well casing itself can be used for the eductor pipe if the diameter of the casing is not much larger than the air line. This provides a practical way to pump sand and mud from the bottom of a well during development and cleaning operations. When test-pumping a well, however, it is better to use a separate educator pipe since the pumping level can then be measured with steel or electronic tapes.

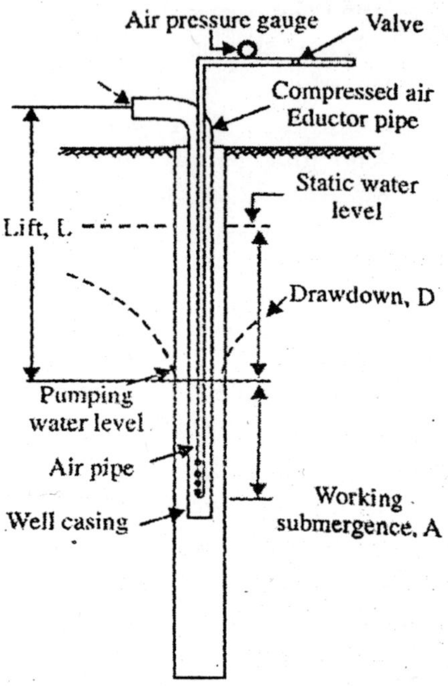

Fig. 8. Principle of operation of an air-lift pump.

The two most important factors in air-lift pumping are the per cent submergence of the airline and the relative sizes of the air and eductor pipes. Generally, the air-lift pump works best with a submergence of 60.6 or more. It is the length of airline that is submerged below the pumping level divided by the distance from the level of water discharge to the lower end of the airline and multiplied by 100, to give the result as percentage. Referring to Fig. 8

$$\text{Per cent submergence} = = \frac{A}{L-A} \times 100$$

The length of air pipe below the static water level is significant only for determining the air pressure required to start the air-lift pump. Before air can discharge from the lower end of the air pipe, the compressed air must push all the water out of the air pipe. To do this, the air pressure must be greater than the water pressure before starting to pump. The depth of water in metres from the static water level to the lower end of the air pipe divided by 10 gives the required air pressure in kg/cm^2.

Thus, starting pressure (kg/cm²) = $\frac{A+D}{10}$

Working presume (kg/cm²) = = $\frac{A}{10}$ in which *A* and *D* are measured in metres.

Pipe sizes for air-lift pumps: The velocity of water in the eductor pipe depends upon the volume of air and water being discharged and the area of the annular space between the airline and the eductor. Table-1 gives the size of airlines for various sizes of eductor used under most conditions. When the water yield is unusually low, the difference between the sizes of eductor and air pipes may have to be less to reduce the area of the annular space and thus increase the discharge velocity enough to prevent excessive air slippage in the eductor. This condition necessitates a larger-than-normal air line when pumping directly from the well casing without a separate eductor pipe.

Table 1. Pipe Sizes for Air-lift Pumps

			Internal diameter	
Pumping rate (litres 1 sec)	**Size of well casing eductor pipe (cm)**	**Size (cm)**	**Size**	**of air line (cm)**
2to4	10.0 or large		5.0	1.25
4to5	12.5 or larges,		7.5	2.50
5 to 6	15.0 or largely		8.7	2.50
6to9	15.0 or larger		10.0	3.13
9 to 16	20.0 or large		12.5	3.75
16 to 25	20-0 or larger		15.0	5.00
25 to 44	25.0 or larger		20.0	6.25

Best operation of an air-lift pump requires good regulation of the amount of air injected. Too much of air causes excessive friction in the pipe lines and waste of air. Too little air results in reduced yield and in surging, intermittent discharge. When developing a well with an air-lift pump, the discharge should be started at a very low rate and brought up gradually. The air flow should be slowly increased in proportion to about what appears to be the increase in water flow into the well from the water-bearing formation.

Air-lift pumping is extensively used in the development and preliminary testing and cleaning of tube wells. It is also sometimes used in crooked tube wells where vertical turbine pumps or submersible pumps cannot be installed. The advantages of air-lift pumps are: simplicity, tube well need not be perfectly straight or vertical, and impure water will not damage the pump. The main disadvantage is its low efficiency. The maximum efficiency usually obtained is about 30 per cent. The initial cost of the pump, including the air compressor, is high and it requires an extra depth of water for proper submergence.

SELECTION OF PUMPS

Irrigation wells and pumps are costly installations which require efficient utilization. A major part of the energy used in agriculture is in pumping. The total energy utilized in irrigation pumping in India in the year 1973-74 was about 8,400 million kilowatt-hours. It was estimated that the requirement would double in the next two decades. Efficient utilization of limited energy resource calls for the selection of the most suitable pump, keeping in view the requirement of irrigation, characteristics of the well or other sources of water, kind of power available, economic conditions of the farmer and other factors.

Criteria for selection of irrigation pumps: The main factors influencing the selection of pumping sets are:

(i) The retirement of irrigation water by the crops to be irrigated,

(ii) Yield of the sources of water (open wells, tube wells, streams rivers, ponds) and

(iii) Availability and cost of the pump and kind of energy.

Determining the Discharge Capacity of Pumps

The data on the safe discharge rate of the well (or other source of water) and the discharge rate required for the crops to be irrigated with a particular copping pattern are estimated. If necessary, the cropping pattern can be adjusted according to the safe yield of the well in different seasons. In many cases, such adjustments may not be possible on one's own farm since the holdings may be smaller than the command area required by a pump set of moderate size. In some cases, the holdings may be fragmented and only a few fragments belonging to the farmer are within the service orbit of the pump set installed in a well. In such cases, any one of the following alternatives could be adopted.

(1) Design the pumping plant according to the safe yield of the well and use it to irrigate the farm of an individual farmer, keeping the spare capacity idle. But it leads to high investment and under-utilization of installed capacity and invested funds.

(2) Design the pumping plant for the safe yield of the well and use the installed capacity to irrigate the land of the well owner and sell the remaining water to irrigate other farmer's land, based on such demands. This alternative is better than the first and leads to the full utilization of the water resource potential developed at a high cost.

Discharge capacity of pump based on crop requirements

The pump discharge should meet the peak demand of water for the selected cropping pattern. The rate of pumping depends on the area under different crops, the water requirement of the crops, rotation period (interval between two successive irrigations of a crop) and the duration the pump is operated each day. It may be computed by the following relationship:

$$Q = \frac{Ay}{RT} \times \frac{1000}{36}$$

$$= 27.78 \frac{Ay}{RT} \tag{1}$$

In which,

Q = rate of discharge of pump (litres/sec)

A = area of land under the crop (hectares)

Y = depth of irrigation (cm)

R = rotation period (days)

T = duration of pumping (hours/day)

Lift Irrigation Systems

In lift irrigation projects, water is lifted directly from rivers or streams or from suitable points at different segments of a canal system. River lift irrigation schemes are feasible in regions where the stream flow is adequate during the peak period of irrigation demand by the crops and where electricity or diesel oil is available at reasonable cost. Canal lift irrigation schemes are beneficial in many areas where sizable land area adjoining a canal, but lying at higher elevations cannot be irrigated through gravity flow from the canal. The water can be pumped to one or more distribution chambers located at suitable points in the command area from where it is distributed through appropriate water distribution systems.

In many tank irrigation projects and small and medium scale reservoir projects, lift irrigation from tanks and reservoirs is sometimes practiced in situations where cultivation in the catchment area adjoining the reservoir may not result in excessive soil erosion and subsequent silting up of tanks/reservoirs.

Assessment of Feasibility of Lift Irrigation Projects

Lift irrigation projects are expensive schemes which have profound influence on river hydrology, riparian rights, the characteristics of the command cropping systems and other site specific requirements. They are usually single purpose projects which have to be sustained by the returns from the increased agricultural production of the command areas. It is, therefore, essential that the techno-economic feasibility of the project should be established before the project is sanctioned or undertaken.

The techno-economic feasibility report of a lift irrigation/water supply project should include the following:

1 Location or the scheme, or

2 Availability of water in the river or the stream, with respect to irrigation seasons.

3 Number of existing irrigation and water supply projects on the same river or stream in both upstream and downstream sections of the proposed scheme.

4 Potential for additional pumping from the river without adversely affecting the water requirement of existing schemes/ riparian users.

5 Suitability of water for irrigation/water supply.

6 Proposed cropping pattern/cropping intensity in the command area/water supply demands of the population.

7 The extent of land development work needed in the command area.

8 On-farm water conveyance system proposed (open channels/ underground pipelines).

9 The method of water application proposed (surface/ sprinkler/ drip).

The economic feasibility of the scheme must be determined by estimating the expected farm income from the command area and/or the returns and social benefits of the scheme. These should be compared with the investment cost of the scheme, including the equipment and accessories and the cost of operation and maintenance. The total cost of the scheme consists of the following:

1 Fixed cost or investment cost.

2 Operating cost.

3 Maintenance cost.

The investment cost or the capital cost of a water lifting scheme includes the cost of the following activities.

1 Preliminary works (investigation, surveying, design of the system, site clearance and construction of approach roads)

2 Land acquisition, if applicable.

3 Civil construction works, including water conveyance and distribution system.

4 Mechanical works, including machinery and equipment, and special tools and plants during the time of construction of projects.

5 Electrification charges.

6 Miscellaneous expenses (contingencies, work charge, and unforeseen expenses) during implementation.

7 Land development work in the command area, if applicable.

The economic viability of a lift irrigation scheme should be determined on the basis of the net incremental income expected to be generated as a result of the implementation of the scheme. The water rate to be charged from the beneficiaries should be estimated on the basis of the annual fixed and recurring costs of the scheme and the investments and returns from the farming enterprise.

Water Availability

The assessment of the amount of river or stream flow which could be safely harnessed for the proposed lift irrigation/water supply scheme is the most important factor in establishing the technical feasibility of a river pumping project. The availability of water in the river should be determined from the observed minimum discharges during the periods when the demand for water is maximum. The river discharge data should be for a sufficiently long period, preferably a ten-year period, if possible. River flow records are usually available from the concerned Executive Engineer of the Government or from local bodies.

If relevant data on river discharges are not available, reasonable estimates are to be made on the quantum of flow during the dry seasons of the year. The number of irrigation and water supply projects existing on the river, upstream and downstream of the project site, should be ascertained. Provision to meet the continued requirements of water for domestic use and other riparian requirements and existing contracts for water supply, like those of industries are to be ensured.

Lift irrigation schemes which are designed and constructed by government agencies usually provide for a certain risk, especially when the water availability is scarce. A confidence level of 75 per cent is usually adequate for schemes in the peninsular region. Lift irrigation schemes depend on the minimum flow (base flow) during the crop period. There is a likelihood of the base flow depleting in certain years. Hence, it is necessary to provide for this contingency.

Design of Lift Irrigation Systems

The design of the various components of lift irrigation systems are highly situation specific. Detailed investigation of the site, including river flow characteristics, topography, and climate conditions, mainly rainfall characteristics will enable the development of the system which best suits the requirements and provides maximum possible economy in installation and operating costs. The main structural components on lift irrigation systems are (i) intake structure, (ii) intake pipe, (iii) jack well/pump well, (iv) pump house, (v) rising main and accessories, (vi) delivery chamber, (vii) distribution chamber, and (viii) on-farm water distribution system. The pumping system equipment comprise of pumps, electric motors/ engines, electrical accessories and fittings, transformer and power connection and protective devices.

Water Lifting from Canals

In canal pumping schemes the capacity of the pumping sets and the periods of pumping have to be fixed after careful consideration of the irrigation requirements of downstream users, including those receiving their supplies from gravity flow and / or are allowed to pump canal flows at specified time periods. The basic requirements of both these categories of users should not be adversely affected by a new pumping scheme. However, in many situations there are ample opportunities to provide adequate supplies for pumping at several points along a canal. They arise due to a change in the cropping system of the command area served by the gravity flow system, in order to combat the problem of excessive rise in ground water table after the introduction of canal irrigation. Canal pumping schemes will help in the effective utilization of canal flows during periods of low demand in the gravity command area. With the present day advances in plant breeding, there is ample scope for selecting crops and their varieties which could be grown successfully during this period. In many regions, the practice of night irrigation is not well established, when surface methods of water application are practiced. Sizable quantities of canal supplies during the night are not effectively utilized. It is desirable to permit pumping during the night to utilize the under-utilized canal flows during the period. There are also situations where the initial reaches of a canal system cannot be brought under gravity flow irrigation due to the unfavorable topography of the surrounding areas. In such cases canal pumping assumes importance as they add to the culturable command area of the canal system. Utilizing canal flows in its upstream reaches results in a major reduction in the total seepage loss of a canal system.

Canal pumping schemes are most economical when the lift command area is close to the canal and the pumping heads are low. In general, lift irrigation results in more efficient use of the canal supplies, as compared to gravity flow systems, as the farmers are more conscious of the cost of pumped water supplied than those obtained from a gravity flow system.

Pump House

A circular pump house could be conveniently located on top of the jack well. A rectangular or square shape of the pump house is more convenient and economical than a circular shape. A rectangular pump house gives more space for locating the pumping sets and ancillary equipment. A rectangular or square pump house can be constructed over the circular steining- by providing girders over the top of the steining and cantilevering them suitably. In case of a sump well, however, much space is not required in the pump house, as the motors are not located at the same level. In such cases, it is more convenient to continue the steining masonry with the circular shape. A lightening conductor should be provided on top of the pump house.

Pump House for Centrifugal Pumps

The pump house for centrifugal pumps shall be rectangular in shape. The pumps, motors, panel box for electrical components, starters and other accessories are located in the pump house separately. The wall of the pump house is about 45 cm thick and the exposed surface is plastered with cement mortor. The dimensions of the pump house, namely length, breadth and height are based on the specific requirements of the installation. The length of the pump house is the total length along the shaft of horizontal centrifugal pumps, the spacing between pumps, the spacing between the pump house walls, the outer ends of the first and the last units. The shaft length of the pumping set (pump with motor coupled to it) it obtained from the specifications of the pumping unit proposed/ The spacing between each pumping unit depends on the h.p. of the units and ranges from 60 cm to lm. For pumps up to 30 h.p.: 60cm, 31 to 60 h.p.: 70cm, 61-90 h.p. : 80cm. 91-120 h.p. : 90 cm and above 120 h.p. : 1 m. The same values may be adopted for the clearance between the wall and the first and the last pump in the row.

The width of the pump house is computed on the basis of the requirements of the following: •

(i) The width of the pump, including the length between the suction side flange and the delivery side flange of the pump,

(ii) The clearance between the wall and the suction flange of the pump is kept 60 cm in all cases. This includes 15cm as the clearance from the wall to the eccentric taper flange and 45 cm as the length of the eccentric taper. When the length of the eccentric taper exceeds 45 cm, the width of the pump house has to be correspondingly increased,

(iii) The length of the non-return valve on the delivery side,

(iv) The length of the sluice valve on the delivery side.

(v) The clearance between the sluice valve flange and the pump house wall is 30 cm for pumps up to 50 h.p., 45cm for those from 50 to 100 h.p., and 60 cm for units above 100 h.p.

The height of the pump house is fixed, taking into account the following factors:

(i) The height of the foundation block of the pumping set above the floor of the pump house is usually kept at 10 cm.

(ii) The thickness of the base plate of the pumping unit is usually kept at about 15 cm.

(iii) The height of the pumping set above the base plate is obtained from the specifications of the pump proposed.

(iv) Clearance required between the top of the pumping unit and the bottom of the gantry hook.

(v) The depth of the I-bean supporting the gantry.

(vi) The depth of the gantry unit: This is the distance between the bottom of the gantry girder and the centre of the hook.

Pump House for Vertical Driven Pumps

Vertical driven pumps for lift irrigation schemes include vertical turbine pumps, submersible pumps, mixed flow pumps and propeller pumps. Proper arrangement of pumps, sluice and reflux valves, delivery manifold, flow training devices in the jack well for ensuring hydrostatic efficiency and gantry arrangement should be designed properly to ensure smooth functioning of the pumping system. Gantry arrangement with chain pulley block is provided at a height of 4m above the floor of the pump house. The capacity of the gantry should be adequate to lift the maximum weight of a single unit which usually varies from 0.5 to 3 tons, depending upon the size and horse power of the pump set. Cost considerations are equally important as the performance requirements. Specific arrangement of the various ancillary components depends on the type of pumps and their capacities.

In case of horizontal centrifugal pumps the sump well and the pump house are generally located separately in order to be economical. However, in the case of vertically operated pumps, the pump house is superimposed on the jack well. The specific requirements in case of vertical turbine pumps, submersible pumps and other vertically operated pumps are as follows:

(i) Provision of suitable hydraulic inflow training devices to ensure hydraulic efficiency.

(ii) Layout of the pumps in the pump house,

(iii) Location of reflux and sluice valves,

(iv) Arrangement of delivery manifold.

(v) Specifications of the pump house and the supporting girders for the pumps,

(vi) Gantry arrangement.

Booster Pumps

While irrigating hilly areas, it is often desirable to provide one or more booster pumps, taking off from a delivery chamber located at an intermediate point. It is highly uneconomical and wasteful of energy to lift water to a hill top and provide field channels and drop structures to irrigate the lower areas with delivery chamber located at the top of the hill. A contour map of a hill area will reveal that the total area between the higher contours is only a small fraction of the area lying between those representing the lower elevations. Hence, for hillside irrigation, intermediate delivery chambers are to be provided, which also serve as reservoirs for locating the inlet of the booster pump. In such a system, high-discharge medium or high head pumps are required to lift water from the intermediate delivery chamber, while one or more low-discharge high-head pumps are required to lift water from the intermediate delivery chamber to the top chamber. Such an arrangement will require a much smaller size prime mover for the pumping system, as well as a smaller diameter rising main between the intermediate and top delivery chambers.

In-line boosters having the downstream section of the rising main connected directly to the suction side of a volute centrifugal pump could also be used in river pumping schemes as per requirements.

References

1. Minor Irrigation Dept, Govt. of Bihar (1978). Outlines of Lift Irrigation Schemes in Chhotanagpur and Santhal Pargana, Mineo, p. 7.

2. National Bank for Agriculture and Rural Development (1995). Lift Irrigation Schemes in Maharashtra-Techno-Economic Evaluation. Study No. 15, Department of Economic Analysis and Research, Mumbai, p. 129.

3. National Bank for Agriculture and Rural Development, (1991). Report of the Working Group on Lift Irrigation Schemes. Mumbai, P. 90.

4. National Sample Survey Organization (1985). Sarvekshana, Vol. 9, No. 1

5. Varshney, R.S., Gupta, S.C. and Gupta R.C. (1979). Theory and Design of Irrigation Structures I. Pub. Nem Chand Brothers, Roorkee, p. 452

Part - B
Strategy Waste and Dry Land Development

Chapter - 20

Wasteland Development in Andhra Pradesh

B. Chandra Sekhar & Prof. K. Govindappa

The term ``Wasteland in common usage means degraded unutilized, uncultivated and common land".

However, different departments at present define the wasteland according to their land use pattern.

Agricultural land lying fallow for more than two years can be treated as agricultural wastelands.

Lands under the control of Revenue department not fit for agriculture lying barren can be termed as Revenue wastelands.

Similarly, the grasslands and lands under the control of Forest Department; which do not have tree cover can be termed as Forest wasteland.

Although, there is no common definition of wasteland, it is clear that, wastelands are the areas, which are under utilised, and which produce less than 20% of its biological productivity. Other areas, which are generally included in the wastelands are, saline and alkaline soils, waterlogged areas, common grazing lands, Panchayat lands, land lying vacant along railway lines, roads, canals, denuded and barren rocky hills, ravine lands, flood plains land infested with thick weeds, and lands lying vacant in the towns, cities and residential colonies.

The wastelands are ecological unstable. These lands have been subjected to different degrees of biotic or ecological interference, as a result, rendering them, degraded, infertile, and un-cultivable. After extensive discussions, with different departments the National Wasteland Development Board (NWDB) defined wastelands, ``that land which is degraded and presently lying unutilized (except as current fallow) due to different constraints". NWDB also suggested that any land, which is not producing green biomass, comparable with the status of the soil, and water, must be treated as wasteland. Wastelands can be divided into cultivable wastelands and uncultivable wastelands. Cultivable wastelands comprise gullies or ravenous lands, undulating upland surface, water logged marshy areas, salt affected lands, shifting cultivation areas, degraded forest land, sandy area, mining, industrial areas, pastures, grazing lands. Uncultivable wastelands comprise barren rocky areas and steep slopes.

1. Ecological status of Wastelands

Before putting the wastelands into proper productive use, it is better to assess their ecological status.

i. Wastelands are situated in most drier parts of the country.

ii. Annual rainfall may not be more than 700 mm.

iii. The soil will be either sterile or alkaline, to saline.

iv. In sloppy topography, the soils must have been subjected to erosion, exposing rocks, or strewn with boulders.

v. The region might be subjected to heavy grazing.

vi. The regions are surrounded by rich population of poor people, and subjected to heavy illicit hacking and encroachments.

vii. Keeping the above in view, the following steps can be taken up.

viii. Provision of complete protection from man and Luis animals.

ix. Existing peoples participation to the maximum.

x. Selection of suitable species for afforestation, or other productive uses.

xi. Adoption of proper techniques for success, and adoption of modern scientific techniques.

2. Classification and Estimation of wastelands

Estimates of degraded land vary considerably, and the extent of land degradation is yet to be determined precisely. Estimates of wastelands differ considerably due to definitional and coverage in consistencies. According to land use statistics for 2002, published by Dept. of Agriculture, the current estimates of cultivatable wastelands are about 13.90 m.ha. However, the information on land use statistics does not clearly indicate the extent of wastelands and degraded and, which could be restored with some interventions.

National Remote Sensing Agency (NRSA) carried out a District wise mapping of wastelands on 1:50,000 scale using satellite data. The total wastelands in the country were placed at 63.85 m. ha. They occur in different agro climatic and soil zones of the country. The following table given the data on total extent of wastelands available in the states.

These wastelands form the core of degraded lands in India. They are in urgent need of attention and have to be accorded the highest priority for treatment.

Some of the most degraded lands in the country are CPRs (Common Property Resources). CPRs are the resources on which people have an equal right of use. These resources include community pastures, community forests, wastelands, and common dumping and threshing grounds. Inspite of concerted efforts to check deforestation, and various afforestation schemes taken up during successive plan periods large tracts of forest continue to be classified as degraded. The 2005 Forest Survey of India report, placed the actual forest cover at only 20.60 percent of total geographical area as against the recorded forest area of 23 percent. In Andhra Pradesh, the Forest Survey of India report, places the forest cover at 16.13 percent, of the total forest area 31 m. ha. suffers from some form of degradation and 14.06 m. ha. of forest suffer from extreme degradation and are part of 63.85 m.ha. of wastelands reported by NRSA.

Table 1. State wise Wastelands of India (Area in sq. km)

S.N.	State	No. of Districts Covered	Total Geog. Area of districts covered	Total Wastelands area in districts covered	% of wastelands to total geog. Area
1	Andhra Pradesh	23	275068.00	51750.19	18.81
2	Arunachal Pradesh	13	83743.00	18326.25	21.88
3	Assam	23	78438.00	20019.17	25.52
4	Bihar	55	173877.00	20997.55	12.08
5	Goa	2	3702.00	613.27	16.57
6	Gujarat	25	196024.00	43021.28	21.95
7	Haryana	19	44212.00	3733.98	8.45
8	Himachal Pradesh	12	55673.00	31659.00	56.87
9	Jammu & Kashmir	14	101387.00	65444.24	65.55
10	Karnataka	27	191791.00	20839.28	10.87
11	Kerala	14	38863.00	1448.18	3.73
12	Madhya Pradesh	62	443446.00	69713.75	15.72
13	Maharashtra	32	307690.00	53489.08	17.38
14	Manipur	9	22327.00	12948.62	58.00
15	Meghalaya	7	22429.00	9904.38	44.16
16	Mizoram	3	21081.00	4071.68	19.31
17	Nagaland	7	16579.00	8404.10	50.69
18	Orissa	30	155707.00	21341.71	13.71
19	Punjab	17	50362.00	2228.40	4.42
20	Rajasthan	32	342239.00	105639.11	30.87
21	Sikkim	4	7096.00	3569.58	50.30
22	Tripura	4	10486.00	1276.03	12.17

23	Tamil Nadu	29	130058.00	23013.90	17.20
24	Uttar Pradesh	83	294411.00	38772.80	13.17
25	West Bengal	18	88752.00	5718.48	6.44
26	Union territories	20	10973.00	574.30	5.23
	Total	584	3166414.00	638518.31	20.17

In addition to the wastelands identified by NRSA, other areas such as deserts, drought prone, flood prone, and tribal areas have been subjected to severe forms of degradation. The capacity of these lands is limited due to environmental factors. Pressures of livestock and human population have further compounded the problems. Table 2 given below, provides estimates the degraded land on the basis of factors that covered the degradation. Attempts were made by NWDB to identify the total wastelands and concluded that an area of 129.57 m. ha. (including degraded forest area) and 93.69 m.ha. (Without degraded forest area) is available. However, the estimate given by NRSA, is fairly realistic.

Table 2. Causes of Land Degradation

Causes of Degradation	**Area (million hectares)**	**Percentage of total area**
Water erosion	107.12	61.70
Wind erosion	17.79	10.24
Ravines	3.97	2.28
Salt-affected	7.61	4.38
Water logging	8.52	4.90
Mines & Quarry wastes	-	-
Degraded land due to shifting cultivation	4.91	2.82
Degraded forest lands	19.49	11.22
Special problems	2.73	1.57
Coastal sandy areas	1.46	0.84
Total	173.64	100.00

3. Land and Environmental degradation and Poverty

It is to be recognized that both poverty and environments are descriptions of states of human and natural resources attributes, and cannot be reduced to simple one-dimensional cause effect relationship. Hence, the wastelands co-relate very strongly with the incidence of poverty in the country. The prevention of land degradation and the augmentation of the carrying capacity of land to provide food, fuel, and fodder requirements have therefore, been a primary concern of the

Government. In our country, for targeting environmentally degraded lands to initiate poverty alleviation programmes, at the behest of ministry of Rural Development, the planners are using the data on wastelands availability extensively. Presently, this data serving as primary input in planning reclamation measures, micro level inventory, and monitoring of wasteland reclamation measures, are being used by the Department of land resources of the ministry of Rural Development, state Forest and Agriculture Departments etc., for various institutional interventions, aimed at poverty alleviation.

Despite more than 70% population in the rural areas in India being dependent on natural resources, the relationship between wastelands and poverty is seem to be complex. In fact, at state level, the occurrence of wastelands does not seem to be connected with the incidence of poverty. Bihar, for example, has just 6% wastelands, but the percentage of population below poverty line is 57%.

There are another set of states wherein the incidence of poverty as well as wastelands both are equally high. Assam with more than 25% wastelands has got more than 45% population, below poverty line. On the other hand Punjab with 4% wastelands, has 11% population B.P.L. followed by Andhra Pradesh, Haryana, Kerala and Gujarat.

At the district level, the relationship between incidence of poverty and wastelands has increased significantly. Similarly, if we move further to micro (village) level, the relationship still increases. Table 3 gives the clear picture.

The linkage between poverty and environmental degradation are, however not just governed by the physical limits of eco systems, but rather, by the income strategies of the poor. Driven by public policies and institutional interventions, economic and spatial integration of markets occurs and several new marginal income earning opportunities become available in the informal sector of economy. Consequently, the dependence of poor on the natural resources base will be decreased.

4. Programmes for Development of Wastelands

DPAP : Drought prone Area Programme was the first major programme aimed at land development and soil and moisture conservation in drought prone areas. It was introduced in 1973-74. Anantapur district is one among the five districts selected under DPAP with World Bank's financial assistance. Later from the year 1985, the Government of India continued the programme with central and state share. Currently it is being implementing in 971 blocks in 16 states.

DDP : The desert development programme, which was introduced 1977-78, is being implemented in 7 states and covers 234 Blocks in 40 districts. Anantapur district in Andhra Pradesh covers all its blocks under DDP scheme.

Table 3. Wasteland statistics and indicators of poverty and food insecurity - in India

S.N.	State	Deficit in food produ-ction	Percen-tage waste-land	Percen-tage below poverty line	Rural in fracture index (%)	Rural literacy (%) 2001
1	Andhra Pradesh	1.16	18.81	15.92	42.30	55.33
2	Assam	1.29	25.52	45.01	74.60	60.92
3	Bihar	1.55	5.90	56.93	99.20	44.42
4	Chhattisgarh	0.42	7.53	44.38	60.00	76.23
5	Gujarat	1.74	21.95	22.18	30.80	58.53
6	Haryana	0.33	8.45	28.02	34.90	64.00
7	Himachal Pradesh	0.76	56.87	30.34	11.80	74.38
8	Karnataka	1.11	10.87	29.88	35.80	60.00
9	Kerala	3.99	3.73	25.76	39.70	90.05
10	Jharkhand	3.70	18.89	62.00	80.00	46.26
11	Madhya Pradesh	1.12	19.31	42.05	57.40	58.10
12	Maharashtra	1.27	17.38	37.93	32.40	71.00
13	Orissa	1.13	13.71	49.72	64.60	66.44
14	Punjab	0.16	4.42	11.95	37.70	65.00
15	Rajasthan	1.25	30.87	26.46	56.90	56.00
16	Tamil Nadu	1.18	17.70	32.48	31.40	67.00
17	Uttar Pradesh	0.94	9.40	44.54	84.10	54.00
18	Uttaranchal	0.85	30.27	24.98	70.00	61.00
19	West Bengal	1.18	6.44	40.80	89.90	64.00

IWDP : Integrated Wasteland Development Programmes started in 1989-90 seeks to develop Government wastelands and CPRs, based on micro plans at village level. The IWDP is aimed at overall economic conditions of resources poor population.

NWDPRA : National Watershed Development Project for Rain fed Areas (NWDPRA) initial in 1990-91, has twin objectives of improving agriculture production in rainfed areas and to restore ecological balance.

NABARD : In order to channelise greater resources, for rainfed areas, the watershed development funds was set up in 2000-01 at the National Bank for Agriculture and Rural Development (NABARD) with corpus fund of Rs. 200 Crores. Andhra Pradesh is one of the recipient states of this fund for rainfed areas development, on watershed basis, through participatory approach.

Table 4. Wasteland statistics and indicators of poverty and marginalization - Andhra Pradesh

S. No.	District	Incidence of poverty (poor %)	Wasteland (%)	Small and Marginal holdings (%)
1	Adilabad	53.10	15.23	54.00
2	Anantapur	50.00	16.90	63.70
3	Chittoor	61.10	38.76	90.00
4	Kadapa	43.10	29.93	84.60
5	East Godavari	28.60	13.41	55.40
6	Guntur	33.00	14.72	87.20
7	Karimnagar	43.20	12.43	84.20
8	Kammam	41.70	10.90	68.30
9	Krishna	29.40	10.56	90.10
10	Kurnool	41.70	21.97	67.00
11	Mahaboobnagar	52.80	13.54	67.20
12	Medak	48.80	11.73	74.20
13	Nalgonda	46.60	12.14	74.00
14	Nellore	49.70	37.61	97.80
15	Nizamabad	41.00	18.70	82.40
16	Prakasham	37.80	21.46	87.80
17	Rangareddy	47.80	19.50	67.90
18	Srikakulam	53.10	18.89	92.20
19	Visakhapatnam	45.90	28.24	74.70
20	Vizayanagaram	51.30	20.63	85.80
21	Warangal	43.20	11.63	80.80
22	West Godavari	35.70	4.38	71.00

5. Scenario of Wasteland development in Anantapur District

Anantapur District is chronically drought affected and receives 552 mm of rainfall, and is situated in rain shadow region. Bulk of rainfall is received from South West monsoon. Failure of South West monsoon in the district leads to drought. The total geographical area of the district is 19.13 Lakh ha. The forestland is about 10% of the geographical area. as per Forest Survey of India, (2005) the area under canopy is hardly 2.16%. Area wise, it is biggest district in Andhra Pradesh. 90% of the agricultural area is under rainfed condition. Irrigation is through bore wells and Thungabhadra Project (High Level Canal) covers an area of 51.771 ha. in 157 villages.

Drought has been a recurring phenomenon in Anantapur district. The adverse effect of drought is felt not only on human beings, animal population but also on ground water table, drinking water, crop, and fodder production. The situation has led to spreading of large chunks of lands to become barren and unproductive. Nearly 16.90% of the total geographical area is classified as wasteland, as per NRSA.

Inspite of all the programmes aiming at land development through soil and moisture conservation and localized water harvesting, the following problems of wastelands still persist.

1. Increase in biotic pressure.
2. Absence of adequate investments and appropriate management.
3. High incidence of poverty in rural areas.
4. Breakdown of traditional institutions for managing CPRs (Common Property Resources) and failure of new institutions to fill the vacuum.
5. Faulty land use practices.

As these problems exacerbated, the consequences are, soil erosion and degradation, depletion of natural resources increase in extent of wastelands, threat to ecological security due to pressure on forest areas.

The following components are suggested to improve the productivity of wastelands.

i. Soil and moisture conservation measures like bunding, trenching, vegetative barriers, and drainage line treatment to check land degradation.

ii. Planting and sowing of legumes and fodder species, promotion of agroforestry and dry land horticulture, Block plantation for meeting fuelwood, fodder, demands, Strip plantations along roads, canal banks, will improve biomass productivity.

N.R.E.G.S. (National Rural Employment Guarantee Scheme): Apart from the land based programmes explained above; NREGS is launched in the district during the year 2006, as per the National Rural Employment Guarantee Act (2005). The main objective of the programme is to provide 100 days of employment per each family, in a financial year in rural areas. 80% of the funds are earmarked for land based programmes like soil and moisture conservation, water harvesting, and development of wastelands, and CPRs, creation of durable assets in rural areas.

Chapter - 21

Wasteland-A threat to Survival and Quality of Human life

Dr. Amrit Patel

Land has been a stock of renewable resources and a source for human survival as well as improving the quality of human life. Since it has competing demand and multiple uses the rate of land degradation far exceeds its natural rate of regeneration. This means the degraded land is not naturally replaced within a human lifetime resulting in loss of opportunities for the next generation.

Extent of degradation

According to National Remote Sensing Agency's district wise mapping of wastelands, using satellite data, the wastelands in India is 63.85 million hectares. Besides deserts, drought prone, flood prone, flood prone and tribal areas have been subjected to severe forms of degradation. Estimates of the cost of soil degradation during 1980s and 1990s ranged from 11% to 26% of GDP. The cost of salinity and water logging is estimated at Rs. 120 billion to Rs. 270 billion. The Working Group on ``Watershed Development, Rain fed Farming and Natural Resource Management" for the 10th Five Year Plan constituted by the Planning Commission had assessed that 88.5 million hectare degraded wasteland including rain fed areas would need development. The Working Group envisaged to cover the entire 88.5 million hectare land in four successive Five Year Plans, commencing from the 10th Plan to 13th Plan at an estimated cost of Rs. 727.5 billion (1994 price).

Causes and Consequences

India has only 2.4% of the world's geographical area and 0.5% grazing area but supports over 16% of the worlds population and over 18% of worlds cattle population. The degradation of environment in the fragile Indian sub-tropical ecosystem is basically attributed to increasing biotic and abiotic pressure; absence of adequate investment and appropriate management practices; high rate of population growth and high incidence of poverty in rural areas; over exploitation of natural resources; the break down of traditional institutions for managing common property resources and failure of new institutions to fil the vacuum and faulty land use practices. All these have resulted into soil and wind erosion; depletion of natural resources; lower productivity; groundwater depletion; shortage of drinking water; reduction in species diversity and increase in the extent of wastelands.

Ninth Five Year Plan

Two centrally sponsored schemes for soil conservation and integrated watershed management in the catchments of flood prone regions introduced in 1961 and 1982 aimed at enhancing productivity of degraded lands, minimizing siltation of reservoirs and chances of floods. During the 9th Plan both these schemes were merged into a new one soil conservation for enhancing productivity of degraded lands in the catchments of river valley projects and flood prone rivers. The scheme is being implemented in 53 catchments having a total area of 113.40 million hectare spread over in 27 States. The scheme for reclamation of alkali soils introduced in 1985-86 was extended to all States in the 9th Plan. It attempts to improve land and crop productivity by taking up production of crops, including horticulture, fuel wood plantation and fodder species suitable to the soil conditions, integrated wasteland development project being implemented, since 1989-90, is based on village/micro watershed plans, which are prepared after taking into consideration the land capability, site conditions and local needs of the people. The scheme also aims at rural employment besides enhancing the contents of people's participation in the development process at all stages, which is ensured by providing modalities for equitable and sustainable sharing of benefits and usufructs arising from such projects. The major activities undertaken are [i] in situ soil and moisture conservation measures like terracing, bunding, trenching, vegetative barriers and drainage line treatment [ii] planting and sowing of multipurpose trees, shrubs, grasses, legumes and pasture land development [iii] encouraging natural regeneration [iv] promotion of agro forestry and horticulture [v] wood substitution and fuel wood conservation measures [vi] drainage line treatment by vegetative and engineering structures [vii] afforestation of degraded forest and non-forest wastelands [ix] development and conservation of common property resources.

Working group

The Working Group on ``Watershed Development' Rain fed Farming and Natural Resource Management constituted for the 10th Plan projected that 107 million hectare of land were subject to degradation, of which 88.5 million hectare would have to be treated under watershed programs during 10th to 13th Plans.

Table 1. Wasteland Development Program During 10th to 13th Five Year Plans

Five Year Plan	Area covered million hectare	Esti-mated Cost Rs/ha	Total cost Rs million	Cost sharing Ratio	By Center Rs million	By States Rs million	By People Rs million
10th Plan [2002-07]	15.0	5000-7000	90,000	50:25:25	45,000	22,500	22,500
11th Plan [2007-12]	20.0	6000-8000	140,000	40:30:30	56,000	42,000	42,000
12th Plan [2012-17]	25.0	7500-9500	21,250	30:30:40	63,750	63,750	85,000
13th Plan [2017-22]	28.5	9000-11000	285,000	25:25:50	71,250	71,250	142,500
Total	88.5		727,500		236,000	199,500	292,000

New approach

The integrated Wasteland Development Program is being implemented since 1st April, 1995 on the basis of new guidelines for watershed development recommended by the Hanumantha Rao Committee, which envisages the bottom up approach whereby the Users Group themselves decide their work program. It aims at creating a scenario where the Government acts as a facilitator and the people at the grass root level become the real executioner of the program. Its strength lies in the flexibility approach followed in the method of release of funds, the area to be covered in each watershed as well as choice of components. It attempts to make the projects sustainable by establishing watershed development fund and involving people in deciding equity issues and usufruct sharing mechanism. It is not just a technical project but encompasses a social program as well. It, *inter alia*, emphasizes on greater flexibility in implementation, well defined role for State, district and village level institutions, removal of overlaps, a greater role for women, an effective role for the PRIs, involving Self-Help-Groups comprising rural poor, especially those belonging to SC/ST categories, seeking credit from financial institutions, transparency in implementation and most effective use of remote sensing data furnished by the National Remote Sensing Agency.

People's participation

Since it is the man who is primarily responsible for degradation of environment, regeneration and conservation can only be possible by creating awareness and seeking participation of the people who inhabit the watershed. The entire watershed community should be involved to implement IWDP and maintain the assets created to ensure sustainability. It emphasizes that the IWDP would have to become a people's movement in order to succeed. It is people's own program, which aims at giving terms of project implementation and fund disbursal. Its implementation seeks to empower people so that sense of collective responsibility can be included among them. It aims at decentralized decision making process by involving local people, PRIs, NGOs Government departments and watershed community at the grass root level and promotion of locally available low cost technology.

The new approach, also, recognizes the need to involve the community as a necessary condition for the sustainability of the program. Activities under the community organization include organizing Self Help Groups and User Groups, Participatory Rural Appraisal exercise, awareness camps, exposure, visits and programs on literacy, family welfare, social services, income generating activities etc. giving small contributions to SHGs or other village institutions like Mahila mandals, Youth clubs, Anganwadis which are considered important for people's participation. Effective community organization is important to establish credibility of the Watershed Development Team and create rapport with the village community who is ultimately going to own and implement the program even after withdrawing the Government machinery. In short, peoples participation and community organization is primarily sought to establish a system under which village people can actually involve themselves in planning, implementation and monitoring of watershed development programs.

Institutional Arrangement

Institutional arrangement has been provided from village to state level for successful implementation of the program and making it sustainable and equitable.

Water development association has a key role to play. It consists of al members of the village whose land is situated in the watershed area called User Group and all those members who derive sustenance from the watershed area called Self Help Group.

Watershed Committee is the key institution at watershed level consisting of about two to three representatives, each of User Group and Self Help Group, Panchayat and women etc. The committee also selects a watershed secretary preferably a local man graduate from the same area.

Watershed development team is a multi-dimensional team responsible for technical and financial supervision of the project activities. The team comprises field level officials drawn from various disciplines like forestry, soil conservation, horticulture, social science etc. These officials are key functionaries for sensitizing SHGs, UGs and villagers.

Performance

There have been 10 schemes for wastelands development implemented by three Ministries viz., Ministry of Agriculture (MOA), Ministry of Rural Development (MORD), and Ministry of Environment & Forest. Wastelands development program was started in each sixties and progressively strengthened during 9th and 10th Five Year Plans. The progress was slow till 8th Plan. It acquired momentum during 9th and 10th Plans as the area treated as well as investment increased sharply from 17,672 million hectare and Rs. 46,388.0 million till 8th Plan to 50.899 [188%] million hectare and Rs. 192.512.2 [315%] million at the end of 10th Plan. The MoRD accounted for 32 million hectare [63%] of the treated area investing Rs. 95,232 million [49.7%] as compared to 18.77 million hectare [37%] at the cost of Rs. 96,804.9 million [50.3%] by the MoA.

National Watershed Development Project for Rainfed Areas [NWDPRA] launched in 1990-91 [7th Plan] on pilot basis was implemented in 28 States during 10th Plan.

River Valley Projects [RVP] and Flood Prone Rivers [FDR] Program is currently implemented in 53 catchments in 27 States.

Watershed Development Project for Shifting Cultivation Area [WDPSCA]. An Area of 4.357 million hectare is affected by jhum/shifting cultivation mainly in seven States of North Eastern region and Orissa. This scheme launched in the 7th Plan [1987-88] has, since 1994-95, been implemented in seven States of NE region.

Reclamation of Alkali Soils [RAS] about seven million hectare is affected by salt problem, out of which about 3,581 million hectare suffers from alkalinity in 11 States.

Drought Prone Area Programme [DPAP] was launched in 1973-74 and currently is implemented in 972 blocks of 182 districts in 16 states.

Desert Development Program (DDP) was launched in 1977-78 and is now under implementation in 235 blocks of 40 districts in seven States.

Integrated Watershed Development Project [IWDP]; a centrally sponsored project was introduced in 1989-90 and since 1st April, 1995, is being implemented through watershed approach under the new guidelines in 443 districts.

Externally Aided Projects [EAP]; The MoRD is servicing externally aided watershed development projects for the development of degraded and wasteland areas.

WHAT TO DO ?

Peoples participation and community organization need to be made very effective to ensure proper planning, implementation and monitoring so as to achieve quantitative and qualitative objectives of the programs.

Table 2. Degraded Lands Developed Since Inception up to the Tenth Five Year Plan [Area : million hectare : Expenditure : Rs. million]

S.N.	Scheme	Year of start	Area Upto 8th Plan	Expend upto 8th Plan	Area upto 9th Plan	Expend Up to 9th Plan	Total Area Up to 10th Plan	Expand upto 10th Plan
	A : MoA							
1	NWDPRA	1990-91	4.233	9679.3	6.979	18777.4	9.309	30255.6
2	RVP & FPR	1962 & 81	3.889	8199.5	5.488	15612.6	6.486	22442.4
3	WDPSCA	1974-75	0.074	937.3	0.258	1662.7	0.393	2955.8
4	RAS	1985-86	0.484	622.9	0.581	763.9	0.711	1217.4
5	WDF	1999-00					0.059	260.2
6	EAPs		1.00	6460.0	1.335	20398.1	1.815	39673.5
	Subtotal				14.641	56764.7	18.773	96804.9
	B : MoRD							
1	DPAP	1973-74	6.86	11099.5	6.895	32847.4	13.727	48425.0
2	DDP	1977-78	0.848	7227.9	3.356	7973.8	7.873	19498.8
3	IWDP	1988-89	0.284	2161.6	3.734	6165.1	9.956	24381.5
4	EAPs				0.140	183.9	0.50	2926.7
	Subtotal				14.125	47170.2	32.056	95232.0
	C : MoE & F							
	NAEP	1989-90			0.070	475.3	0.070	475.3
	Total		17.672	46388.0	28.836	104410.2	50.899	192512.2

Program wise comprehensive evaluation by an independent professional team needs to be conducted at the end of each plan period to assess improved productivity of wastelands, improved availability of fuel wood and fodder, increase in water table, reduction in migration, improvement of economic status of the people, cost benefit ratio and rate of return on investment.

There is need to develop policies, which would result in the best use and sustainable management of land and water resources so as to prevent and becoming degraded and waste.

In order to formulate an appropriate plans for treatment of degraded ands a complete census of degraded or wasteland, its location, extent of area, ownership, the vegetative cover currently available and the biological physical and chemical properties of land needs to be done.

Since there are a number of schemes for the development of wasteland/degraded lands, a MIS with clearly defined benchmarks needs to be designed to get a realistic picture of the targets assigned, progress achieved and the tasks ahead.

The corporate sector may need to be involved in restoring wastelands and reclaiming degraded lands for which financial resources be channeled through involvement of financial institutions.

In collaboration with the NRSA, the Department of Land Reforms has released a wasteland Atlas of India in March 2000 indicating district level information on 13 categories of wastelands, which should be updated periodically with annual status of land records.

Proper technical advice on the reclamation of wasteland and on improving biological potential should be indicated in the Soil Health Passbook issued to farmers.

Conclusion

The United Nations has designated year 2008 as the international Year of Planet Earth and is celebrating it during the Triennium 2007-09. The event coincides with the launching of 11th plan.

Let all of us commit to protect and preserve our precious land, water and environment through all possible preventive and curative measures.

Chapter - 22

Wasteland Development Programme

Dr. A. Chelladurai & Dr. B. Manihar Sharma

Agriculture is the most important business of Indian economy and the natural resource base of the economy sustains performance of the Agriculture sector. So Agricultural activities have direct and immediate relations with the natural resources. The term natural resource includes land, water, rain, forest Land use pattern. In term of area and fertility of soil etc. India ranks seventh in the world and in terms of population, it ranks second with a total geographical area of 328.7 million hectares and reporting area at 306.03 million hectares. The net cropped area, which was 41.7 percent in 1950-51 has increased to 46.1 percent in 2000-2001 and it is believed that the area under net cropped area has reached a saturation point.

Table 1. Land utilisation pattern in India.

S.N.	Particulars	Area	Percent in million hectares
1	Total geographical area	329	-
2	Total reporting area	306	100
3	Barren land not available for cultivation	42	14
4	Area under forests	69	23
5	Permanent postures and grazing land	11	3
6	Culturable waste lands etc.	18	6
7	Fallow lands	25	8
8	Net area sown	141	46
9	Area sown more than once	49	16
10	Total cropped area	190	62

Source : Statistical Abstract, India 2002 (CSO)

An attempt has been made in this paper to understand the need for effective implementation of integrated wasteland Development Programme in India. When the present proportion of area under cultivation in

India is higher than in most of the countries in the world. It is not advisable to bring more area under plough in India as it may endanger the space for other uses. For instance in China it is 10 percent and the world average is only 14.7 percent. Table 1 describes the land utilisation pattern in India for the year 2000-2001.

Eighteen million hectares of land or 6 percent of the total land area is under culturable waste land i.e., land available for cultivation but not cultivated during the previous 5 or more years. And also total cropped area in India in 2000-2001 was 190 million hectares (it was only 185 million hectares in 1990-91).

What is wasteland?

Land and water are of critical importance for Agriculture development. Vast tracts of the land are however degraded but can be brought under plough with some effort. Such lands are known as wastelands. The productivity of wastelands is very low and people owning these lands are poor and are forced to earn a living from wage employment. Therefore, waste lands is regarded as a powerful tool and attacking the issues of poverty and backwards.

NATIONAL WASTELANDS DEVELOPMENT BOARD (NWDB)

The Central Government set up NWDB in 1985 to bring 5 million hectares of wasteland per year under fuelwood and fodder plantations. The setting up of this Board was the Governments response to the continuous deforestation in the context of the exploding population, on the one side and the tremendous suffering of the weaker sections in the rural areas in their search for fuelwood and fodder on the other NWDB works under the overall guidance of the National Land use and wastelands. Development Council Minister and attempts to reverse the trend of rapid deforestation and to conserve the already depleted forests by bringing wastelands under tree cover. Former Prime Minister Rajiv Gandhi had said, ``We shall develop a people's movement for afforestation".

INTEGRATED WASTELANDS DEVELOPMENT PROGRAMME (IWDP)

Further, the Government of India have launched the Integrated Wastelands Development Programme (IWDP) in 1989, throughout the country to improve the productivity of these lands and there by improve the living standards of the rural poor who own these lands. The IWDP is a 100 percent centrally sponsored scheme. The development of wastelands is taken up on watershed basis. Watershed is a geographical until where rain falling in the area drains through a common point.

OBJECTIVES

i. To arrest rainwater run off and conserve it in situ where it falls.

ii. Control of soil erosion which is usually caused by rainwater run off.

iii. Water and soil conservation also leads to improved green cover in the project areas leading to improved productivity of land.

iv. Wastelands are sought to be developed in an integrated manner based on village micro wasteland plans.

v. These plans are prepared after taking into consideration the land capability and site conditions and in consultation with the local people in regard to their needs.

vi. The Watershed projects are executed by the local people using low cost technologies locally available.

We are aware that the natural resources like the forests and agricultural lands have gradually been devastated and degraded to gradually been devastated and degraded to a great extent which is responsible for the cause of a number of problems. Vast areas have been depleted of the fertile top soil due to the siltation and salinization mainly attributed to floods. Crops yield have been much reduced. Besides we have come across the various effects caused due to the fast removal of the vegetational cover and massive deforestation in all almost regions of the world. The consequences are well known to all of us. Besides depletion in forest production, a number of ecological changes such as reduction in the quantity of water in the Groundwater (Aquifer) as well as drastic changes in the environmental quality and climate are concomitant. The reason which is quite obvious of the reduction in the production of the crops and tress is the degradation and destruction of the soils. The erosion and depletion of the fertile top layer of the soil are caused by a number of factors such as unscientific agricultural practices, construction of large dams, massive deforestation and number of man's activities concerned with over cultivation, urbanization and industrialization etc. All these activities denude the land of its protective green cover, hastening the process of soil erosion, degradation water logging and salinity. This affects both the cultivated and uncultivated lands. When these problems are not checked in time by taking proper remedial measures, there is every likelihood that the soil might lost its fertility forever turning the same into a barren and desert.

Desertification has been progressing in many states of India, which reduces the crops, and tree yields to a great extent. Generally such lands, which do not give economic, return or in the other word, which is useless, are termed as `wasteland'. Such lands are ecologically degraded and unstable as the topmost fertile layer of the soil (rich in organic matter and humus) has nearly or completely lost.

In India, the National Land Use and Wasteland Development Council (NLWDC) and the national Wasteland Development Board (NWDB) which are under the Ministry of Environment and Forests respectively have undertaken the tasks of preparing programs for conserving and protecting the lands from turning into wastelands.

A Technical Task Group was formed by the planning Commission and NWDB. This group has defined the wasteland as `the land which is degraded and is presently lying unutilized except as current fallows due to different constraints' (CSIR 1990). Depending upon the casual factors, the wastelands may be grouped under (i) Water erosion, (ii) Wind erosion and (iii) Salinity and Alkalinity. The wastelands comprise of three groups of land of viz. (i) cultivated land affected by soil erosion, (ii) degraded forest land (iii) degraded land with special.

Problems or problem soils. Depending upon the extent of degradation, the wastelands have been categorized by NWDB into (i) Cultural wastelands which are not being used currently due to different constraints but have the potential for the development of vegetative cover after appropriate treatments and (ii) Unculturable wastelands which are not being used currently and cannot be developed for vegetative cover under any circumstances. The culturable wastelands comprise many land areas such as gullied and ravinous lands, water logged lands and marches, salt affected lands. Areas under shifting cultivation (jhumming cultivation), degraded forest lands, degraded pasture and grazing lands, degraded non forest plantation land, striplands, sandy areas, mining/ industrial wastelands. The uncultural wastelands include barren rocky/ stony waste areas, steep sloping areas and snow covered areas.

Wastelands in India

Out of the total geographical area of 329 million hectare (mil. ha) in India 175 mil.ha has become wastelands. An area of 150 mil.ha has been degraded due to wind and water erosion which constitutes 85.7% of total wastelands. The remaining 25 mil.ha (14.3%) has been degraded due to water logging, salinization alkalinity, shifting cultivation etc. The degradation of top-soil in India has been estimated as 141 mil.ha constitutes 11% of the total loss of top soil of the world.

Deforestation which is one of the potent factors for turning fertile.

Lands into wastelands require immediate attention. India loses its forests at the rate of 1.5 mil.ha per year while the total rate of deforestation for the world accounts for 17 mil.ha per year. Based on the satellite imagery data Forest Survey of India reports that the total forests area of India stands at 637.293 sq. km by 1999 which constitutes only 19.4% of the geographical area though the New Forest policy of India states that 33% of the geographical area of India should necessarily be forest area (The Hindu, 2001). The present demand of fuel wood in India is about 235 million cum whereas our forests can supply only 40 million cum per annum. As India would require about 350 million cu.m. of fuel wood by 2010 A.D. There would occur a deficit of 305 million cum of fuel wood per year by then (Sagreiya, 2000).

Besides, the rate of annual afforestation is very meager as the Afforestation : Deforestation stands at 1 : 14 as against the world ratio of 1 : 10 (Newman, 1990; Sharma, 2001). Besides causing a number of ecological factors which are known to us, the removal of trees from our forests at a massive scale remove 60-75% of organic matter and other important nutrients thus rendering the soil nutrient depleted. Hence there is need to enhance massive planting of trees of overcome the various problems discussed herewith. In India, Rajasthan has largest area of wastelands (37 mil.ha) constituting 21.1% of the total wastelands of India. The percentages of wastelands in the North East constitute 16-25% of their respective geographical areas.

Wastelands in Manipur : Strategy

Let us assess the extent of wastelands in the state and analyze in what way the same can be improved. According to the reports of the Ministry of Agriculture, Govt. of India and CSIR (1990) the area of wasteland in the state is estimated at 7,340 sq. km out of the total geographical area of 22,366 sq. km. This constitutes 30.4% of the geographical area of the state and 0.42% of the wastelands of India. The area in the state stands at the 4th position of the largest wasteland areas in India. The high percentage of wastelands in the state clearly justifies why the crop and forests yields become reduced significantly.

The growth rate of rice in the N.E. region including Manipur was reduced from 5.6% in 1980's to 2.1% during 1991-95 while the growth rate for rice for India was reduced from 3.7% in 1980 to 3.28% in 1990 and 1995 respectively (The Hindu 1996). The fuel wood production from our forests has significantly reduced 1980-1990.

The causes of the conversion of fertile lands into wastelands in the state are multifarious. The degradation of the soils finally converting them into sterile soils without the top-soil has been noticed both in the plains and hilly areas alike. These are primarily attributed to Man's activities. Some of the important causes are highlighted below :

(i) The rapid rate of deforestation in the state has been the most potent factor for converting the land into wastelands. About 4,937 sq. km constituting 22.1% of the total area of Manipur represent only the dense of true forest as per the reports of the concerned department. But the area of the dense forest has been estimated at about 719.3 sq. km in 1990's based on the satellite imagery data given by the National Remote Sensing Agency (NRSA) and this constitutes only 3.2% of the total area of

Manipur. The heavy depletion of forest in state due to wide practice of shifting cultivation (Jhumming or pamlou in Manipur) as about 83,526 families in hilly areas have been practicing this system of cultivation. The total area cleared due to jhumming has been estimated as 4,905 sq km in 1990's Senapati and Ukhrul districts have the largest areas forests (25-28%) under the jhoom practice. Of many consequences of burning during jhum practice the most important effect is degradation of the fertile soil due to removal of the organic matter layer of the soil which depletes the important nutrient in the soil. The fuel wood production has decreased from 70,264 cu m in 1986 to 29,078 cu m by 1998. The per capita availability of fuel wood has been reduced from 0.03 in 1981 to 0.015 cu m in 1998. The average rate of annual tree plantation in the state is very less the annual ratio of afforestation : deforestation is found to be 1 : 13 : 5 (Sharma, 2001) which far exceeds the ratio of 1 : 10 for the forests of the world as given by Newman (1990). Besides degrading the soils, one of the serious effects is the threatening to extinction of over 65 plants species comprising of shrubs, bamboos, canes, orchids, aromatic, and medicinal plants from our forests (Table).

(ii) Removal of vegetational cover in the catchment areas of important rivers and lakes are responsible for the fast rate of siltation.

(iii) Lack of road side plantations especially in those areas facing the hill slopes. Even if such plantations are there, the right or appropriate plant species have not been used.

(iv) Heavy removal of brown earth from hill slopes for earth work which thereby exposes the hill slopes to wind and water erosion.

(v) Over use and misuse of inorganic chemicals, fertilizer as well insecticides to the fields without organic manure which thereby leads to the sterility and destruction of the soils.

(vi) Non practice of crop rotation in many fields keeping the soil as fallow areas exposes the field to wind and water erosion which finally leads to gully erosion.

Suggestion for Improvement

It may be noted that out of the total area of wastelands in the state, about 880 sq km has been treated and reclaimed by the government and this constitutes only 12% of the wasteland in the states and hence the major areas are yet to be reclaimed. Besides the efforts taken up by the Government a joint co-operative effort from the public, NGOs and organizations and remedial measures is called for some suggestions may be helpful towards implanting measures for reclamation of the wastelands in the states.

(i) Measures for the effective massive planting of trees in the forest areas with the right and appropriate selection of tree species.

(ii) Planting of sand binge species of sedges shrubs etc. on the naked hill slopes whish are exposed to wind direction. Some species like Pennisetum, Chrysopogon, Hetoropogon, Themeda Arundenella as well as Vitex and Sida species may be useful.

(iii) Planting of appropriate medium size trees in the fields and foot hills of hills towards the leeward side to protect the areas from wind and storm erosion.

(iv) Discouraging planting if Eucalyptus and Acacia confusa species around the fields, roadsides and hill slopes etc and substitution by appropriate species.

(v) Use of crop rotations in the fields as far as practicable using

Crop (paddy) legume - paddy alternation for restoring natural fertility in the fields.

(vi) Reduction of burning in the steep areas of hills stoppage of the burning phase during shifting cultivation and substitution of shifting cultivation by the terrace cultivation etc.

(vii) Mulching should be done in the fields as soon as the crop season is over.

This can be done by covering the fields using paddy stalks, corn stalks, paddy husks, saw dust etc. so that the soil is not exposed to wind and at the same time the soil gets enriched in organic matter.

(viii) Checking of flooding and water logging in the fields

(ix) Minimizing the use of inorganic chemicals and insecticides in the fields. Use of such chemicals be made upon advice from experts. Inorganic chemicals should always be used along with compost, FYM and any other organic manure.

(x) Watershed should be managed properly in the foothills.

1. Geographical area of Manipur : 22,366 sq.km
2. Area of wastelands : 7,340 sq. km percentage of geographical Area : 30.4%

   Percentage of wastelands of India : 0.42%
3. Category of wasteland in the state : Culturable.
4. Area so far reclaimed till 1990's : 880 sq. km

   Percentage of treated areas : 12%

   Percentage area remaining untreated : 6460 sq. km (88%)
5. Area degraded due to practice of shifting cultivation : 4,905.2 sq. km

Senapati Districts	1,388 sq.km	(28.3%)
Churachandpur	1,232 sq.km	(25.1%)
Ukhrul	718 sq.km	(14.6%)
Tamenglong	675 sq.km	(14.0%)
Chandel	655 sq.km	(13.3%)
Imphal	168.8 sq.km	(3.4%)
Thoubal	51.1 sq.km	(1.0%)
Bishenpur	17.3 sq.km	(0.3%)

6. Annual Afforestation : Deforestation ratio : 1 : 13.5
7- Forest Loss Index of the state (After Sharma, 2001) 0.04
8. No. of threatened plant species in the forests : 65 sp. (After Sharma, 1997 & 2001).

Some important threatened species

(i) Trees

Phoebe hainisiana, Junglans regia, Rhus hookeri, Ghrewia elastica, Flacourtia cataphracta, Butea frondosa, Aphanomixi polystachys, Magnolia griffithii, Cycas pectinata and *Podocarpus wallichii*

(ii) Orchids

Cymbidiumgiganteum, Arundina graminifolia, Ascocentrum ampellecum, Dendribium arachnites, Dendrobium densiflorum, Dendrobium nobile, Kalimpongia narajitii, Paphiopedilum insigne, Paphiopedilum villosum, Spiranthes sp. and *Vanda cuerulea.*

(iii) Aromatic and medicinal plants

Aquilaria agalocha, Dioscorea deltoidea and *Rauwolffia serpentina.*

Chapter - 23

Micro Irrigation and Drought Management

I. Satya Sundaram & Dr. K. Srinivasa Rao

In India, many challenger face water planners, managers, users and generally policy makers. Of course, we have not been able to properly exploit rain, surface and ground water. One third of the country's geographical area is always under threat of drought mainly because of uneven distribution of rainfall.

Given the serious water shortage, proper management of water has assumed special importance. The efficient water management is a complex issue, and it calls for joint efforts by engineers, irrigation agronomists and economists. While scarcity of water may retard agricultural productivity in a region, excess water supply too creates problems.

This paper has been divided into two parts : (i) Importance of Micro Irrigation : and (ii) Drought Management.

Micro irrigation is the term used to describe the method of irrigation which is characterized by the following irrigation features.

- Water is applied at low rate.
- Water is applied over a long period of time.
- water is applied at frequent intervals.
- Water is applied directly into the plant root zone, and
- Water is applied via a low pressure delivery system.

Micro irrigation ensures 30 to 70 per cent savings in water 25 to 100 per cent increase in yields and 15 to 30 per cent reduction in operating and crop production costs. It economists energy usage by around 50 per cent by reducing pumping hours and frictional losses. Experts say micro irrigation can double the area under irrigation, and also improve quality of end product. Micro irrigation can address the problem of grave disparties arising out of lopsided distribution of canal water.

Watershed management is receiving due attention. The unit of planning may be a village or a group of villages covered by a watershed. Depending on the local situation, water harvest structures should be constructed for harnessing rainwater. The surplus runoff, if any, will be stored in medium/minor dam(s). They serve the more needy land have shorter gestation period and are easier to manage.

We need to begin with watershed grid in the small watershed of individual village and integrate minor projects into the scheme and further use available water in dams and canals only through micro irrigation system.

Macro watershed will represent backward integration of the present drip irrigation practices and micro irrigation will be its forward integrating plank. Together they will form a modern irrigation package. We have to work out a proper balance amongst mini watershed developments, minor dams and micro irrigation methods.

The watershed management in dryland regions is integrated use of land, water and vegetation in an area for providing an answer to alleviate drought moderate floods prevent soil erosion, improve water availability and increase food, fuel, fibre and fodder production on sustained basis.

Drip irrigation is an advance method of applying water and fertilizer near the root zone of the crop with the help of low cost plastic pipes and emitters.

In spite of its high initial cost, drip irrigation is having an edge other methods, particularly in areas having saline soil, saline water and high evaporation rates. Drip irrigation also eliminates the need for construction of irrigation and drainage channels. The system can be used even by illiterate farmers. It has an efficiency up to 95 per cent.

In sprinkler irrigation, water is sprayed evenly on the crops through revolving nozzle-fitted emitters. Water is applied once in four to seven days. This reduces its moisture stress of the crop area to some extent. However, the water application being controlled, only the needed water can be regulated in this system. This has an efficiency of up to 7 per cent. In the conventional method of flooding the fields, losses from evaporation and percolation down to below root zones are high.

While drip irrigation is suitable for horticultural crops, where the plants are spaced apart, the sprinkler irrigation is commended for field crops where delivery of water to individual plant is not feasible.

Some problems

The micro irrigation technology is yet to become popular in India. It was introduced in the 1980s. But, be now, it could cover only 1.2 million hectares in 12 states. About half of the acreage is in Maharashtra alone. Of course, Karnataka, Tamil Nadu, Andhra Pradesh, Gujarat, Madhya Pradesh and Haryana are showing keen interest.

One reason for slow spread is that the initial cost of installation of micro irrigation system is high. The system requires equipment like pipes, tubes, filtration equipment, water emitters and so on. The equipment is also heavily taxed whereas the subsidy was 90 per cent in 1996, it came down to 25 per cent in the Tenth Plan.

Other problems include lack of technical knowledge and trained human resource. The scheme lacks sufficient institutional and credit support.

In the irrigation sector, there is great untapped potential. Micro irrigation can ensure more crop per drop.

Drought Management

Lining of water courses has been recognized as an effective measure to minimize the water losses in the transit and to avoid damage to the crops.

The warabandi system of irrigation aims at ensuring assured, adequate, dependable, timely supply of water in proportion to the land holding to all the beneficiaries in rotation under an outlet, irrespective of location. The system has the district advantage of ensuring equal distribution of water per acre in tune with size of land holding, cropping pattern, crop water requirement, nature of soil, rainfall etc.

The cropping pattern is an important determinant of water use efficiency. We have to select crop varieties and crop rotations in such a way that the available water is fully utilized.

Dryland farming requires action at two levels :

- an intensive approach aimed at integrated development of selected micro watershed, and
- an extensive approach for breeding water stress tolerant varieties, fertilizer dosage, soil and moisture conservation.

The order to provide adequate tree cover and promote subsidiary occupation, horticulture, afforestation and pasture development form an integral part of dryland farming programme.

In a drought prone area the emphasis should be on.

- on farm development in canal and tank covered areas;
- conjunctive use of surface and ground water;
- crop diversification wherever necessary;
- introducing sprinkler system for closely spaced crops in canal and tank command areas;
- large scale adoption of drip irrigation especially in well irrigated areas and for wide spaced commercial and high value crops;
- farmers managements/participation in irrigation;
- reuse of waste waters, both sewage water from municipalities and effluent water from industries, after proper reclamation.

The long term strategy for drought initiation should aim at :

- rainwater harvesting drought watershed development and efficient use of available water through micro irrigation systems such as drip and sprinkler irrigation :
- development of common property resources for water and fodder such as water ponds and grazing lands along canals and other water sources;
- promotion of less water requiring crops and other farm activities including silvipastures, in drought prone areas;
- evolution of drought resistant varieties of crops such as legumes (pulses) oilseeds and fodders.

In mitigating drought, village ponds and farm ponds have to play an important role. The village ponds can be used for livestock and to meet daily needs of villagers. Of course the extent of help depends on catchments characteristics, runoff volume and its utilizations. The natural farm ponds are for pre-sowing, protective irrigation an livestock consumptions.

A pilot scheme in the state sector for repair, renovation and restoration of water bodies directly linked to agriculture is proposed to be taken up during the remaining period of the Tenth Plan. The objectives of the scheme are :

- to restore and augment storage capacity of water bodies, and
- to recover and extend their lost irrigation potential.

As micro irrigation aims at water conservation, it is the most important instrument to bring down the severity of drought.

Drought connotes a situation of water shortage for human, cattle and agriculture consumption resulting in economic losses, primarily in agriculture sector. India, being an agriculture driven nation and farmers majorly relying on monsoon, they stand handicapped if the rains are not on time. Agriculture is determined by the climate. In other words, climate is the key factor in any operation of agricultural production right from field preparation to marketing. The success and failure of farming is closely associated with the prevailing weather conditions. In India, drought essentially occurs due to failure of south-west monsoon (June-September). Although the onset of South-West Monsoon 2009 was in time, the advancement of monsoon delayed in Central and Northern parts of the country resulting in delay in advancement of monsoon to other parts of the country. It was deficient by 39% as on 10th June, 2009, and was -54% as on 24th June, 2009 which has resulted in drought like conditions in some parts of the country.

The declaration of drought was primarily based on the quantum of rainfall, damage to kharif crops and lesser availability of drinking water and less moisture in the soil. Monsoon failure results in crop failure, shortage of drinking water as well as undue hardship to the rural and urban community. Indeed, the Southwest monsoon accounts for about 80 per cent of the country's rainfall in a year. With only about 40 per cent of country's sown area irrigated, the monsoon becomes crucial in determining agricultural output. There is no gainsaying the fact that too much or too little rain can prove disastrous. Monsoon rain generates food and provides labour; and creates cash flow in the market. Bad rain could even result in dipping stock market and falling corporate investment, Further, a great deal of the country's electricity requirement is generated by water power provided by the monsoon rain. A severe drought could reduce gross domestic product (GDP) in a given year by about 2 to 5 per cent. This is why the progress of monsoon - its onset and performance - is followed with such a keen interest in India. The present paper is an attempt to explore the drought conditions that are prevailed in the country and its effect on the proposed economic growth.

A historical snapshot

Historically, a severe drought caused a major economic crisis. When rainfall became below normal, by say, 15 per cent or more, growth rates of both GDP from agriculture and aggregate GDP declined, and in extreme cases, both became negative. The incidence of unemployment and under-employment increased, especially in the rural areas. Inflation rate shot up to double-digits. The food imports necessitated by the fall in domestic production led to foreign exchange crisis. Finally, the burden of drought relief, borne mainly by the government of India, increased the fiscal deficit. The drought, therefore, became the most important factor behind the economic crisis in 1965-66, 1966-78, 1972-73, 1974-75 and 1979-80.

A Concern on Agriculture and Economic growth

The fact is that agriculture remains the basis of economic growth for most South Asian countries. The economies are therefore affected directly by the vagaries of monsoon, which has far reaching implications for agricultural productivity and commodity prices. Around 65% of food grain production is undertaken during the monsoon period and approximately 60% of the population in South Asia depends on crop husbandry, animal husbandry, fisheries and forestry for food, income and employment.

Since agriculture is also the basis for industrial development, supplying the raw materials for manufacturing industries and stimulating industrial output, the monsoon can directly affect government savings, public investment and foreign exchange reserves. Agricultural operations have been adversely affected in several parts of the country causing distress to farmers and their families. a deficit of more than 6 million hectares has been reported in paddy, which is the worst affected crop. If we see the overall distribution of rainfall, the northern region has not received adequate rainfall for its kharif sowing. Crops which stand at risk are paddy, pulses, bajra, cotton, soybean and sugarcane. If the overall rainfall in the country remains deficient in the current season, we shall expect the prices of these commodities volatile and may touch higher levels. Scanty early rains would mean a definite drop in the kharif crop and a lower agriculture output. With the consequent rise in prices of agricultural products, a larger proportion of the household income would be devoted to its purchase. This, together with lower disposable income with farmers, would press down industrial demand and hence growth.

Farm output may be affected due to late monsoon

Delayed and deficient monsoon sowing of the kharif crops which account for nearly 57% of the country's total agricultural output. It could adversely affect farm production this year impacting food prices. The progress of the south west monsoon has been slow and halting. While Kharif sowing has picked up in June 2009. As monsoon continues to play truant and the possibility of a drought gets real, the performance of the Indian economy is under threat. assuming a normal monsoon, most research agencies and organizations had forecast a 6 per cent plus real growth in gross domestic product (GDP) for 2009-10. India's central bank, the Reserve Bank of India (RBI), had forecast 6 per cent, National Council for Applied Economic Research (NCAER) 6.5 per cent, the Prime Minister's Economic Advisory Council 6.5 to 7 per cent, Crisil 6.3 per cent, ICRA 6.5-7.5 per cent and Fitch 5 per cent. Agricultural production is likely to fall 5 per cent and GDP with service sector estimated to grow at 7 per cent and industrial growth at 4 per cent, is likely to be around 4-4.5 per cent against an estimated over 6 per cent. Agriculture contributes around 25 per cent to GDP. Most hit would be the agro-based industries (edible oil, textiles, sugar), big ticket consumer durable items and the postponable FMCGs.

Drought years have brought negative growth rates for the agricultural economy. During the past 60 years, there was a drought every few years. And almost always, the agri economy witnessed a negative growth. For instance, in 1965-66, when agriculture contributed to 43 per cent of overall GDP, a massive 4.8 per cent of the GDP growth rate got knocked off due to drought. In 2002-03, a notable 1.55 per cent of the GDP growth got wiped off because agriculture then comprised 21 per cent of overall GDP.

How drought impacts the overall GDP growth rate of the country

Drought years	Agriculture GDP growth rate (%)	Impact on overall GDP growth rate (%)
1951-52	1.49	Nil
1965-66	-11.04	-4.78
1966-67	-1.42	-0.6
1968-69	-0.16	-0.07
1972-73	-5.02	-2.06
1974-75	-1.52	-0.63
1979-80	-12.77	-4.64
1982-83	-0.28	-0.1
1985-86	0.31	Nil
1987-88	-1.59	-0.5
2002-03	-7.24	-1.55

Source: CMIE, National Remote Sensing Centre, ISRO, BW Research

Economic effects of drought

- Reduced production of food and allied food products
- Decreased availability of dairy and livestock products
- Loss to industries directly dependent on agricultural production (e.g. machinery and fertilizer manufacturers, food processors, dairies, etc.)
- Cost of water transport or transfer
- Cost of new or supplemental water resource development.
- Increased commodity prices
- Revenue losses to state, and local governments
- Increased demand for monetary assets and increased interest rates
- Reduction of economic development
- Decrease of gross national product and economic growth

Drought conditions curtail economic growth

Economic growth and employment will suffer in a drought. The share of in GDP is now just 18%, and barely 12% relates to crops (the rest is animal husbandry, horticulture, etc). Despite extensive irrigation, a drought will turn agricultural growth negative, as against average growth of 4.3% in the last five years. That will directly reduce GDP growth by at least 0.5%. Its indirect impact - reduced employment and reduced rural spending on good and services - may be greater. So, a drought could pull GDP growth well below the 6.7% achieved last year. But this drought may hurt growth more than past monsoon failures if it prompts rural consumers to slash spending on domestically produced items like clothing and medicines, spilling the drought's impact into the industrial sector.

Drought Situation in Andhra Pradesh

After four years of bountiful kharif seasons, the farmers in Andhra Pradesh are going to witness an unprecedented situation, with at least 1,000 mandals out of the 1,128 reporting deficit rainfall. While vast stretches are yet to see sowings, crops on lakhs of acres face the threat of withering or wilting due to severe deficits in rainfall. The State shows a deficit of 52 per cent, with the South-west monsoon playing truant. Against the normal as on date figure of 322.20 mm, the State received only 153.80 mm. The problem seems to be more serious than what is reported officially.

The hopes of the farmers and officials that the first week of August might see showers crashed, with seven districts reporting deficits of 60 per cent of 99 per cent. The remaining districts, barring the three North coastal districts, registered deficits between 20 per cent and 59 per cent. This has impacted almost all crops. Paddy and groundnut were sown in just 26-56 per cent of the normal area. Sugarcane, maize, onion and tur dal were sown in 51-75 per cent, while cotton is slightly better off with 76-100 per cent sown area.

Crop area : Same story across regions

(In lakh hectares)

Region	Normal for the day	Same day last year	As on date
Coastal Andhra	14.91	13.45	9.57
Raylaseema	14.40	13.71	5.74
Telangana	26.77	24.39	22.73
Total	56.09	51.56	38.06

Source : Ministry of Agriculture (*Paddy growing districts*)

The three key paddy growing districts of East Godavari, West Godavari and Krishna districts reporting deficits of 41 per cent, 47 per cent and 65 per cent shows the enormity of the problem. ``Jowar, maize, cotton, tur dal, and groundnut show withering or wilting symptoms. The growth could be stunted if there are no rains in the coming days. Yields too will drop significantly."

Short & Long Term Measures to overcome

Government of India has taken a decision recently to reimburse 50% of the cost that the State would incur in providing diesel subsidy to those farmers who are affected. A few states have been provided additional power from the Central pool.

Short term Measures

1. Providing drinking water by better management of water sources in the affected areas by -

   Collecting available water and supplying it either by pipes or by tankers including railway tankers/wagons to the affected areas; or

   By exploiting the possibility of new sources of water such as tube-wells, deepening of existing tube-wells, etc.
2. Providing fodder for cattle and other livestock affected by drought.
3. Providing employment to the drought affected people so that they may have some income in lieu of the loss of income caused by the loss of crops. (At least one person in a family should benefit out of the employment oriented schemes during the drought period).
4. Ensuring that the required amount of food grains is made available through the Public Distribution System.
5. By optimizing irrigation support to the crops out of existing facilities.
6. By cultivation of short duration varieties of crops after the drought situation is relieved by some rainfall.

Long term measures

1. Developing drought resistant varieties of crops and propagating their spread.
2. Completing quick maturing irrigation schemes and ensuring water flow in the medium-term, say, within 3 years.
3. Taking up watershed development activities and developing a cropping system based on such watershed development.
4. Rain water conservation including roof-top harvesting.
5. Afforestation of all exposed hill slopes as per Ridge to Valley Watershed Development approach and increasing forest canopy cover in order to ensure conservation of rain water.
6. All works in drought affected areas should ensure sustainability.

7. The State Governments should interact with local/state level NGOs in drought related activities.
8. The State Governments should try to construct check dams in the drought affected areas with actual participation of beneficiaries/NGOs so that the stored water could be used for various purposes during drought.

Chapter - 24

Dryland Conservation

Altaf Hussain

Drylands are limited by rainfall, high evapotranspiration and show a gradient increase in productivity from hyper arid to arid and semi-arid to dry sub-humid areas, on decreasing aridity or moisture deficit. Drylands cover about 41 per cent of earth's available land surface and three quarters of world food supplies come from drylands (FAO, 1999). The challenges for global agriculture in 21st century is to produce 7 per cent more food to feed a projected population of 10 billion by 2050 by making sustainable use of existing resources and responding climate change (FAO, 2009). Drylands span over 41 per cent of earth's available land surface will need to contribute their share to this yield increase. So improving dryland crop yield is important, both to maintain food security and to improve livelihoods of the poor.

Drylands in India contribute 70 per cent total cultivated area and about 50 per cent of the geographic area is affected by desertification) Food insecurity, extreme poverty and environmental nexus are the most challenging in the drylands. Improving crop productivity is important both to maintain food security and to improve livelihoods of the people in drylands. Investments are needed for soil and water conservation in order to improve soil fertility and soil moisture. Conservation and efficient utilization of natural resources are two key components to achieve sustainability in drylands. Land degradation and over exploitation of resources prompted researchers and policy makers to evolve innovative technologies which halt degradation and restore productivity. A number of technological innovations are used which include cultural practices, engineering methods, sustainable agricultural practices, precision conservation and agroforestry. Hence, transforming drylands is necessary to achieve second green revolution.

Every continent contains dryland regions. Drylands are most extensive in Africa (13 M km^2) and Asia (11 M km^2). About three quarters of the world food supplies consisting of wheat, maize, sorghum, pulses, oilseeds, potato and fruits are grown of drylands (FAO, 1999). According to the Millennium Assessment (MA) report there are 2.3 billion people living in the drylands, out of which 1 billion are below poverty line accounting half of the world's poor (MA, 2005). Millions of rural dryland dwellers are directly dependent on local dryland ecosystem services for their daily survival. Therefore, any shortfall in any one of such services will create food insecurity, famines, conflicts and vulnerability of millions of rural poor. Climate change will have a disproportionate effect of dryland areas, contributing to desertification and increasing the vulnerability of people in drylands. We need to put the conservation of dryland ecosystem services at the heart of development policy, if we want to reduce poverty and achieve the millennium development goals.

Background

Definition and characteristics

Drylands are generally defined as lands with limited rainfall. Mainly their dryness is due to the negative balance between precipitation and evapotranspiration rates. Drylands are thus been defined in terms of water stress as areas where mean annual precipitation (P) is less than half of the potential evapotranspiration (PET). According to the FAO (1993), drylands are agroclimatic zones having short growing periods, which is defined as the period when both water and temperature permit crop growth. So drylands are zones falling between 1-74, 75-119 and 120-180 growing days representing arid, semi-arid and dry sub humid lands, respectively.

Drylands are characterised by low (100-160 mm annually) erratic and highly inconstant and unreliable rainfall levels. Precipitation is low concentrated during short periods, resulting much of the rainfall to be lost in evaporation and the usual intensity of storms ensures that much of the rainfall runs off in floods. Fragile environments and unpredictable drought and floods are common features of drylands ecosystems.

Classification of drylands

Dryland ecosystems are mainly categorised into four subtypes according to aridity index and annual rainfall levels into hyper arid, arid, semi-arid and dry sub-humid areas as shown in Table 1.

World drylands

Dryland ecosystems occupy over 41 per cent of the earth's land surface. Desertification affects 70 per cent of the world drylands, amounting to 3.6 billion ha or one fourth of worlds land surface (IFAD, 1995).

Asia possesses the largest land area affected by desertification, 71 per cent of which is moderately to severely degraded. In Africa two thirds of which is desert or drylands. 73 per cent of agricultural drylands are moderately to severely degraded (IFAD, 1995). Africa is under greatest desertification threat, with a rate of disappearance of forest cover of 3.5 to 5 million ha per year bearing down on both surface and ground water resources and with half the contents farmland suffering from soil degradation and erosion.

Table 1. Total dryland categories according to FAO (1993), classification and extension (UNEP, 1992)

Subtypes	Aridity Index	Current area		Dominant Biome	Current Population	
		Mk m²	%global		*1000	% global
Hyper arid		9.78	6.6	Desert	101,615	1.7
Arid		15.66	10.6	Desert	222,204	3.7
Semi arid		22.59	15.3	Grassland	828,341	13.9
Dry sub humid	0.50-0.65	12.87	8.7	Forests	909,273	15.3

[Source : World resource Institute, 2002]

Causes of dryland formation

Limited rainfall, poor soil quality, fragile environments are the main factor behind dryland formation. There is always water scarcity in drylands. The dryness of drylands is due to negative balance between mean annual precipitation and potential evapotranspiration rates. Besides, limited rainfall, the soils are of poor quality, low in organic matter, hence less fertile. Harsh climates are another important issue which limits crop diversification in drylands. What makes the drylands a difficult environment is not only less rainfall, but also its erratic distribution. Inter annual rainfall can vary from 20-100 per cent and periodic draughts are common (Zurayk and Haider, 2002).

Problems of drylands

Water scarcity due to limited rainfall, low soil fertility, mostly deep sandy soil with poor water holding capacity, shallow and rocky soils with low organic matter content. Fragile environments with unpredictable floods and droughts are other factors limiting drylands to become productive ecosystems. Lack of technologies limitation of resources and biotic pressures contribute further in conversion of drylands into deserts.

Loss of resources

In the world as a whole about 25000 million tones of soil are being washed away from land every year. In India, the figure is 6.25 thousand million tones. Due to erosion and degradation, the world is losing between 5-7 million ha of cultivated land every year which is nearly the same as the new land brought under cultivation, which means that the extent of cultivated land remains more or less same (Lazarus, 1992). According to Millennium Ecosystem Assessment (MA) Report about 1020 per cent of world drylands are degraded accounting about 6-12 million km^2 (MA, 2005). Drylands in India contribute to over 70 per cent of the total cultivated area and about 50 per cent of the total geographic area is affected by desertification (Hegde, 2006). Land degradation is particularly problematic for both environmental sustainability and poverty reduction in dryland areas. The UN Convention to Combat Desertification (UNCCD) and others use ``desertification" to describe dryland degradation which is caused due to several factors including climatic variations and human activities. Depending on the level of aridity, dryland biodiversity is relatively rich, still relatively secure and is critical for the provision of dryland services. Of 25 global ``biodiversity hot spots" identified by Conservation International, 8 are in drylands. So to conserve dryland are very important to ensure food security, conserve rich biodiversity of drylands and improve livelihoods of dryland people. To conserve the scarce resources of drylands a number of practices or methods are used which constitute dryland conservation technologies. These technologies are agronomic or cultural practices like conservation tillage, mulching, organic manure application, contour faming, strip cropping, use of wind breaks, allay cropping, vegetative barriers etc. and mechanical or engineering methods which include basin listing, sub-soiling, terracing, contour bunding, contour trenching, use of gully plugs, check dams and water harvesting structures like community tanks, intra terrace water harvesting and roof top water harvesting etc. In spite of these practices or methods there are several other measures which can be applied for dryland conservation. These approaches are :

i. Sustainable farming practices

ii. Precision conservation

iii. Integrated watershed approach, and

iv. Use of agroforestry

Table 2. Drylands of the world

Continents	**Land mass (M ha)**	**Hyper-arid (<0.05)**	**Arid (0.5 < 0.20)**	**Semi-arid (0.20-0.50)**	**Dry sub-humid (0.50-.65)**	**% of world drylands**
Africa	2965.6	672.0	503.5	513.8	268.7	31.9
Asia	4255.9	277.3	625.7	693.4	352.7	31.7
Australia	882.2	0.0	303.0	309.4	51.3	10.8
Europe	950.5	0.0	11.0	105.2	183.5	4.9
North America	2190.9	3.1	81.5	419.4	231.5	12.0
South America	1767.5	25.7	44.5	264.5	207.0	8.8
Total	13012.6	978.1	1569.2	2305.3	1294.7	100.
% of world drylands		16	26	37	21	

[Source : Reynolds and Smith, 2002]

Agronomic or cultural practices

Agronomic or cultural practices for soil and water conservation in drylands help to intercept rain drops and reduce the splash effect, help to obtain a better intake of water by the soil by improving the organic matter content and soil structure; help to retard and reduce the surface runoff through the use of mulches, strip cropping, mixed cropping and contour cultivation. Use of vegetation on mechanical structures such as gully checks and water harvesting structures etc. enhance their strength and extend their life span.

Mechanical and engineering methods

These are permanent structures used to supplement the agronomical practices, when the later alone are not adequately effective. These measures play a vital role in controlling soil erosion and reducing runoff. These are used mostly in drylands where the slope of the soil is more than permissible limit. The main objective of the mechanical methods for controlling soil erosion are : (i) to increase the time of concentration by intercepting the runoff and thereby providing an opportunity for the infiltration of water and (ii) to divide a long slope into several short ones so as to reduce the velocity of the runoff and thus preventing erosion. These measures are basin listing, sub-soiling, terracing, contour bunding, contour trenching, gully plugging, check dams and water harvesting structure for hilly areas.

Water harvesting structures for dry hilly areas

Water harvesting is a prominent and technically feasible technology in arid hilly areas. It helps in runoff harvesting and ground water recharging. Different types of water harvesting structures are used for efficient utilization of rainfall. Such as community tanks, inter-terrace runoff harvesting, hill spring outflow harvesting and rooftop harvesting structures. Runoff utilization is increasingly becoming a common practice in dryland conservation agriculture.

There are other approaches which can be adopted for conservations of dryland ecosystems. These research based approaches are as :

(i) Sustainable farming practices

(ii) Precision conservation

(iii) Integrated watershed approach and

(iv) Use of agroforestry

1. Sustainable farming practices

The past decades have witnessed a dramatic change in agriculture with food production soaring due to green revolution. The green revolution entailed the use of improved technologies like high yielding crop verities, expansion of irrigation, mechanization and the use of chemical fertilizers and pesticides. Sustainable agricultural practices are not new, but drawn on traditional knowledge and practices, adopted to ensure food security and maintaining productivity of dryland ecosystems on sustainable basis. These practices are conservation tillage, integrated nutrient management, agroforestry, water harvesting, livestock integration, use of FYM and mulches, green manuring and integrated pest management etc. to maximize productivity without compromising the needs of the future generations.

2. Precision conservation

Precision conservation offers an alternative to integrate the use of spatial technologies such as global position system (GPS), remote sensing (RS) and geographic information system (GIS) and the ability to analyze spatial relationship within and among mapped data to develop management plans that account for the temporal and spatial variability of flows in the environment. Hence precision conservation practices helps to maintain maximum production by improving soil and water conservation by developing efficient land use management plans.

Precision conservation is an innovative three tier approach comprising a set of spatial technologies and procedures linked to mapped variables, which is used to implement conservation management practices that take into account spatial and temporal variability across natural and agricultural systems (Berry *et al.,* 2003; 2005).

3. Integrated watershed approach

An approach towards dryland conservation. Basically a watershed is a basin like landform defined by high points and ridge lines that descend into lower elevations and stream valleys. A watershed carried a water ``shed" from the land after rainfalls and snow melts. Drop by drop water is channeled into soils, groundwater, creaks and streams making its way to rivers and eventually the sea. In other words a watershed is a geohydraulic unit or piece of land that drain at a common point. The aim of watershed management is to ensure that every drop of water and every square foot of land is best utilized.

Integrated watershed approach is not only anti erosion and anti-runoff approach but also a comprehensive integrated approach of land and water resource management. This approach is preventive, progressive, corrective as well as curative.

4. Role of agroforestry in soil and water conservation in dryland ecosystems

Agroforestry is the science of developing integrated self sustainable land use systems in which trees are grown on farm lands along with field crops. It includes the introduction and/or retention of tree crops for timber and fodder, fruit trees, shrubs bamboos, canes and palms along with cultivated filed crops including pasture simultaneously or sequentially on the same piece of land and at the same time to meet the ecological and socio-economic needs of the people. A well planned and properly managed agroforestry programme substantially increase the yield of the land and maintains sustained productivity.

The following are the major agroforestry systems :

(1) Agrisilviculture (trees + field crops)

(2) Boundary plantation (trees on boundary + field crops)

(3) Block plantation (sequential blocks of trees and field crops).

(4) Energy plantation (trees + field crops during trees establishment period).

(5) Allay cropping (hedges of economic value + field crops).

(6) Agrihorticulture (fruit tree + field crops)

(7) Silvipasture (trees + pasture/animal husbandry)

(8) Forage forestry (fodder trees + pasture).

Besides above mentioned systems, two main practices are adopted with the object of intensifying farming on slopes alongwith reducing soil erosion and increasing moisture conservation. These are (i) sloping agriculture land technology (SALT), (ii) Biomass transfer technology (BTT).

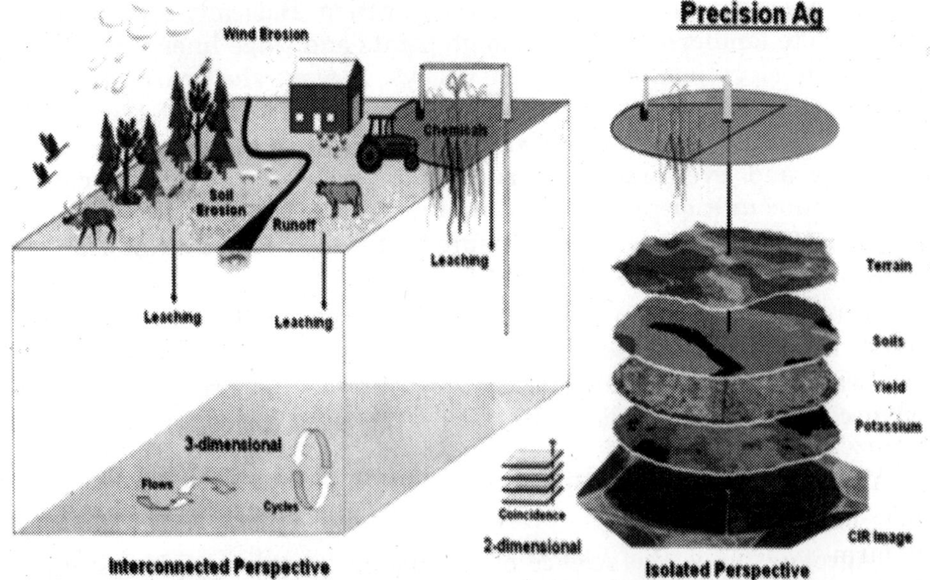

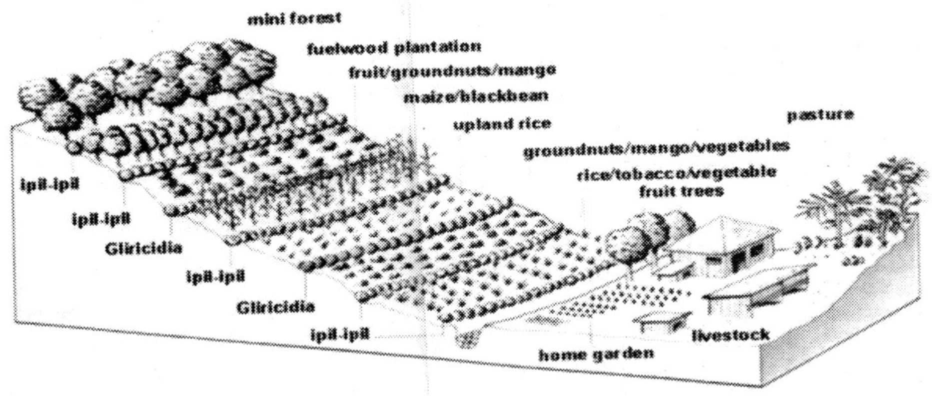

Sloping Agriculture Land Technology (SALT)

The sloping agricultural land technology (SALT) is a farming system developed by the Mindanao Baptist Rural Life Centre in the southern Philippines during the 1970's. Basically attuned to the production needs of small scale hill farmers. This agroforestry technology has gained wide popularity in Asia because it is culturally appropriate, economically sound and is designed to limit soil erosion. SALT is a technology package of soil conservation and food production that integrates several soil conservation measures (Tacio, 1989; Evans, 1992). Basically, the SALT method involves planting field crops and perennial crops in bands 3-5 m wide between double rows of nitrogen fixing shrubs and trees planted along the contour. These minimize soil erosion and maintain the fertility of the soil. SALT helps considerably in the establishment of a stable ecosystem, the double hedge rows of leguminous shrubs or trees prevent soil erosion. Their branches are cut every 30-45 days and incorporated back into the soil to improve its fertility (Palmer, 1991). The crops provide permanent vegetative cover which aids the conservation of both water and soil.

Biomass transfer technology (BTT)

Various agroforestry technologies are finding enormous application in the east and central African (ECA) region and are lifting many out of poverty and mitigating declining agricultural productivity and natural resources. One such example is biomass transfer in which trees that are rich in mineral elements (fertilizer trees), when integrated with inorganic fertilizer can double or triple crops yields in degraded lands.

Biomass transfer technology involves the growing of trees/shrubs along boundaries or contours on farms or the collection of the same from off farm niches such as roadsides and applying the leaves on field at planting. In western Kenya, *Tithonia diversifolia* become the preferred species used by farmers to grow maize, beans or kale etc.

Conclusion

Drylands cover about 41 per cent of land surface. Characterised mostly by low, erratic and highly inconsistent rainfall, water scarcity, soil erosion and climate change are its prominent features. About half of the world food supply comes from drylands and host over half of the world's poor. Keeping in view the rich biodiversity of drylands and home land for millions of rural poor people which are directly dependent on scarce resources of drylands, different innovative technologies for conservation of drylands area adopted to ensure food security, improve productivity and maintain environmental stability. By this way dryland resources and biodiversity reserves are conserved.

Apart from soil and water conservation measures, enhancement of soil fertility is also a vital component of dryland conservation. From a high cost external input oriented agricultural production, to an integrated nutrient management approach, soil fertility can be thought of with inputs like biofertilizers, organic manures and composts green manures and use of mulches etc.

Chapter - 25

Dryland and Wasteland Farming

K. Srivani & K.K. Tripathy

India has about 108 million hectares of rain fed area which constitutes nearly 75 per cent of the total 143 million hectares of arable land. In such areas crop production becomes relatively difficult as it mainly depends upon the intensity and frequency of rainfall. The crop production, therefore, in such areas in called rain fed farming as there is no facility to give any irrigation, and even protective or life saving irrigation is not possible. These areas get an annual rainfall between 400 mm to 1000 mm which is unevenly distributed, highly uncertain and erratic. In certain areas the total annual rainfall does not exceed 500 mm. The crop production, depending upon this rain, is technically called dry land farming and areas are known as dry lands.

India has about 47 million hectares of dry land out of 108 million hectares of total rain fed area. Dry lands contributes 42 per cent of the total food grain production of the country. These areas produce 75 per cent of pulses and more than 90 per cent of sorghum, millet, groundnut and pulses from arid and semi-arid regions. Thus, dry land and rain fed farming will continue to play a dominant role in agricultural production.

Dry land, besides being water deficient, are characterized by high evaporation rates, exceptionally high day temperature during summer, low humidity and high run off and soil erosion. The soils of such areas are often found to be saline and low in fertility. as water is the most important factor of crop production, inadequacy and uncertainty of rainfall often cause partial or complete failure of the crops which leads to period of scarcities and famines. Thus the life of both human being and cattle in such areas becomes difficult and insecure.

Despite all these improvements in agriculture, we have yet not been able evolve an appropriate package of practices for our dry land areas. The income of farmers of dry land regions is still very low land areas.

Characteristics of Dry land Agriculture

Dry land areas may be characterized by the following features:

1. Uncertain, ill- distributed and limited annual rainfall;
2. Occurrence of extensive climatic hazards like drought, flood etc;
3. Undulating soil surface;
4. Occurrence of extensive and large holdings;
5. Practice of extensive agriculture i.e. prevalence of monocropping etc;
6. Relatively large size of fields;
7. Similarity in types of crops raised by almost all the farmers of a particular region;
8. Very low crop yield;
9. Poor market facility for the produce;
10. Poor economy of the farmers; and
11. Poor health of cattle as well as farmers.

Problems of Dry Farming in India

The major problem which the farmers have to face very often is to keep the crop plants alive and to get some economic returns from the crop production. But this single problem is influenced by several factors which are briefly described below.

Moisture stress and uncertain rainfall According to definition the dry farming areas receive an annual rainfall of 500 mm or even less. The rains are very erratic, uncertain and unevenly distributed. Therefore, the agriculture in these areas has become a sort of gamble with the nature and very often the crops have to face climatic hazards. The farmers also take up farming halfheartedly as they are not sure of being able to harvest the crops. Thus, water scarcity becomes a serious bottleneck in dry land agriculture.

Effective storage of rain water According to characteristics of dry farming, either there will be no rain at all or there will be torrential rain with very high intensity. Thus, in the former case the crops will have to suffer a severe drought and in the latter case they suffer either flood or water logging and they will be spoilt in case of very heavy downpour, the excess water gets lost as run off which goes to the ponds and ditches etc. This water could be stored for providing life saving or protective irrigation to the crops grown in dry land areas. The loss of water takes place in several ways namely runoff, evaporation, uptake through weeds etc. The water could be stored for short period or long period and it can be preserved either in soil, pond or ditches based on situation and utilized fr irrigation during dry periods.

Dry land agriculture cannot compete with conventional standards and definitions of productive agriculture. If such a competition is attempted, it will only turn out to be an economic and environmental disaster for the dry lands. Therefore, norms, standards and definitions have to redraft for dry land agriculture separately and realistically.

The burgeoning population growth of India coupled with rapid urban development has led to an increasing demand on the country's land resources. An indication of this burden on the natural resources is a simple comparison between India's share in total world land area and in the total world population. While the former is a meagre 2 per cent of the world geographical area, the latter constitutes 16 per cent of world's population. Land resources provide livelihood to two thirds of India's population. The increasing pressure on land, relentless exploitation of this valuable resource for agricultural and allied, housing, industrial and manufacturing activities has made the productive farm lands less productive, leading to its constant degradation.

The total geographical area of the country is around 329 million hectares out of which only 264 million hectares (80 per cent) are fit for vegetation. While 142 million hectares are covered under all types of crops, 67 million hectares of land are under forest cover and 68.35 million hectare area of land is lying as wastelands in India. The Government of India (GoI) defines wastelands as the degraded land which is currently under utilised and can be brought under vegetative cover, with reasonable effort by resorting to effective and appropriate water and soil management.

It is estimated that approximately half of the wastelands in India which are not covered under forests of any kind can be made productive if treated properly. It is the unprotected and unpreserved non forestlands, which are subjected to constant degradation. The tremendously increasing biotic pressure on the land resources, in the last six decades, have promoted deforestation and done irreversible damage to the soil and environment. Land degradation is not only impacting the livelihoods of the land dependent communities but also disrupting the ecosystem as a whole. Keeping this in view the government created the Department of Wasteland Development (presently renamed as Department of Land Resources) in July 1992 under the Ministry of Rural Development to restore ecological imbalance through development of degraded non forest wastelands.

Status of Wasteland in India

The status of wastelands in India between 1986-2000 and 2003 is highlighted in Table 1. During 1986-2000, 6.38 lakh square KMs of land was categorised as total wasteland. This fell by 2.71 per cent by 2003. As can be seen from the Table, there were 5.52 lakh square KMs of land which required treatment to become productive. While wastelands under the category of sands (either in the coastal region or inland), shifting cultivation, degraded notified forestland witnessed a sharp fall, the wastelands in the category of mining and industrial and steep sloping areas increased.

Table 1. Wastelands of India (1986-2000 to 2003)

S.N.		Category Wastelands (sq.km)		(Area Change in Waste land 1986-2003 (per cent)
		1986-2000	2003	
1	Guilled and/or Ravenous land	20,553.35	19,039.34	-0.05
2	Land with or without scrub	194,014.29	187,949.49	-0.19
3	Waterlogged and Marshy land	16,568.45	9,744.97	-0.22
4	Land affected by salinity/ alkalinity-coastal/inland	20,477.38	12,024.05	-0.27
5	Shifting Cultivation Area	3,5142.2	18,765.86	-0.52
6	Under utilised/degraded notified forest land	140,652.31	126,551.81	-0.45
7	Degraded pastures/grazing land	25,978.91	1,9344.3	-0.21
8	Degraded land under plantation	5,828.09	2,138.24	-0.12
9	Sands-Inland/Coastal	50,021.65	3,3984.2	-0.51
10	Mining/Industrial wastelands	1,252.13	1,977.35	-0.02
11	Barren rocky/stony waste/sheet rock area	64,584.77	57,747.11	-0.22
12	Steep sloping area	7,656.29	9,097.38	0.05
13	Snow covered and/or glacial area	55,788.49	54,328.16	-0.05
	Total Wasteland Area	638.518.31	552,692.26	-2.71

Source : Dept. of Land Resources, M/o Rural Development, GoI (www.dolr.nic.in)

Government Intervention

In 1985, the government created the National Wasteland Development Board (NWDB) under the Ministry of Forests and Environment with a view to tackle the problem of degradation of lands, restoration of ecology and to meet the growing demands of fuel wood and fodder at the national level. In 1992, the NWDB was reconstituted and placed under the Ministry of Rural Development, where emphasis was laid on treating wastelands in non-forest areas with active involvement of the community. The programmes designed and implemented by this Board aimed at improving productivity of waste and degraded lands.

Integrated Wastelands Development Project (IWDP) scheme, designed specially to develop wastelands has been under implementation in the country since 1989. Besides taking up the development of non forest wastelands, this Scheme provides for the development of an entire micro watershed in a holistic and integrated manner. The basic objective of this scheme is an integrated wastelands development based on village/micro watershed plans which are prepared after taking into consideration the land capability, site condition and local needs of the people. The scheme also aims at rural employment besides enhancing the contents of people's participation in the wastelands development programmes at all stages, which is ensured by providing modalities for equitable and sustainable sharing of benefits and usufructs arising from such projects.

Under the Five Year Plans a lot of wasteland development related activities have been designed and implemented by various Ministries/ Departments of the government. Table 2 highlights the status of the development of degraded lands during the Plan periods till the Tenth Five Year Plan (2002-2007).

Table 2. Degraded Lands Development under Various Watershed Development Programmes up to Tenth Five Year Plan (2002-07)

S.N.	Ministry/ Scheme and Year of Start	Progress since inception		Progress in Tenth Plan (2002-07)		Total since inception up to Tenth Plan	
		Area	Expend-iture	Area	Expend-iture	Area	Expend-iture
(A) Ministry of Agriculture (D/o Agriculture & Co-operation)							
1	National Watershed Development Project for Rainfed Area (1990-91)	69.79	1,877.74	23.30	1,147.82	93.09	3,025.56 (3,250)
2	River Valley Project & Flood Prone River (1962 & 1981)	54.88	1,516.26	9.98	727.98	64.86	2,244.24 (3.460)
3	Watershed Develo-pment Project for Shifting Cultivation Area (1974-75)	2.58	166.27	1.35	129.31	3.93	295.58 (7,521)
4	Reclamation of Alkali Soil (1985-86)	5.81	76.39	1.30	45.35	7.11	121.74 (1,712)
5	Watershed Develo-pment Fund (1999-00)	0.00	0.00	0.59	26.02	0.59	26.02 (4,410)

6	Other Externally Aided Projects	13.35	2,039.81	4.80	1,927.54	18.1	53,967.35 (21,858)
	Sub Total	146.41	5,676.47	41.32	4,004.02	187.73	9,680.49 (5,157)
(B) Ministry of Rural Development (D/o Land Resources)							
1	Drought Prone Area Programme (1973-74)	68.95	3,284.74	68.32	1,557.76	137.27	4,842.50 (3,528)
2	Desert Development Programme (1977-78)	33.56	797.38	45.17	1,152.50	78.73	1,949.88 (2,477)
3	Integrated Waste-land Development Programme (1988-89)	37.34	616.51	62.22	1,821.64	99.56	2,438.15 (2,449)
4	Other Externally Aided Projects	1.40	18.39	3.60	274.28	5.00	292.67 (5,853)
	Sub Total	141.25	4,717.02	179.31	4,806.18	320.56	9,523.20 (2.971)
(C) Ministry of Environment & Forests							
1	National Affores-tation and Eco Deve-lopment project (1989-90)	0.70	47.53	0.00	0.00	0.70	47.53 (6.790)
	Grand Total (A + B + C)	288.36	10,441.02	220.63	8,810.20	508.99	19,251.22 (3.782)

Source: 11th Five Year Plan (2007-2012), Planning Commission, Government of India

Conclusion

Growing population, unsustainable land use practices, extensive deforestation, increased demand on land based agro activities are leading to the fast degradation of world's scarce land resources and affecting productivity in agriculture. While over exploitation of natural resources like land has a direct bearing on agricultural productivity and food security, the treatment of wastelands and protection of farm lands from constant degradation through integrated land management procedures is the need of the hour. In spite of large public investments, the innovative and integrated land management practices have not been able to show satisfactory results in bringing degraded lands of the country into the cultivable land fold. Integrated watershed management, due to its inbuilt emphasis on social mobilization and community involvement, have been successful in somewhat preventing wastelands from further degradation and have also been successful in developing barren lands for productive agricultural use in various arid and semi arid areas of the country. Thus, there is an immediate need to identify and list land treatment processes and procedure already in vogue under various land development programmes and to build awareness on these techniques amongst the community in villages so as to ensure sustainable wasteland management.

Chapter - 26

Dryland Farming : Issues and Strategies

Y.S. Ramakrishna & B. Venkateswarlu

Covering 90 m.ha of net cultivated area, rainfed areas contribute 41% to the national food basket and support 40% of human and 60% livestock population. Although dryland farming has been used historically to represent drought prone (<750 mm) unirrigated areas in the country for planning purposes, more recently rainfed farming is used as an appropriate term to encompass all unirrigated areas of the country which are some times interspersed within the irrigated plains.

Rainfed farming is used to represent this area rather than dryland farming. Within the rainfed areas, 15 million ha. area falls in the arid zone receiving less than 500 mm rainfall; another 15 million ha. in dry semi arid zone with rainfall between 500-750 mm. 42 million ha. in wet semi arid zone receiving 750-1150 mm rainfall and the remaining 25 million ha. in high rainfall (>1150 mm) areas. Though crop production is the main stay of livelihood in rainfed areas, livestock play an important role, particularly in arid and dry semi arid regions. Most often, livestock protect the farmers from distress during drought years when crops fail.

Maintaining food security in the country for growing population is a major challenge. Since the per unit area productivity in irrigated areas is reaching a plateau, it is argued that the bulk of the future increases in food production have to come from rainfed areas. In fact, the available technologies can enhance the productivity from 0.8 t/ha to 1.5 to 2 t/ha. This doubling of productivity can contribute another 40 m t food grains from rainfed regions. However, to achieve this, a multi pronged approach involving technology transfer, resource and risk management and market support is needed.

The production and productivity of major rainfed crops since independence show variable trends. In oilseeds, the per ha. productivity has doubled in nearly 50 years : part of it was aided by increased area under irrigation (15%), whereas in pulses, the productivity moved from 441 kg/ha in 1950-51 to 623 in 2003-04,, an increase of just 182 kg (42%) in 52 years. This is a major area of concern. In coarse cereals (maize, sorghum and millets), the productivity rose steadily from 408 kg/ha to

1228 kg/ha, a jump of 300% in the same period. This can be considered as an appreciable achievement despite a steady fall in area. The availability of hybrid seed in maize, sorghum and pearlmillet and its spread made this possible to a large extent. Overall, the productivity of food grains in rainfed areas is around 1.0 t/ha. now which is substantially lower than the over 3 t/ha from irrigated areas. These figures show that the rainfed areas have clearly lagged behind in their contribution to the national food basket. As already stated, there is an urgent need to improve this productivity upto 2 t/ha to meet the growing food needs, particularly due to the yield plateaus in irrigated areas and a steady diversion of agricultural land to non agricultural uses all over the country including rainfed zones.

However, during the same period, the contribution of livestock to the farmer's income improved significantly. In most dryland and hill regions, livestock farming is a major support to family income. Besides livestock, contribution of small ruminants (sheep and goats) to the rural economy is estimated to be around Rs. 240 million per annum. About 5 million families in India are engaged in rearing small ruminants which provide gainful employment of 184 to 437 man days per annum depending on the size of the flock. In other words, the contribution of livestock to the farmer's income in rainfed areas is quite significant.

Key Constraints

The major constraints limiting the productivity in rainfed farming are :

- High risk in cropping due to weather aberrations :
- Land degradation due to water wind erosion leading to depletion fertility.
- Slow rate of adoption of new technologies due to inherent risk.
- Low public investments on farm related infrastructure.
- Poor socio-economic base of the farmers.

Due to these constraints, the yield gap between the research station and farmers fields remains high. Most of the improved rainfed farming technologies are presently centered around resource management (efficient use of soils and rainwater) and best results from such technologies can come through community adoption. Therefore, technologies for improvement of rainfed areas are implemented through watershed concept. The early successful models of watershed development could not be upscaled on large scale due to several weakness in the adoption strategy. Though a number of expert reviews, these weaknesses are being overcome and from Xth plan onwards, a new strategy based on increased people's participation has emerged which is likely to get a boost in XI plan through the proposed National Rainfed Areas Authority (NRAA).

However, climatic risks, poor credit availability and price fluctuations avenue to constrain the progress in rainfed farming.

Goals and Strategies

In order to minimize the disparities in incomes of farmers in irrigated and rainfed areas, there is an urgent need to improve the household income of farmers in dryland either through improved productivity or higher profitability through farm level value addition, or providing off farm employment opportunities; to minimize risk through crop and livestock insurance. A multi pronged strategy of adopting new technologies, creation of enabling mechanisms for the farmers to absorb improved technologies and sustain them, investments on infrastructure and appropriate policy initiatives is called for in a mission mode approach. Important issues that need focused attention are summarized.

Watershed Development

Watershed development continues to be the key strategy in rainfed farming. Rainwater conservation, improved crop production technologies and income generating options for landless are integrated into the watershed programmes. Despite the huge effort by various ministries, the self replication of this approach still remains a challenge. The returns from water conservation technologies depend on the amount and distribution of rainfall and the end use of harvested water. For eg. in relatively high rainfall areas of eastern India, water harvesting and recycling was successfully demonstrated on an operational scale under the National Agricultural Technology Project (NATP). Within 3 years the entire cost on capital expenditure could be recovered and rainfed rice could be saved from severe drought during kharif 2002 (NATP, 2005). While watershed approach is an accepted strategy, it has to be region specific. For example, areas like deccan plateau, fragile coastal rainfed areas, saline patches and large parts of Indo Gangetic plains require a significant modification of the conventional ridge to valley approach (Samra, 2005). A meta analysis 311 pilot watersheds across India (Joshi *et al.,* 2005) revealed that crop based farming systems are more relevant in 700 to 1100 mm rainfall zone, livestock based activities in <700 mm zone and fish based production systems in >1100 mm regions (Reference). Overall, watershed projects gave a BC ratio of 2.41, IRR of 22% and employment generation of 181 persons/ha/year. Higher benefits were realized by low income group farmers.

In low rainfall semi arid and arid areas, *in situ* moisture conservation is of extreme importance. Simple practices like ridges and furrows, conservation furrows and compartmental bunding have improved yields between 25-40%. Significant gains in productivity and income from these areas can not come from any revolutionary adoption of new technology which gives quantum jump in yield, but from wider adoption of low cost but proven technologies which may individually cause small improvements but translate in to larger gains at the national level. However, to encourage individual farmers to adopt these practices, such conservation practices may be included in the national rural employment guarantee scheme (NREGS) so that the incentive of wage alone could catalyze its wider adoption.

Improved Seeds

The adoption of HYVs of major rainfed crops continues to remain low. Availability of quality seeds is a major constraints. The rainfed farmer has to depend more on the public sector seed production agencies which are presently in weak condition as the private sector is more based towards seed production of cash crop which gives them more profit. Thus, a major thrust on promoting local seed production through farmer's participation can alone make a significant impact on the productivity of rainfed crops. This is particularly relevant in pulses and oilseeds. Concerted efforts are needed particularly to increase the pulse production which remained stagnant over the last two decades at 12-14 million tons. Though there has been no major varietal break through in pulses, other production technologies like IPM, INM and water conservation have potential to increase the yields by 30-40%. What is needed is a major development push to the adoption of these technologies by strengthening the activities of the mini missions under pulses.

Soil Health

Declining soil health is stated to be one of the major causes of yield stagnation across the country. There is a wide spread deficiency of micro nutrients which is limiting the response of crops to other management practices. A mission mode strategy on improvement of soil health by undertaking diagnostic surveys at farmers field level and formulation of a correction strategy that could involve balanced application of nutrients, adoption of INM practices, emphasis on organic manures, biomass generation and redesigning of cropping systems with emphasis on legumes in the rotation. Innovative approaches to encourage biomass production for soil health improvement can be thought of. Small farmers who produce biomass for improving their soil fertility need to be encouraged just like farmers using chemical fertilizers or helped by way of subsidies. This can be done for generation of biomass either on farm (during or after the season) or off farm through an approved list of the annual or perennial plant species either for a running length (in case of bunds) or area. The incentive may be given in the form of coupons by the Department of Agriculture, which can be exchanged by the farmers for purchase of agricultural inputs from designated private or cooperative stores. The expenditure incurred on such redemption of coupons by these establishments may be reimbursed by the State Governments. However, a mission oriented, whole village adoption is required to have an impact at the district or state level. Integrated village level bio centers need to be set up to produce the bio agents and bio fertilizers through rural youths under the agri-clinic scheme, which can also provide new livelihood options.

Bio energy Crops

At the current level of consumption of petroleum products, energy shortages is going to impact agriculture as seriously as water. With proper energy farming strategy, agriculture can become a net contributor of energy rather than a hopeless consumer. The potential of tree born oilseeds like Jatropha, Pongamia and Mahua to produce bio diesel is quiet high considering the extensive areas of culturable wastelands in the country. Energy farming alone can contribute millions of jobs in rural areas, through primary and intermediate level processing units. However, there are critical gaps in our knowledge on the actual potential of these crops under real farm conditions and the problems that might arose while scaling up. The Government of India has already initiated a mission on bio fuel crops. CRIDA is participating actively in the network projects on bio fuels both at the national and regional level and clear answers to some of the questions on viability of biofuel crops may be available soon.

Farm Mechanisation

Farm mechanization is another area which can make a significant impact on production and cost reduction in rainfed agriculture. With declining animal energy, there is an enormous scope for introduction of planting and post harvest machinery. Timely planting and weeding which are most crucial in drylands to capitalize on the limited moisture availability can only be ensured with good planting equipment. Here again, innovative institutional mechanisms like custom hiring enabling small farmers to have access to large farm implements for crop management, harvest and pest control have to be widely adopted to have a significant impact at the regional and national level.

Livestock and Pasture

In addition to large tracts of arid and semi arid areas which are chronically drought prone where cropping is a gamble despite all efforts, the country has vast stretches of wasted lands which can be put under silvipasture and energy farming. During drought years, it is only the livestock farming which provides succor to the farmers in dry areas and any major effort towards enhanced fodder production will certainly contribute to better livestock production and additional income at the farm level. A deliberate shift in priorities through appropriate incentives and subsidies is required at the national level in favour of silvi pasture and horti-pasture for such lands and agroclimatic sub regions, which can also help improve soil health, instead of continuing under unviable arable cropping.

Organic Farming

Rainfed agriculture also offers opportunities in the areas of organic farming. Coarse (nutritious) cereals, beans, medicinal oils and herbs can be grown in identified niche areas for export purpose. The strategy here should be to realize more value from relatively lower yields and capitalize on the inherent advantage of low input use by small farmers in drylands (many areas are by default organic) rather than go for more expensive and complicated methods of organic production.

Change in food habits leading to declining consumption of coarse cereals is affecting farmers profitability. Coarse cereals are critical not only the nutritional security of the poor but also to support the livestock in dry areas. Besides promoting them as major fodder crops, value addition into health foods and bio fuels can contribute to improved profitability to farmers. One of the recent project under NATP showed that alcohol can be produced from stalks and grains of sweet sorghum on a comparative scale to molasses but with less use of water and fertilizers; more importantly providing higher returns to rainfed farmers who grow sorghum. Use of coarse cereals as livestock/poultry feed can be promoted by linking farming groups with food industry. Similarly, significant value can be added at the farm level to large number of rainfed horticulture products which are grown by poor farmers through simple methods of drying and primary level processing (for supply to industry) which can be done by women.

ICT and Farm Services

Better climate management through integration of information from short, medium and long term weather forecasts into location and farming systems specific action plans for better crop management, providing drought resilience and sustained and enhanced productivity.

Issues

Despite the new technologies and enabling mechanisms, making small holders farming viable is a continued challenge. The livelihoods of the families depending on small farms are at threat due to continued forces of globalisation where economy of scale becomes prime driver of profitability. In this context, research on diversified farming system with market linkages are critical. There is an urgent need to identify a variety of off farm livelihood opportunities in rural areas that can strengthen the land resources, agriculture and livestock sector or which support the rapidly growing service sector in semi urban areas.

New Policy Initiatives

In order to effectively translate the above strategies into reality, a number of new policy initiatives are required at panchayat, district, state and national level. The watershed strategy has to be modified with more focus on livelihood enhancement of the entire village community. New institutional innovations and enabling mechanisms have to be put in place so that the community can sustain the gains made through new technology and derive long term benefits. Watershed has to be the nucleus on which all public and private funded farm development activities have to converge. This new approach may be tried in the 200 backward districts identified for the National Rural Employment Guarantee Scheme. The proposed National Rainfed Areas Authority (NRAA) hopefully will address this convergence issue urgently.

Considering that the rainfed areas have to play an important role for achieving the next quantum jump in food production, there is a need to set up a technology mission on rainfed agriculture which can bring together the research and development efforts on pulses, oilseeds which are currently under different missions and also include the related issues of fodder production and livestock farming. Only a mission mode approach can help in realizing the target of 2 t/ha of productivity of major rainfed crops set out in this road map from the current level of 1 t/ha, which is attainable looking to the current yield gap between attainable yields and the current yield levels of farmers with proper management strategy.

Chapter - 27

Development of Dryland Agriculture in India

B. Hemalatha and Y.V.R. Reddy

About 65 per cent of arable land of 143 m ha of total cultivated/arable land even today depends upon monsoon rains. The productivity of dryland crops is not only low but also highly fluctuating depending upon the rainfall and its distribution. Government of India has taken several measures to improve the position of dryland farmers in India, but all measure have not helped farmers to adjust to drought spells, late arrival and early seizure of monsoons. Farmers in dryland have been facing peculiar situation in India during this year. Government has initiated watershed programme for development of agriculture in drylands in 1983-84 and has been allotting funds annually to change scenario of lands but this year drought proved that this is a futile exercise. Hence there is a need to develop technologies and carry out works addressing to drought spells, low productivity, developing measures for the benefit of farmers during drought/famine periods, etc. Thus this paper deals some of the issues pertaining to dryland farming and dryland farmer so as to ameliorate the conditions of rural people in arid and semi-arid regions in India.

Research in Agro-meteorology

Indian Council of Agricultural Research and State Agricultural Universities have established agro-meteorology research centres/ departments to focus on the research pertaining to agro-meteorology in relation to crop improvement. The research has become routine and repetitive rather giving warning. to farmers much in advance regarding seasonal variation in rainfall and its distribution, climatic conditions and its variability, pest/insect problems etc. so that farmer can decide about his choice on crops, application of nutrients, strategy to overcome other problems. This is only possible for the farmer to adjust to any climatic conditions if he has sufficient knowledge on agro climatic conditions. Department of agro meteorology should focus research on such aspects to benefit the farmers, through there are many problems to predict weather conditions much in advance. Hence dynamic growth processing techniques may be evolved to forecast weather/climatic conditions and to suggest farmers risk aversion processes in dryland agriculture.

Watershed Programme

Though Government of India have been spending annually about Rs. 10,000 crore on this programme in India in rainfed areas, the desired results have not been achieved due to non/low participation of farmers in this programme. The programme is common in India irrespective of rainfall and its distribution, soil conditions, cropping programme, socio economic conditions of farmers etc. This programme has been found to be beneficial during normal rainfall years but has no solution during the drought years and particularly it has not given any solution during this prolonged drought spell year during sowing period itself. Hence there is a need to change the programme in relation to structures etc. It generally requires more emphasis on *in situ* moisture conservation techniques in low and normal rainfall regions. In slightly more rainfall areas, farm ponds, percolation tanks/wells, soil conservation methods etc. are to be suggested. In high rainfall areas irrigation tanks are preferred in addition to soil conservation techniques.

In addition, suitable varieties of crops based on different climatic conditions should be evolved and made available to farmers. Whenever confirgency plan is prepared, farmer should be supported with resources for following suitable and profitable cropping programmes. Hence research should be focused to address to meet the aberrant weather situation in dryland agriculture. Soil and water conservation structures should be planned based on soil, rainfall, cropping programmes and socio-economic conditions in the area.

Depletion of Ground Water

Though it has been claimed that the ground water table has been raising due to watershed programme in India, in reality, the ground water has been declining very fast due to over exploitation through increased bore wells and water use efficiency is also low due to adoption of age old practices in irrigation system rather than adopting drip/ sprinkler irrigation. Moreover, the electricity charge/rate is subsidised on flat rate basin irrespective of quantum of water pumped out. Hence Government may come out with suitable policies/measures to arrest over exploitation of ground water.

Agro-forestry Systems

In arid/semi arid regions, suitable agro-forestry system including agro-horticulture, agri-silvi-pastoral system etc. may be developed for the benefit of farmers so as to balance the eco system and prevent degradation leading to a threat to population to live harsh/unfavourable climatic zones. This will lead to socio-conflicts due to socio-economic imbalance among the regions. Hence income should be generated through different plantations, horticulture, silvipastoral systems including livestock management etc. so as to improve economic condition of people. Thus plantation/orchards have been affected due to drought spells and over exploitation of ground water which touched beyond 400 ft. depth. Even deep-rooted plants are drying/dying due to non-availability of water to plants due to over exploitation of ground water.

Crop Improvement programme

Evolving suitable improved genotypes of different crops/plantations suiting to different agro-climatic conditions is required to improve the productivity at par with advanced countries so as to meet the competition in international market, as the position of India in international market for agricultural products is precarious due to high cost of production.

If the productivity of crops were increased, unit cost of produce would decline accordingly. Hence crop improvement programme should be given more prominence. The breeding work should receive more attention in crops, plantation, orchard/fruit plants, livestock etc. to improve the productivity through proper breeding methods/policies. Though biotechnology is not a substitute for breeding to evolve improved genotypes, but biotechnology helps in reducing the cost through gene modification techniques and some other changes. This is a continuous process. As crops, animals, soils etc. in India are low productive factors due to various reasons, there is a need to improve these stock through different breeding procedures/policies. Top priority is to be given to bring another green revolution in India.

Fertilizer

The use of chemical fertilizers has been increasing due to non-availability of sufficient Farm Yard Manure, costly affair in using green manure, easy to apply etc. Though soils are fed with nutrients and micro-nutrients through chemical fertilizer, farmers report that soil texture, structure, water holding capacity, etc. have been changing from poor to poorer conditions and even these chemical fertilizer is harmful to beneficial bacteria in the soil. The taste in vegetable differs based on use of Farm Yard Manure and chemical fertilizers. Moreover, chemical fertilizers are costly requiring only cash. Thus Indian farmers are financially handicapped. Hence alternate methods should be suggested.

Bio-fertilizers can be applied at a cheaper rate. These fix nitrogen bacteria in the soil. There are different bio-fertilizers even for crops in drylands. Hence there is a need to suggest and popularize suitable bio-fertilizers to different crops, the application methods and cost of bio-fertilizers is cheaper but improves the nutrient position in the soil. Thus natural process to improve soil productivity is also possible through bio-fertilizers. Hence there is a need to develop bio-fertilizers and popularize even in drylands through extension education/training.

Vermin-compost is very good source of manure when compared to Farm Yard Manure, compost etc. Hence farmers raise economic factors.

Thus enrichment of soil fertility through different techniques is more important as our drylands are not only `thirsty' but also `hungry'.

Plant Protection Measures

Farmers have been indiscriminately using pesticides/insecticides. This causes more harm/damage than help. There are some pesticides/ insecticides banned by international organizations/advanced countries, but such dangerous plant protection chemicals are still used in India. Though pesticides/insecticides are required to control pests/disease problem is improved/hybrid varieties of different crops for protecting and harvesting a good crop, optimum and appropriate chemical vegetables, fruits are rejected in international market due to residual content of such chemicals on these items. Thus valuable market is lost for our crops. There is a need to develop resistant varieties like B.T. cotton etc. to reduce the cost on plant protection chemicals substantially. Research in biotechnology is required to solve pest problem for reducing cost.

For certain diseases, even juice extracted from Neem. Pongama, Jatropa, bitterguard, castor and some locally available plants leaves can be mixed in water and sprinkled/sprayed to crops for preventing diseases. Similarly soil from diseases can be prevented through application of Neem/Pongamia/castore cake which even improves the soil fertility. There is strong belief in pest incidences. Farmers are to be trained in this direction.

Alternate Land Use Farming Systems

As soil in drylands are varied and even marginal land are brought under plough, there is a need to grow crops based on land capability classification, nutrient position, water holding capacity, natural factors such as wild animals, socio-economic conditions of a farmer and group of farmers etc. Thus suitable arable crops, orchards plantations, pasture etc. are to be decided/suggested considering all factors for feasible proposition so as to harvest highest returns from same piece of land, as the scope to expand the area under arable crops diversified farming systems can be followed based on resources available.

Soil and Water Conservation

As degradation of soil is a crux of the problem in India and it is severe in dryland farming areas, there is a need to suggest suitable soil and water conservation techniques based on rainfall and its distribution, intensity of rainfall, runoff, soil type and condition, participatory attitude of farmers socio-economic conditions of farmers and so on. It should be made as people's programme through extension education methods. These methods should be cost-effective. Vegetative barriers may be cheaper and sustainable than mechanical measures which are not only costly investment but also costly in maintenance. Investment on water grass channel, plantations, growing pastures/grass on waste/barren lands etc would reduce the runoff. Thus greenery may be encouraged to minimize soil erosion in addition to minimum and low cost soil and water conservation structures in arid and semi-arid regions.

Policies

Though share of agricultural income in national income may be low, agricultural development is the indication of national prosperity and peace maintenance in the society. For the past 10-15 years, the agriculture are stagnant due to many factors including drought spells, lack of break through in seed improvement, high cost of inputs, stagnant prices for agricultural products, increased wage rates, exploitation by traders, market policies and so on. The cost of production per unit produce (say quintal or ton) is higher in drylands compared to irrigated areas but it is higher in India compared to advanced countries. Hence Indian agricultural products are not competing in international market under liberalization, privatization and globalization of World Trade Policy. Thus Indian agriculture suffers from all angles from production, price, marketing, processing etc. Hence Government may evolve such policies for benefiting farmers who can be made to compete in the world market.

As in drylands same crop is grown every year, it reduces the yield and this has been observed in Andhra Pradesh in case of groundnut. Hence crop rotation is a must in drylands for effective utilization of resources.

Green-leaf manuring, which has been in vogue, has been loosing its importance due to costly in application because higher wages of labour etc. Suitable and fast growing bushy leafy plants may be developed and planted on bunds in drylands so as to cut and incorporate in the soil for enrichment of soil fertility in drylands.

The above methods are to be popularized due to dwindling Farm Yard Manure position in India gradually because of decline in maintenance of livestock by the cultivators based on different socio-maintaining present ecosystem. To minimize the risk of crop failure, different crops in different combinations can be grown. Crops such as intercropping/ sequence cropping, agro-forestry, horticultural plantations/ crops may be decided and grown accordingly. Based on the resources availability and utilization, different farming systems such as dairy, goatry, poultry, sheep rearing, rabittry, duckery, piggery, vegetable farming, floriculture, sericulture, horticulture plantations etc. are to be practiced for effective utilization of resources so as to maximize profits under dynamic situation, as aged practices which is static and stagnant profit maximization technique gave a path to dynamic profit maximization to withstand in ever growing technology process and needs of people. Thus revolutionary changes are required in Indian agriculture and more so in dryland agriculture where.

Even rural/cottage industries can also be developed in rural areas through marketing of such products viz., making of soaps, pickels, detergents, ropes, squashes, juices, jam, jelly and so on. Thus employment can also be provided through alternative enterprises in rural area. This will stop the migration of people from rural areas to town or cities or greenery areas.

It can be conducted that agriculture can be developed through evolving and supplying improved on hybrid seeds educating farmers in bio-fertilizers. Vermi-compost, indigenous plant protection measures, supplying chemical fertilizers and pesticides at cheaper rate, providing service community centres for machanization of agriculture, development of market infrastructure and trade policies, making the farmers self-reliant through evolved and acceptable policies or techniques and so on.

Thus age - old and stagnant growth of profit maximization should be given to path to dynamic growth profit maximization process so as to make agriculture sustainable and competitive in the world.

Chapter - 28

Land Management can Improve Rural Economy

Nivedita Thapliyal

Land degradation indicates temporary or permanent long term decline in ecosystem function and productive capacity. It may refer to the destruction or deterioration in health of terrestrial ecosystems, thus affecting the associated biodiversity, natural ecological processes and ecosystem resilience. It also considers the reduction or loss of biological/ economic productivity and complexity of croplands, pasture, woodland, forest, etc.

Land degradation is increasing in severity and extent in many parts of the world, with more than 20% of all cultivated areas, 30% of forests and 10% of grasslands undergoing degradation (Bai *et al.,* 2008). Millions of hectares of land per year are being degraded in all climatic regions. It is estimated that 2.6 billion people are affected by land degradation and desertification in more than a hundred countries, influencing over 33% of the earth's land surface (Adams and Eswaran, 2000). This is a global development and environmental issue highlighted at the United Nations Convention to Combat Desertification, the Convention on Biodiversity, the Kyoto protocol on global climate change and the millennium development goal (UNCED, 1992; UNFP, 2008).

The decline in land quality caused by human activities has been a major global issue since the 20th century and will remain high on the international agenda in the 21st century (Eswaran *et al.,* 2001). The immediate causes of land degradation are inappropriate land use that leads to degradation of soil, water and vegetative cover and loss of both soil and vegetative biological diversity, affecting ecosystem structure and functions (Snel and Bot, 2003). Degraded lands are more susceptible to the adverse effects of climatic chance such as increased temperature and more severe droughts.

Land degradation encompasses the whole environment but includes individual factors concerning soils, water resources (surface, ground), forests (woodlands), grasslands (rangelands), croplands (rain fed, irrigated) and biodiversity (animals, vegetative cover, soil) (FAO, 2005).

On the other hand the NRC (1994) stressed that land degradation is complex and involves the interaction of changes in the physical, chemical and biological properties of the soil and vegetation. The complexity of land degradation means its definition differs from area to area, depending on the subject to be emphasized.

Dryland

The phenomenon is most pronounced in the dryland, which cover more than 40% of the earth's surface (Dobie, 2001). Around 73% of rangelands in dryland areas are currently degraded, together with 47% of marginal rain fed croplands and a significant percentage of irrigated croplands (UNCCD Agenda 21, 1992; UNCCD, 1994). Overgrazing has damaged about 20% of the world's pastures and rangelands (FAO, 1996).

The degradation of land may be the result of numerous factors or a combination thereof, including anthropogenic (human related) activities such as unsustainable land management practices and climatic variations. Note that degradation processes e.g. erosion do occur naturally, and are generally balanced by the rate of soil formation. However, accelerated degradation is typically associated with human modification of the environment. The underlying causative factors of land degradation and environment are poverty and undervaluing of natural resources. In both cases people focus on immediate economic gain irrespective of damage to the same resources they are dependent on. The latter in particular promotes inefficient use and wastage of resources.

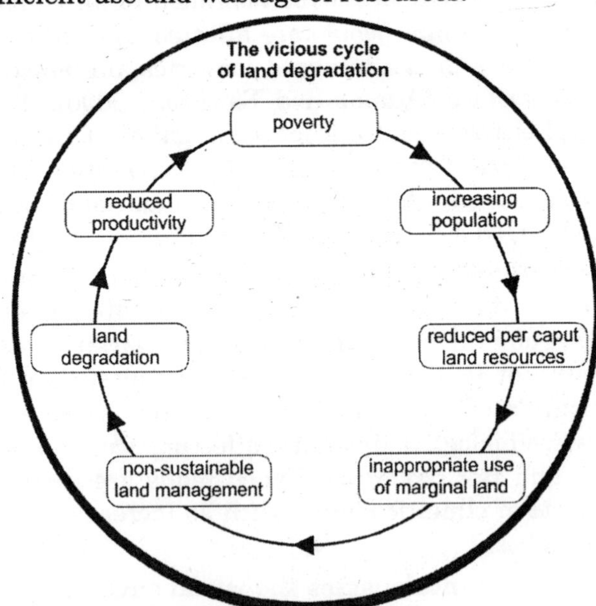

The cycle of processes leading to and perpetuating land degradation
(Source : FAO)

Land degradation, resulting from unsustainable land management practices, is a threat to the environment as well as to livelihoods, where the majority of people directly depend on agricultural production. There is a potentially devasting downward spiral of overexploitation and degradation, enhanced by the negative impacts of climate change leading in turn to reduce availability of natural resources and declining productivity : this jeopardizes food security and increases poverty. Sustainable land and Ecosystem management (SLEM) project is rooted in the rational that food security through enhanced agricultural productivity can be achieved through sustainable management of the country's natural and agro ecosystem.

Poverty in India

Poverty is one of the main problems which have attracted attention of sociologists and economists. It indicates a condition in which a person fails to maintain a living standard adequate for his physical and mental efficiency.

According to 2010 data from the United Nations Development Programme, an estimated 37.2% of Indian live below the country's national poverty line. A recent report by the Oxford Poverty and Human Development Initiative (OPHI) states that 8 Indian states have more poor than 26 poorest African nations combined which totals to more than 410 million poor in the poorest African countries.

According to a new UN Millennium Development Goals Report, as many as 320 million people in India and China are expected to come out of extreme poverty in the next four years, while India's poverty rate is projected to drop to 22% in 2015. The report also indicates that in Southern Asia, however, only India, where the poverty rate is projected to fall from 51% in 1990 to about 22% in 2015, is on track to cut poverty half by the 2015 target date. The latest UNICEF data shows that one in three malnourished children worldwide are found in India. 42 percent of children under five were underweight. It also showed that a total of 58 percent of children under five surveyed were stunted. The 2011 Global Hunger Index (GHI) Report ranked India 45th, amongst leading countries with hunger situation. It also places India amongst the three countries where the GHI between 1996 and 2011 went up from 22.9 to 23.7, while 78 out of the 81 developing countries studied, including Pakistan, Nepal, Bangladesh, Vietnam, Kenya, Nigeria, Myanmar, Uganda, Zimbabwe and Malawi, succeeded in improving hunger condition.

Initiatives by SLEM Projects in India

With the support of the Sustainable Land and Ecosystem Management (SLEM), Technical facilitation Organization and its partners have taken up socio economic as one of the key issues which include Poverty,

Malnutrition, Agricultural GDP and livestock. The main focus of the SLEM projects has always been to the support the target groups in taking up activities for land management and subsequently improving the socio-economic condition. The partners have developed alternative land management strategies, often based on land use practices that rest on local knowledge and local traditions that have stood the test of time. Many of these approaches have achieved noteworthy successes. However, these successes are often not published and need to be brought to the attention of colleagues in other countries as well as policy makers. SLEM-CPP is devoted to the analysis and publication of these cases and supports the exchange of learning experiences amongst the national as well as global context. The project Partners of SLEM-CPP have documented various case studies which has improved the living condition with the introduction of improved sustainable livelihood.

SLEM has apply ``options analysis" for land management where different possible solutions are explored for their effectiveness in addressing the causes and impacts of land degradation as well as improving the standard of living of the poor. Key questions are : why do land users employ inappropriate practices, or what inhibits them from applying more appropriate technologies? Frequently, resource users are aware of degradation but are not in a position to rectify it, often due to political and economic circumstances e.g. insecure land tenure, misuse of subsidies and incentives, market price distortions, etc. These complementary paths help to form solutions from political, technical and economic perspectives. The complex interrelated causes of or contributors to land degradation must be identified to effectively design remedial interventions. Activities to be considered must also include those which support training and education; improve knowledge, local planning procedures and land management skills; create awareness; enhance institutional development; and address pertinent policy issues. Such measures would ensure that the work done to combat land degradation is not reversed because people and governments continue in their old practices, but that they would acquire new knowledge and skills, and make policy improvements.

Several tools are available to assess the costs and impacts of land degradation and the changes and benefits of implementing SLM. These would aid more informed decision making and strategic planning regarding the approach to SLM that should be taken. These include assessing ecosystem services and economic valuation. A major component common to all the projects working on SLM is the emphasis on capacity building and inter agency integration of functions and activities to address several of the barriers listed above.

- A framework for action
- Improving land use

- Involving the people participation
- Developing local and national programme integrated with land management and socio-economic parameters
- Strengthening State/Regional/District level Institution dealing with land management issues.
- Coordinating international action

Some of the case studies included below which has focused on the land management and social and economic improvement by changing the pattern of practice and knowledge.

Chapter - 29

Strategy to Develop Degraded Land

Dr. Gopal Kalkoti

India has world's 2% of geographical area and 1.5% of forest and pasture lands to support 18% of world's population and 15% of livestock population. The increasing human and animal population has been instrumental in the reduction in the availability of land over the decades. While the per capita availability of land has declined from 0.89 hectare in 1951 to 0.37 hectare in 1991 and is projected to decline to 0.20 hectare in 2035, per capita agricultural land has declined from 0.48 hectare to 0.16 hectare and likely to decline to 0.08 hectare in respective years.

Extent of land degradation

Agencies that have so far estimated land degradation include National Commission on Agriculture (1976), Society for Promotion on Agriculture (1976), Society for Promotion of Wasteland Developments (1984), National Remote Sensing Agencies (1985), Ministry of Agriculture (1985), National Bureau of Soil Survey and Land Use Planning (1985 & 2005). The estimates on the extent of land degradation in India vary widely from 63.9 million hectares to 187.0 million hectares due to different approaches, methodologies, defining degraded soils, adopting various criteria for delineation, among others. However, one cannot underestimate the challenging nature and extent of land degradation in India. The National Bureau of Soil Survey & Land Use Planning (NBSS & LUP) of the ICAR, Nagpur in 2005 has reported that out of 328.60 million hectares of geographical area in India Net Cultivated Area is about 141 million hectares (42.9%) of which irrigated area is about 57 million hectares (40.4%) and about 84 million hectares (59.6%) are rainfed. Area of around 146.82 million hectares (44.7%) out of 328.60 million hectares is suffering from various kinds of land degradation. In absence of comprehensive and periodic scientific surveys, the figures reported by NBSS & LUP based on studies and several estimates (2005) for various land degradation have been considered as logically concluded and are being used for various purposes.

Land degradation is caused by several factors viz. water and wind erosion, water logging, salinity/alkalinity, soil acidity, among others. India has been experiencing a very high degree of land degradation as 44.7% of its geographical area is classified as degraded. Of this 93.68 million hectares (63.8%) are affected by water erosion, 16.03 million hectares (10.9%) by soil acidity, 14.30 million hectares (9.7%) by water logging, 9.48 million hectares (6.5%) by wind erosion, 5.94 million hectares (4.1%) by salinity/alkalinity and 7.38 million hectares (5.0%) by complex problems.

Across regions, all six regions had very high percentage of geographical area as degraded ranging from as high as 56.3% for Central region to 35.4% for Northern region and even 29.5% for Delhi and Union Territories. Among States, 11 states had extremely high percentage of geographical area degraded above mean value of 44.7% ranging from 52.0% to 89.2% and other 15 states too had significantly high percentage of geographical area degraded varying from 25.4% to 43.9%. In particular Mizoram (89.2%) Himachal (75.0%) Nagaland (60.0%) Madhya Pradesh and Chhattisgarh combined (59.1%) were states with very severe intensity of degradation.

Policy and Programs

Acknowledging the acute problem of land degradation, the Government, in its efforts to sustain ecological environment, agricultural productivity and production, has initiated from time to time several policies and programs to prevent land degradation on one hand and take remedial measures to improve the quality of degraded land on the other.

Institutional Support

In order to facilitate the understanding of the problems, nature and magnitude of land degradation and initiate measures to remedy the situation by formulating national policy and programs the Government has created institutional infrastructure, viz. [i] Soil and Land Use Survey of India [SLUS] was established in 1958 at IARI, New Delhi with seven centers located at Noida, Kolkata, Bengaluru, Nagpur, Hyderabad, Ahmedabad and Ranchi, SLUSI has a mandate to provide detailed scientific database on soil and land characteristics to various States for planning and implementation of programs relating to soil and water conservation and natural resource management. During 2011-12, SLUSI had targets of 101 lakh hectares of Rapid Reconnaissance Surveys, 3.62 lakh hectares of Detailed Soil Surveys and 131 lakh hectares of Soil Resource Mapping against which it has completed soil surveys of 82.15 lakh hectares [81.3%], 1.65 lakh hectares [45.6%) and 52.25 lakh hectares [43.2%] respectively till January 2012. [ii] It has also established Remote Sensing Center in 1982 for application of advanced technologies in soil survey [iii] Detailed Soil Survey of ``very high" and ``high" priority watersheds to provide detailed soil data base for planning and execution of soil conservation projects and for scientific land use planning using large scale base map [iv] District wise Soil Resource Mapping to create repository of soil data base in the country [v] Development of Digital Spatial Data Base on Hydrologic Units, Soil and Land Information using Geographic information System, Ration Data Base Management System for GIS based Web Services [vi] Development of State wise Micro watershed Atlas of India [vii] Creation of Platform Free State wise Micro watershed Atlas for dissemination of watershed information [viii] Organization of short term training Courses on soil and Land Resource data Base for integrated watershed Development planning. Soil Conservation Training Center, DVC, Hazaribagh, Jharkhand organizes medium and short term training courses for field functionaries and project staff of State Governments engaged in implementation of soil and water conservation programs.

Table. Region wise extent of Land Degradation in India [Area in '000 ha]

Region	Water erosion	Wind erosion	Water logging	Salinity/ alkalinity	Soil acidity	Complex problem	Degraded area	Geographical area	8 as % of 9
1	2	3	4	5	6	7	8	9	10
Northern	12002 [47.4] [12.8]*	8828 [34.9] [93.1]*	2040 [8.1] [14.3]*	1962 [7.7] [33.0]*	157 [0.6] [1.0]*	324 [1.3] [4.4]*	25313 [100] [17.2]*	71472 [21.7]*	35.4
North-East	4146 [32.8] [4.4]*	00	522 [4.1] [3.6]*	00	5534 [43.9] [34.5]*	2422 [19.2] [32.8]*	12614 [100] [8.6]*	26219 [8.0]*	48.1
Eastern	9249 [61.0] [9.9]*	00	3392 [22.4] [23.7]*	474 [3.1] [8.0]*	1848 [12.2] [11.5]*	194 [1.3] [2.6]*	15157 [100] [10.3]*	41833 [12.7]*	36.2
Southern	22330 [73.1] [23.8]*	00	5031 [16.5] [35.2]*	723 [2.4] [12.2]*	1179 [3.9] [7.3]*	1302 [4.3] [17.6]*	30565 [100] [20.8]*	63576 [19.3]*	48.1
Western	16446 [77.0] [17.5]*	443 [2.1] [4.7]*	599 [2.8] [4.2]*	1350 [6.3] [22.7]*	519 [2.4] [3.2]*	1993 [9.3] [27.0]*	21350 [100] [14.5]*	50743 [15.4]*	42.1
Central	29275 [70.5] [31.2]*	212 [0.5] [2.2]*	2709 [6.5] [18.9]*	1416 [3.4] [23.8]*	6796 [16.4] [42.4]*	1126 [2.7] [15.2]*	41534 [100] [28.3]*	73786 [22.4]*	56.3

Delhi + UTs	242 [84.3] [0.3]*	00	06 [2.1] [0.1]*	19 [6.6] [0.3]*	00	20 [7.0] [0.3]*	287 [100] [0.2]*	973 [0.3]*	29.5
Total	93680 [63.8] [100]*	9483 [6.5] [100]*	14299 [9.7] [100]*	5944 [4.0] [100]*	16033 [10.9] [100]*	7381 [5.0] [100]*	146820 [100] [100]*	328602 [100]*	44.7

Figures in parentheses indicate percentage share of concerned category of degradation to the total degraded area. Figures in parentheses with * indicate percentage share of the region in the total

Programs and Performance

Three ministries viz. Ministry of Agriculture, Ministry of Rural Development, and Ministry of Environment and Forest are implement in various watershed development programs for development Plan (2002-07) 50.89 million hectares have been developed at a costs of Rs. 19251.22 crore (Rs. 3783/ha). Parts of such developed lands are brought under cultivation to maintain balance in different types of land uses. Following are among a few ongoing programs under implementation with physical performance data for 2011-12.

Soil Conservation in the Catchment of River Valley Project and Flood Prone River scheme : This was launched in 1961-62 and from November 2000 onwards is being implemented through Macro Management of Agriculture (MMA) scheme in 60 selected inter state catchments spread over all States [except Goa]. Its objectives are (i) prevention of land degradation by adopting a multidisciplinary approach to soil conservation and watershed management in catchment areas (ii) improvement of land capability and moisture regime in watersheds (iii) promotion of land use to match land capability (iv) prevention of soil loss from catchments to reduce siltation of multipurpose reservoirs and enhancing in situ moisture conservation and surface rainwater, storage in catchments to reduce flood peaks and volume of runoff. To assess impact of soil and water conservation measures, system of continuous monitoring of rainfall, runoff and sediment parameters (prior to, during and after treatment) is followed by establishing Sediment Monitoring Stations at outlet of watershed. Since inception till 2010-11, an area of 78.85 lakh hectares [26.1%) against priority area of 301.50 lakh hectares needing urgent treatment have been treated. During 2011-12, 1.78 lakh hectares are targeted for treatment against which 1.26 lakh hectares [70.8%] have been treated till January 2012.

Reclamation and Development of Alkali and Acid Soils (RADAS) : This program was launched in 1985-86 and was restructured during 2007-12 for development of alkali and acid soils. Currently this program is being implemented through MMA scheme in seven States of

Arunachal Pradesh, Mizoram, Gujarat, Haryana, Punjab, Karnataka and Rajasthan. It aims at improving physical conditions and productivity status of alkali and acid soils for restoring optimum crop production. Under the program, up to 2010-11, 8.41 lakh hectares have been developed at the cost of Rs. 166.49 crore. During 2011-12, 22000 hectares were targeted for reclamation and development against which 16000 hectares (72.7%) have been reclaimed up to January 2012.

Watershed Development Projects in Shifting cultivation areas: These projects are being implemented from 1992-93 in seven States of North Eastern Region with objectives to (i) protect hill slopes of jhum areas through soil and water conservation measures on watershed basis and to reduce further land degradation (ii) encourage and assist jhumia families to develop jhum land for productive uses with package of practices leading to settled cultivation (iii) improve socio economic status of jhumia families through household/land based activities (iv) mitigate ill effects of shifting cultivation by introducing appropriate land use according to land capability and improved technologies. Under the scheme, arable and non arable land is treated through various measures. Rehabilitation components include improvement in production system of households with land and enhancing income of households without land/asset through provision of income generating activities and assets. Since inception up to 2010-11, an area of 5.49 lakh hectares has been developed at a cost of Rs. 455.79 crore. During 2011-12, 38000 hectares [90.5%] of jhum land has been developed up to January 2012 against target of 42,000 hectares.

World Bank Aided Sodic Land Reclamation and Development project : In June 2009, technical and financial assistance was sought from World Bank for reclamation and development of 1.35 lakh hectares of degraded land comprising 1.30 lakh hectares of sodic land and 5000 hectares of ravine area at estimated cost of Rs. 1,224 crore for six years. During 2010-11, 20,000 hectares were reclaimed at the cost of Rs. 85.18 crore. During 2011-12, 26000 hectares have been developed up to January 2012 as against target of 25,000 hectares (104%).

Strategic Actions

In order to mitigate the threat of the severity of 146.82 million hectares of degraded land (44.7% of country's geographical area) to environment, agricultural productivity and human survival, need is to (i) formulate a Vision 2025 document detailing comprehensive strategy to develop at least 110 million hectares (75%) of degraded land (ii) review current status of land degradation by end of 2013 (iii) formulate strategic action plans with sharp focus, inter alia, on (a) disseminating proven and demonstrated technology among farmers (b) understanding local constraints inhibiting acceptability of existing technologies through action research program (iii) invest adequately to strengthen State wise research institutions, human resources, training facilities and mechanism to effectively transfer technology to users (iv) institute comprehensive survey once in five years to scientifically assess the status of land degradation (v) initiate policy and programs based on local requirements in the districts of each State in light of the nature and magnitude of land degradation caused by factors, viz. water and wind erosion, water logging, salinity/alkalinity, soil acidity, among others (vi) design robust Management information system to provide quarterly progress districtwise to understand gaps between planning and implementation (vii) install effective monitoring and review system at district level to quarterly monitor the implementation process (viii) undertake once in three years comprehensive evaluation State-wise through independent professional team to sharply bring out the inadequacies in the policy, planning and implementation process that could not yield expected results and suggest measures to improve performance and arrest land degradation assigning annual targets with measurable performance indicators (ix) train program implementers and users to meticulously implement programs on the basis of new guidelines for watershed development recommended by the Hanumantha Rao Committee, emphasizing the bottom up approach whereby the User Groups themselves decide their work program the Government acts as a facilitator and the people at the grass root level become the real executioner of the program.

Chapter - 30

Initiative to Sustain Land Resource Development

Arpita Sharma

Land includes benefits to arise out of land, and things attached to the earth or permanently fastened to anything attached to the earth". Land is one of the most critical resources for the rural poor dependent on farming for their livelihoods. Today, about 2 million hectares of rainfed and irrigated agricultural lands are lost to production every year due to severe land degradation, among other factors. This degradation is a critical link in a downward spiral with respect to poverty. Poor land quality compromises farm incomes, resulting in ongoing poverty and a lack of resources to invest in increasing land and labor productivity, condemning farmers to repeat the cycle often worsening degradation. Inappropriate land management, particularly in areas with high population densities and growth rates, further increases loss of productivity. This in turn affects food security and the potential for rural on and off farm income generation. The challenge for developing countries is to develop land management programs to increase the availability of high quality fertile lands in areas where population growth is high, poverty is endemic, and existing institutional capacity is weak.

Land laws in post-Independent India :

1956 : Before 1956 devolution of both acquired and inherited property was governed by the personal laws of the community. Although equal rights were granted to women in acquired property through the Hindu Succession Act of 1956, rights in inherited agricultural land were specifically exempted from the Act, and were made subject to tenancy and land reform laws of the states.

1950 : In India, agrarian reforms through the 1950s took place at a time when gender equality was marginal to the policy agenda and women's organisations lacked their current visibility. Hence, in most government land reform programmes and land transfers, women's land rights remained a non-issue.

1980 : From the 1980s onwards gender equality was talked about, but restricted only to land distributed by government. The plans called for titles to spouses in productive assets, houses, house sites and directed state governments to register government allotted wasteland/ceiling surplus lands in joint names, but remained silent on the inequities in devolution laws as regards women. However, the potential of wasteland distribution in future is extremely limited, as the cultivable waste has already been allotted or encroached. Hence the main source of land title in the years to come is not through distribution of government land or leasing, but through inheritance.

The main source of tenure has always been through inheritance and will be more so in future and therefore we need to examine the tenancy laws and the extent of discrimination inherent in such laws.

Other laws : As already stated, the Hindu Succession Act left the question of devolution of inherited agricultural land and property to be decided by the respective state tenancy laws. For example, in the tenurial laws of Haryana, Himachal Pradesh, Jammu and Kashmir, Punjab, Delhi and Uttar Pradesh, the specified rules of devolution show a strong preference for agnatic succession, with a priority for agnatic males. In all these states the tenancy develops in the first instance on male lineal descendants in the male line of descent. The widow inherits only in the absence of these male heirs. In addition, in the first four states mentioned, daughters and sisters are totally excluded as heirs. In Delhi and Uttar Pradesh, daughters and sisters are recognised but come very low in the order of heirs.

States where the tenurial laws explicitly mention that the devolution of tenanted land will be according to personal law are very few and include Rajasthan and Madhya Pradesh where the personal law applies for all communities. Also in the Telangana region of Andhra Pradesh, the commentary following Section 40 of the relevant Act clarifies that for Hindu tenants the Hindu Succession Act will apply. In practice, however, even in Rajasthan daughters have been recognized as heirs only in some judgments, while in others male heirs alone have received recognition. In addition, there are states which do not specify the order of devolution in their laws dealing with tenancy land, such as Gujarat, the Bombay region of Maharashtra, West Bengal, Karnataka, Kerala, the Andhra region of Andhra Pradesh and Tamil Nadu. In these states we can presume that the personal laws automatically apply. Then there are states such as Bihar and Orissa for which the tenancy acts specify that occupancy rights shall developed in the same manner as other immovable property. ``subject to any custom to the contrary". This leaves open the possibility of admitting gender - in egalitarian customs if established, especially for the tribal communities in these regions.

According to the Hindu Personal Law, sons and daughters are entitled to equal shares in the deceased man's ``notional" share in Mitaksara joint family property. But sons, as coparceners in the joint family property additionally had a direct birth to an independent share; while female heirs (e.g. daughter, widow, mother) had claims only in the deceased's ``notional" portion. This meant that if a man had four acres of land and a son is born, he is left only with two acres and the rest has notionally gone to the new born son. But if a daughter is born she gets nothing unless her father dies, that too from the remaining two acres of land of which the son will also get his share in addition to two acres that was his since birth. Also, sons could demand partition; daughters could not. In actual practice, daughters get nothing, as mutation of land is generally done in favour of male heirs. In some cases they are asked to give a letter in favour of the sons.

2005 : Little effort was made unitll 2005 to do away with these discriminatory laws. Finally after 50 years of the 1956 Hindu Succession Act (HSA), the Government addressed some persisting gender inequalities in the HSA by bringing in the Hindu Succession (Amendment) Act, 2005. Once of the most significant amendments in the 2005 Act is deleting the gender discriminatory Section 4 (2) of the 1956 HSA. Section 4 (2) exempted from the purview of the HSA significant interests in agricultural land, the inheritance of which was subject to the devolution rules specified in state level tenurial laws. The 2005 Act brings all agricultural land at par with other property and makes Hindu women's inheritance rights in land legally equal to men's across States, overriding any inconsistent State laws. This can benefit millions of women dependent on agriculture for survival. The second major achieve-ment lies in including all daughters, especially married daughters, as coparceners in joint family property. They can also demand partition in the life time of their father just as sons could. Third, the Act deletes section 23 of the 1956 HSA, thereby giving all daughters (married or not) the same rights as sons to reside in or seek partition of the family dwelling house. Section 23 did not allow married daughters (unless separated, deserted or widowed) even residence rights in the parental home. Unmarried daughters had residence rights but could not demand partition. Fourth, the Act deletes Section 24 of the 1956 HSA, which barred certain widows, such as those of predeceased saons, from inheriting the deceased's property if they had remarried. Now they too can inherit.

The Central Government persuades and incentives the States through schemes or policy initiatives. India faces tremendous challenges on the issues related to land governance. The following data will make it clearer : India has approximately 2.16 million sq. km. of cultivable area (1) India has about 18 per cent of world's population; (2) 15 percent of world's live stock population is to be supported from this land (3) India has about 2 percent of world's geographical area and 1.5 percent of forest and pasture land (4) The per capita availability of land has declined from 0.89 hectares in 1951 to 0.37 hectares in 1991 (5) The average agriculture land holding has declined from 0.48 hectares in 1951 to 0.16 hectares in 1991 (6) 95.65 percent of the farmers are within small and the marginal category owning about 62 percent of the land, while the medium and the lage farmers who constitute 3.5 percent own about 37.72 percent of the total area (7) Most of the cases pending in the Courts relate to land disputes; 7.9 million persons are without dwelling units to live in (8) In the rural areas alone, there are more than 140 million land owners, owning more than 430 million records (9) There are approximately 55 million urban households (10) In most of the States last cadastral survey was done around 70 to 80 years ago. In fact in some States, e.g., North Eastern States this survey has not been done till now. The issues related to land may be described in following five divisions.

Land Management : Land figures as Entry 18 in the State list of the Constitution as ``Land, that is to say, right in or over land, land tenures including the relation of landlord and tenant, and the collection of rents; transfer and alienation of agricultural land; land improvement, and agricultural loans; colonization." Entry 45 in the State list is ``Land revenue, including the assessment and collection of revenue, the maintenance of land records, survey for revenue, the maintenance of land records, survey for revenue purposes and records of rights, and alienation of revenues". So, the land and its management fall in the exclusive domain of the States. Each State has a different set up for land and land records management. In most of the States Revenue Department handles the land records along with the other issues related to land management. Survey Department deals with the survey of lands, Consolidation Department deals with the consolidation of the lands, and Gram Panchayats do undisputed mutations in some States. The change of the land records by any one of them makes the records of another obsolete. So, the records are out of date in most of the States and they do not reflect the ground reality. Before independence, the revenue from the lands was a major consideration for the proper management of land and land records. But after independence as revenue from the lands dwindled, the land and land records management was also neglected. In fact in some of the States the land revenue has been abolished altogether. The surprising fact is that the States hardly give any priority to this subject and most of the initiatives have been taken at the Central Level.

Government laws and rules for land development

(1) Watershed Development Project (1989) : The Watershed approach has conventionally aimed at treating degraded lands with the help of low cost and locally accessed technologies such as in situ soil and moisture conservation measures, afforestation etc. and through a participatory approach that seeks to secure close involvement of the user communities. The brad objective was the promotion of the overall economic development and improvement of the socio-economic development and improvement of the socio-economic conditions of the resource poor sections of people inhabiting the programme areas. Many projects designed within this approach were, at different points of time, taken up by the Government of India. The Drought Prone Areas Programme (DPAP) and the Desert Development Programme (DDP) were brought into the watershed mode in 1987. The integrated Wasteland Development Programme (IWDP) launched in 1989 under the aegis of the National Wasteland Development Board also aimed at the development of wastelands on watershed basis. All these three programmes were brought under the Guidelines for Watershed Development with effect from 1.4.1995.

Other major programmes now being implemented through this approach are the National Watershed Development Project in Rainfed Areas (NWDPRA) and the Watershed Development in Shifting Cultivation Areas (WDSCA) of the Ministry of Agriculture (MoA). The objectives of Watershed Development Projects will be : (1) Developing wastelands/degraded lands, drought prone and desert areas on watershed basis, keeping in view the capability of land, site conditions and local needs. (2) Promoting the overall economic development and improving the socio-economic condition of the resource poor and disadvantaged sections inhabiting the programme areas. (3) Mitigating the adverse effects of extreme climatic conditions such as drought and desertification on crops, human and livestock population for their overall improvement. (4) Restoring ecological balance by harnessing, conserving and developing natural resources i.e. land, water, vegetative cover. Encouraging village community for : (1) Sustained community action for the operation and maintenance of assets created and further development of the potential of the natural resources in the watershed. (2) Simple, easy and affordable technological solutions and institutional arrangements that make use of and build upon, local technical knowledge and available materials. (3) Employment generation, poverty alleviation, community empowerment and develoment of human and other economic resources of the village.

(2) Drought Prone Areas programme (1994) : The basic objective of the programme is to minimise the adverse effects of drought on production of crops and livestock and productivity of land, water and human resources ultimately leading to drought proofing of the affected areas. The programme also aims to promote overall economic development and improving the socio-economic conditions of the resource poor and disadvantaged sections inhabiting the programme areas. Upto 1994-95, DPAP was in operation in 627 blocks of 96 districts in 13 States. Prof. C.H. Hanumantha Rao Committee recommended : (1) Exclusion of 245 existing blocks; (2) Including of 384 new blocks; and Transfer of 64 blocks from DPAP to DDP. The Government did not agreed for exclusion of existing DDP blocks. However, inclusion of new blocks and transfer of blocks from DPAP to DDP was agreed to. Thus, from 1995-96 total blocks covered under DPAP became 947. These 947 blocks were in 164 districts in 13 States. Subsequently, with the re-organization of States, Districts and Blocks, the programme is now covered in 972 blocks of 183 districts in 16 States. These States are Andhra Pradesh, Jammu & Kashmir, Jharkhand, Karnataka, Madhya Pradesh, Maharashtra, Orissa, Rajasthan, Tamil Nadu, Uttar Pradesh, Uttaranchal and West Bengal. The identified dry sub humid area under the programme is about 7.46 lakh sq.kms (74.6 million has.).

(3) Desert Development Programme (DDP) : The basic object of the programme is to minimise the adverse effect of drought and control lesertification through rejuvenation of natural resource base of the identified desert areas. The programme strives to achieve ecological balance in the long run. The programme also aims at promoting overall economic development and improving the socio-economic conditions of the resource poor and disadvantaged sections inhabiting the programme areas. Upto 1994-95, Desert Development Programme was under mplementation in 131 blocks of 21 districts in 5 States. The Hanumanntha Rao Committee recommended : (1) Inclusion of 32 new blocks; and (2) Transfer of 64 blocks from DPAP to DDP. Inclusion of new blocks and transfer of blocks from DPAP to DDP was agreed to. Thus, rom 1995-96 total blocks covered under DDP became 227 in 40 districts f 7 States. Subsequently, with the re-organization of Districts and Blocks, the programme is now covered in 235 blocks of 40 districts in 7 States. The corresponding physical area under the programme is about .57 lakh sq. kms.

(4) Technology development extension and training for vastelands development in non forest areas : The Department of Vastelands Development was set up in July 1992 and placed under the Ministry of Rural Development. The restructured National Wastelands Development Board (NWDB) was given the specific responsibility to volve mechanisms for integrated development of non forest wastelands hrough systematic planning and implementation, in a cost effective nanner, specially to meet the needs for the people in the rural areas in espect of fuel wood and fodder. As part of its activities in fulfillment of s mandate the NWDB sponsors research and extension of research ndings to disseminate new and appropriate technologies for wastelands evelopment.

(5) Hariyali (2003) : To involve village communities in the nplementation of watershed projects under all the area development rogrammes namely, integrated Wastelands Development Programme WDP), Drought Prone Areas Programme (DPAP) and Desert Development Programme (DDP), the Guidelines for Watershed Development were adopted w.e.f. 1.4.1995, and subsequently revised in August 2001. To further simplify procedures and involve the Panchayat Raj Institutions (PRIs) more meaningfully in planning, implementation and management of economic development activities in rural areas, these new Guidelines called Guidelines for Hariyali are being issued.

Rural Land Resources Management (LRM) Program : The rural Land Resources Management (LRM) Program, at the World Bank develops and promotes knowledge based technical social institutional and policy choices for our clients, which improve management of this critical resource. These choices focus on : (1) Developing sustainable land management through improved land tenure systems and community natural resources management; (2) Raising the Profile of the risk and vulnerability impacts of climate change on communities natural resources, (land/water) and promote appropriate adaptation mechanisms (3) Mainstreaming of integrated approaches to Land and Water resources management for food security and poverty reduction (4) Creating and strengthening an enabling environment, which will enhance national regional, and global capacities to implement the convention to combat desertification and restore degraded lands.

Sustainable Land Management (SLM) : SLM is defined as a knowledge based procedure that helps integrate land, water biodiversity and environmental management (including input and output externalities) to meet rising food and fiber demands while sustaining ecosystem services and livelihoods. SLM is necessary to meet the requirements of a growing population. Improper land management leads to land degradation and a reduction in the productive and service (biodiversity niches, hydrology, carbon sequestration) functions of watersheds and landscapes. In layman's terms, SLM involves : (1) Preserving and enhancing the productive capabilities of land in cropped and grazed areas - that is, upland areas, down slope areas, and flat and bottom lands; sustaining productive forest areas and potentially commercial and noncommercial forest reserves; and maintaining the integrity of watersheds for water supply and hydropower generation needs and water conservation zones and the capability of aquifers to serve farm and other productive activities. (2) Actions to stop and reverse degradation - or at least to mitigate the adverse effects of earlier misuse which is increasingly important in the uplands and watersheds especially those where pressure from the resident populations is severe and where the destructive consequences of upland degradation are being felt in far more densely populated areas "downstream".

Chapter - 31

Unirrigated Agriculture : Problems and Prospects

A. Gayathri & P. Veerachamy

Introduction

This paper examines the problems and prospects of unirrigated agriculture in Tamil Nadu. Agriculture continues to be the main economic activity in rural areas of the developing world in spite of a steady diversification of their economic base during the preceeding decades. Likewise, agriculture is the backbone of the rural India and the largest industry in the country. The role of agriculture is important in terms of food security, international trade and economic development. India ranks first among the countries that practice unirrigated agriculture both in terms of extent and value of production, India has 143 million hectares of agricultural land and about 108 million hectares are unirrigated area, which constitutes nearly 75% of the total land (Kumar *et al.,* 2009). Unirrigated agriculture is largely practiced in arid, semi arid and sub-humid regions of our country. With about 68% of rural population, these regions are also home to 81% of rural poor (Rao *et al.,* 2005). In such areas, crop production has become difficult as the intensity and frequency of rainfall is low.

The unirrigated agriculture refers to crop production in a farming system which depends entirely on rainfall but may include supplementary irrigation from small dams or tanks fed from rainfall and associated run off on a particular land holding. However, all unirrigated areas are not of the same characteristics. Unirrigated areas are highly diverse, ranging from assured rainfall and resource rich areas with good agricultural potential to erratic rainfall and resource poor areas with much more restricted potential. Some resource rich unirrigated areas potentially are highly productive and already have experienced widespread adoption of improved seeds. In drier, less favorable areas, on the other hand, productivity growth has lagged behind, and there is widespread poverty and degradation of natural resources (Bhatia, 2005). However, nearly 50% of the total food grains are grown under unirrigated agriculture and millions of rural poor depend on unirrigated agriculture. In addition, 85% of the cereals, 83% of the pulses, 70% of the oilseeds and

65% of the cotton are predominant unirrigated crops grown in India. Nearly 50% of the total rural workface and 60% of livestock in the country depend on unirrigated agriculture (CRIDA, 2011). It emphasizes the crucial role played by unirrigated agriculture in food security and livelihood of the rural households.

By considering the above facts, the policy makers given much importance to the unirrigated agriculture in order to meet the rising demand for food, basic staples, non food grains, and exports. At the same time, the productivity of irrigated land is being utilized at the maximum level. The growth in total factor productivity in irrigated agriculture has declined slightly in major crops (Singh and Rathore, 2010). As a result, the opportunity for continued expansion of irrigated agriculture is limited and the need for Unirrigated agriculture has always been an important part of the agricultural sector. However, the state of unirrigated agriculture is precarious and the problems associated with it are multifarious. To name the more striking ones; low cropping intensity, high cost of cultivation, poor adoption of modern technology, uncertainty in output, low productivity, increasing number of suicides among farmers, lack of institutional credit, inadequate public investment and high incidence of rural poverty (Anon, 2009).

With this background, this paper explores and identifies the major problems of unirrigated agriculture and opportunities for stimulating agricultural growth in Tamil Nadu.

Unirrigated Agriculture in Tamil Nadu

Agriculture continues to be the mainstay of livelihood for more than 50 percent of the population in Tamil Nadu. It contributes 12% of the Nest State Domestic Product. Agriculture is the single largest sector providing job opportunities for rural people, besides being the source of supply of food grains and other dietary staples and serving as the chief source of raw material for industries. In Tamil Nadu, out of 7 million hectares of cultivable area, around 3.1 million hectares comes under unirrigated agriculture. The major segment of the pulses and oilseeds are produced by the unirrigated agriculture (Season and Crop Report 2010).

The existing studies in the area of unirrigated agriculture have used the gross cropped area under irrigation as an indicator to identify unirrigated agricultural areas. They consider that the predominant rainfed agriculture as ``unirrigated areas" and predominant irrigated agriculture as ``irrigated area". ``However, several previous studies have faced this conceptual issue in categorizing unirrigated agriculture. Therefore, the studies in the area of unirrigated agriculture have followed both average rain fall and gross cropped area under irrigation (Rangasamy, 1981; Bapna *et al.,* 1984; Jodha, 1985; Subbarao, 1985; Shah and Sah and 1993; Throat, 1993).

As a result, the Statistical Hand Book of the Tamil Nadu has classified the districts according to the range of actual rain fall (see Table 1). The average actual rain fall is categorized below 800 mm, 801 to 1000 mm, 1001 to 1200 mm, 1201 to 1400 mm, 1401 to 1800 mm and above the 1800 mm. This categorization may be useful to identify the gross cropped area under irrigation and unirrigated districts of Tamil Nadu state.

Table 1. Distribution of Districts by Range of Average Actual Rainfall 2009-10.

Sl.No.	Range of Rainfall	Distribution of Districts by Range of Rainfall
1	below 800 mm	Namakkal, Erode, Tiruchirapalli, Karur, Perambalur, Madurai, Virudhunagar and Thoothukudi
2	801 to 1000 mm	Vellore, Salem, Dharmapuri, Krishnagiri, Thiruvannamalai, Pudukkottai, Dingugul, Theni, Ramanathapuram, Sivagangai, Tirunelveli and Ariyaliur
3	1001 to 1200	Kancheepuram, Thiruvallur, Villupuram, Coimbatore and Kanniyakumari
4	1201 to 1400	Chennai, Cuddalore, Thanjavur and Thiruvarur
5	1401 to 1800 mm	Nagapattinam
6	1801mm and above	The Nilgiris

Source : Tamil Nadu Statistical Hand Book 2010, Department of Economics and Statistics, Chennai

Table 2. Details of Percentage of Gross Cropped Area under irrigation and Annual Actual Rain Fall in Districts of Tamil Nadu State

Sl.No.	Districts	Percentage of Gross Cropped Area under Irrigation	Total Annual Rain Fall (Actual in mm.)
1.	Kancheepuram	88.42	1156.80
2.	Thiruvallur	85.59	1062.00
3.	Cuddalore	59.03	1351.40
4.	Villupuram	72.02	1096.90
5.	Vellore	53.89	814.80
6.	Thiruvannamalai	74.75	957.70
7.	Salem	50.29	860.20
8.	Namakkal	48.59	592.30
9.	Dharmapuri	43.83	812.70
10.	Krishnagiri	28.65	920.50

11.	Coimbatore	61.53	1177.80
12.	Thiruppur	61.27	-
13.	Erode	69.52	708.60
14.	Tiruchirappalli	59.19	757.30
15.	Karur	60.31	637.10
16.	Perambalur	30.99	760.20
17.	Ariyalur	31.98	823.70
18.	Pudukottai	74.27	813.70
19.	Thanjavur	82.21	1217.00
20.	Thiruvarur	68.42	1325.80
21.	Nagapattinam	57.00	1666.90
22.	Madurai	63.68	713.30
23.	Theni	57.46	821.40
24.	Dindigul	48.44	820.00
25.	Ramanathapuram	35.56	866.40
26.	Virudhunagar	46.64	503.20
27.	Sivagangai	73.81	892.50
28.	Tirunelveli	77.38	901.10
29.	Thoothukudi	24.94	634.70
30.	The Nilgiris	0.52	2368.60
31.	Kanyakumari	41.70	1142.40
	Tamil Nadu	58.12	937.80

Source : Tamil Nadu Statistical Hand Book 2010, Department of Economics and Statistics, Chennai

Mentioned above, the average actual rain fall and gross cropped area under irrigation are considered for identifying the unirrigated agricultural districts. In Tamil Nadu, Namakkal, Erode, Tiruchirappalli, Karur, Perambalur, Madurai and Virudhunagar districts are identified as low rain fall district and their actual rainfall is below 800 mm. Among the low rain fall districts, percentage of gross cropped area under irrigation is relatively lower in Thoothukudi (24.94), Perambalur (30.99) and Virudhunagar (46.64) districts.

In general, unirrigated agriculture is considered as a gamble with monsoon. in this context, Government of Tamil Nadu has initiated a Missio9n on Rain fed Farming to increase the productivity and income of the farmers in unirrigated agricultural areas by adopting integrated watershed approaches of International Crop Research Institute for Semi Arid Tropics (ICRISAT) with the assistance from TNAU and Central

Research Institute for Dryland Agriculture (CRIDA). However, the problem of unirrigated are unresolved and those are discussed in the forthcoming sections.

Cropping pattern and crop diversification in unirrigated agriculture

Historically unirrigated farmers practice high diversity in cropping systems with livestock integration which is an inbuilt risk management strategy. The cropping patterns have evolved based on the rainfall, length of the growing season and soil types. However, due to changed consumer preferences and market demand, farmers are now rapidly shifting to crops and cropping patterns which are more remunerative. But the change in cropping pattern not towards the food crops to commercial crops and other high remunerative crop. The change in cropping pattern shows sharp increase in area under maize and cotton took place in few years at the cost of coarse cereals like sorghum and pearl millet mainly due to higher returns. Such changes will have implication on fodder availability to livestock. However, it is viable only in unirrigate where the miner irrigation sources are possible.

The change in cropping pattern will have implications on the resource use. Continuous mono cropping increase vulnerability of farmers to weather risks depletes soil fertility, ground water and leads to build up of pests and diseases. This issue has to be dealt both through technology and policy. In general, the India agrarian structure is dominated by marginal and small farmers not only in terms of number but also in terms of area cultivated. In this context, type of crop diversification and extent of crop diversification may differ among the different land holders (Gupta and Tewari, 1985; Kalpana, *et al.,* 2009). The cost of cultivation in Unirrigated agriculture includes plough, manure and harvesting. Hence, the cost and return of the Unirrigated agriculture may differ according to farm character, farmer character, type of crop and other factors.

The Problems of unirrigated agriculture

The important problems in unirrigated agricultural are explored by reviewing the relevant literature of the subject. In addition, it examples the role of economic and social policies, area development programs, infrastructural investments and provides the measures for promoting sustainable unirrigated agricultural development.

Unstable Production

Unirrigated agriculture is often characterized by high variability of production outcomes or, production risk. Unlike most other farming systems, farmers are not able to predict with certainty the amount of output that the production process will yield due to external factors such as weather, pests, and diseases. Farmers can also be hindered by adverse events during harvesting or threshing that result in production losses. However, unirrigate agricultural farmers have developed various coping strategies to insulate themselves from income risk, at least to a certain degree. As a result, even if individual crop yields vary greatly across year's farmers incomes may not, so increased yield variability of HYVs is not necessarily a deterrent to adoption. Due to the frequent crop failure, the farmers in unirrigated agriculture need to involve plough, seeding and manure practices within the single season.

Unstable market

Input and output price volatility is important source of market risk in agriculture. Prices of agricultural commodities are extremely volatile. Output price variability originates. It causes increase in cost of cultivation in the mid of great fluctuations on crops yield at from both endogenous and exogenous market shocks. Segmented agricultural markets will be influenced mainly by local supply and demand conditions, while more globally integrated markets will be significantly affected by international production dynamics. In local markets, price risk is sometimes mitigated by the ``natural hedge" effect in which an increase (decrease) in annual production tends to decrease (increase) output price (though not necessarily farmers' revenues). In Integrated markets, a reduction in prices is generally not correlated with local supply conditions and therefore price shocks may affect producers in a more significant way. Another kind of market risk arises in the process of delivering production to the marketplace.

Lack of Infrastructure

The inability to deliver perishable produce to the right market at the right time can impair the efforts of producers. Lack of infrastructure and well developed markets make this a significant source of risk in one hand, and the large fluctuations in input output prices which restrict the reliability on price predictions on the other hand. Unstable farm income resulting from business and financial risk coupled with lack of infrastructures in the area may affect production decisions, delay adoption of the new technology, prohibit long term investment in agriculture and hence delay the agricultural development in this sector (Hazal and Ramasamy, 1991).

Lack of drought management of strategies

Implementations of farmers drought management strategies fail, however, in the event that widespread drought causes crop failure over a wide area and depresses the rural economy so much that all sources of income are affected. Such aggregate level, covariate risk calls for government intervention to help stabilize incomes and prevent famine. In this context, the effective implementation of rural employment and food subsidy programs deserve credit for reducing drought related hunger in India in the last two decades. On the other hand, government sponsored rainfall insurance schemes have probably not contributed to increased adoption of improved seeds, but they have done a great deal to drain public funds.

Failure of Institutional finance

Credit is well known to play an important role in facilitating investment in improved agricultural technology. Commercial Banks, Regional Rural Banks (RRBs) and Cooperatives are the three main rural financial institutions that provide credit to the agricultural sector at the village level. In most of India, weak formal banking and cooperative systems provide subsidized credit, but defaults are extremely high and funds are provided disproportionately to relatively large farmers. In addition, occasional interference by politicians to forgive farmers debts only serves to weaken the banking system. In addition, some of the studies estimates that the excluding Kerala, the ratio of credit supply to farmers short term credit requirements in India is abut 1 :10. Meanwhile, informal village moneylenders provide coverage to a wider range of clients but a very high rates of interest (Desai, 1988). Whereas, many village moneylenders borrow from the formal sector at concenssional rates in order to relent to their poorer neighbors at higher rates (Hanumantha Rao and Gulati, 1994). These phenomenons are frequent and problems are acute in unirrigated agriculture. In some extreme cases, these unfavorable events become one of the factors leading to farmers suicides which are now assuming serious proportions (Raju and Chand, 2007).

Lack of Crop Insurance

As stated earlier, production process in unirrigated agriculture is entirely different than in other kind of farming. It has been observed through the variability and instability in production and productivity and high cost of cultivation. These factors lead severe impact negatively on most rural households simultaneously and are therefore difficult to manage through traditional risk sharing and coping strategies. However, the risk bearing capacity of the average farmer in the unirrigated agriculture is very limited. As a result, government policy may sometimes play an important role in helping farmers manage risk. In this context, the Government Policy of Crop Insurance was established to manage risks in agriculture (Walker and Jodha, 1986; Rao *et al.*, 1988).

Conclusion and Policy Suggestions

In Tamil Nadu, low rainfall districts are Namakkal, Erode, Tiruchirappalli, Karur, Perambalur, Madurai and Virudhunagar. Therefore, the authorities need to evolve management practices for farmers of the districts. Besides, choice of remunerative crops without degradation of the natural resource base has to be suggested and also to define agro ecological zones where such cropping systems can be adopted sustainably. Simultaneously, need based policy incentives are required to encourage farmers adopt agro ecology compatible cropping systems so that the farmers income is maintained and the natural resource base of the country is not degraded.

Financial assistance to the farmers in the low rainfall area is highly imperative. But, formal sector funds often are not available due to rationing and bureaucratic hassles. The neediest farmers would be made better off if concessional lending were abandoned and bank managers were given more autonomy and protection against political interference. Banking operations could be made simpler and more decentralized in order to reduce transactions costs of both banks and their clients. Higher interest rates would help banks become viable credit institutions rather than merely a means for channeling concessional funds. Under these circumstances, banks could attract deposits, and they would have more incentive to develop better loan portfolios. In short, this step would help develop greater professionalism in the banking sector (Hanumantha Rao and Gulati, 1994).

Crop insurance is provided by the public sector in many countries. However, it fails to reach its target in unirrigated agricultural areas. The primary reasons are follows : Majority of the unirrigated holdings is in small and marginal farm categories and these farms have poor access to institutional credit. Since crop Insurance was linked to crop loans, many small and marginal farmers could not participate in the crop insurance scheme. The threshold yield was fixed on the basis of the average of the preceeding 10 years whereas the trend in the growth of yield levels for most of the crops was positive. Further, the threshold yield or level of non indemnifiable yield was very high even for low risk areas and the high risk areas in unirrigated agriculture are exclusion of from crop insurance scheme. As a result, the farmers in unirrigated agriculture not prefer to adopt the crop insurance policies. In particular, unawareness among the farmers about the crop insurance scheme and non availability of insurance coverage for the major commercial crops like cotton and others were excluded from the crop insurance scheme. Therefore, the agencies for crop insurance need reform the policies and insurance coverage to include the different type of farmers and crops for managing the uncertainty and risks in unirrigated agriculture (Bhende, 2005).

As a whole, the policy makers have to look into the above mentioned problems and suggestions thereby to improve the agricultural production, farmer's livelihood and sustainable agricultural development.

REFERENCES

Anon. 2009. ``Agricultural Statistics at a Glance." Ministry of Agriculture, New Delhi.

Bapna, S.L., H. B Inswanger, and J. Quizon 1984. ``Systems of Output Supply and Factor Demand Equations for Semi-arid Tropical India." *Indian Journal of Agricultural Economics*, 39 (2) : 179-202.

Bhatia, M.S. 2005. ``Viability of Rainfed Agriculture in Semi Arid Regions" NABARD Occasional Paper No. 40.

Bhende, M.J. 2005. ``Agricultural Insurance in India : Problems and Prospects." NABARD Occasional Paper No. 44, Mumbai.

CRIDA 2011. Vision 2030. Central Research Institute for Dry land Agriculture, Hyderabad, Andhra Pradesh.

Desai, D.K. 1988. ``Institutional credit requirements for Agricultural Production - 2000, *Indian Journal of Agricultural Economics*, 43 (3).

Government of Tamil Nadu 2010. Tamil Statistics Hand Book 2010. Department of Economics and Statistics, Chennai, Tamil Nadu.

Gupta, R.P. and Tewari, S.K. 1985. ``Factors Affecting Crop Diversification : A Critical Analysis : ``*Indian Journal of Agricultural Economics,* 40 (3) : 304-309.

Hanumantha Rao, C.H. and Gulati, Ashok 1994. ``Indian Agriculture : Emerging Perspectives and Policy Issues." New Delhi : ICAR and Washington, DC : IFPRL.

Hazell, Peter, B.R., and Ramasamy, C. 1991. ``Baltimore : Johns Hopkins University Press.

Jodha, N.S. 1985. Development strategy for rainfed agriculture: Possibilities and constraints. Economics Program, ICRISAT.

Kalpana, R., Devasenapathy, P. and Kaleeswari, R.K. 2009. ``Crop Diversification for Increasing Productivity and Profitability in Irrigated Uplands of Tamil Nadu." *Indian J. Agricultural Research* 43 (1) : 73-76.

Kumar, P., Joshi, P.K. and Birthal P.S. 2009. ``Demand Projections for Food Grains in India." *Agricultural Economics Research Review,* 22 (2) :237-243.

Raju, S.S. and Ramesh Chand 2007. ``Progress and Problems in Agricultural Insurance in India". *Economics and Political Weekly,* 1905-1908.

Rangaswamy, P. 1981. ``Economics of dry farming in drought prone areas : A case study of Hissar in Haryana." Research Study No. 81/1. Agricultural Economics Research Centre, University of Delhi.

Rao, K.P.C., Bantilan, M.C.S., Singh, K., Subramanyam, S., Deshingkar, P., Rao P. Parthasarthy and Shiferaw B. 2005. ``Overcoming Poverty in

Rural India : Focus on Rainfed Semi Arid Tropics." Patancheru 502 324, Andhra Pradesh, India : International Crops Research Institute for the Semi-Arid Tropics, 96 pp.

Rao, C.H.H., Ray, S.K. and Subbarao, K. 1988. ``Unstable Agriculture and Droughts Implications for Policy." New Delhi : *Vikas Publishing House Pvt. Ltd.*

Shah, Amita and Sah, D.C. 1993. ``Dry Land farming under the Changing Source Environment : A Case Study of Gujarat." *Artha Vijnana* 35 (3) :

Singh, S. and Rathore, M.S. 2010. Rain fed Agriculture in India - Perspectives and Challenges, Rawat Publications, Jaipur.

Subbarao, K. 1985. ``Institutions, Infrastructure and Regional Variations in India's Input Delivery System." In Agricultural Markets in the Semi arid Tropics : Proceedings of the International Workshop, October 24-28, 198. Patancheru, India : ICRISAT.

Thorat, S.K., 1993. ``Technological Change and Regional Differentiation." Khama Publishers, New Delhi.

Walker, T.S. and Jodha, N.S. 1986. ``How Small Farm Households Adapt to Risk. In Peter Hazell, Carlos Pomareda and Alberto Valdes" (eds.), *Crop Insurance for Agricultural Development.* Baltimore and London. The Johns Hopkins University Press.

Part - C

Land Use and Interlinking of Rivers for Irrigation

Chapter - 32

Value Added Fertilizers for Enhanced Prdouctivity

Surinder Sud, Y.S. Shivay & Dinesh Kumar

The Fertiliser sector in India is the third largest in the world in terms of both production and consumption. But the average per hectare use of these chemical nutrients is confined largely to a few states and even fewer crops. After stagnating for several years, the overall fertiliser consumption rose perceptibly by about 10 per cent in 2004-05 to reach 18.47 million tonnes (in terms of nutrients), against 16.8 million tonnes in the previous year. But the average per hectare consumption was only around 97 kg in India, against over 157 kg in Bangladesh. 109 kg in Pakistan and 120 kg in Sri Lanka.

Significantly, over half of the total fertiliser consumption is accounted for the five states Uttar Pradesh, Andhra Pradesh, Maharashtra, Punjab and Karnataka and virtually three crops - paddy, wheat and sugarcane - with some nutrients going to cotton and groundnut. The fertiliser use in unirrigated crops, be they coarse cereals, pulses, oilseeds or others, is rather limited.

Indeed, the history of the country's fertilizer industry can be traced back to 1906 when a super phosphate production factory with an annual capacity to produce 6,400 tonnes of active ingredient P205 was set up at Ranipet in Tamil Nadu. However, the subsequent expansion of fertilizer industry, as also growth in consumption, remained rather sluggish for several decades for want of awareness about the benefits of the plant nutrient application, lack of fertiliser responsive crop varieties and inability of the farmers to invest in costly farm inputs like fertilisers.

The turning point in the fertiliser sector came with the availability of the high yielding varieties of wheat and paddy which essentially required fertilisers and water for their optimal performance. This not only triggered off the green revolution in the 1960s but also led to a spectacular jump in fertiliser demand, necessitating higher production and imports. The emergence of new fertiliser factories and the field programmes launched by them to educate the farmers on the efficient use of fertilisers, coupled with availability of crop production credit to the farmers, gave further impetus to fertiliser consumption.

Major investment is setting up of fertiliser plants came in the 1980s and the early 1990s thanks to the HBJ pipeline which made natural gas available for urea production. However, further investments dried up subsequently as the gas availability did not increase further and the overall pricing, supply and demand scenario eroded returns on fertiliser production. In fact, even the installed plants began feeling the pinch of paucity of gas, adversely affecting their capacity utilisation.

But, nevertheless, the total installed capacity of the country's fertiliser industry has risen to 12.27 million tonnes of nitrogen and 5.5 million tonnes of phosphates in 2005-06. The capacity utilisation in the fertiliser industry in 2004-05 was estimated at 92.1 per cent in nitrogen and 74.9 per cent in phosphates. But, the single super phosphate industry has continued to be in doldrums for the past few years with its capacity utilisation dropping in 2004-05 to 47.2 per cent from about 50.2 per cent of the previous year due to unfavourable pricing policies and inordinate delays in price declarations.

Notably, the response of the crops to applied plant nutrients is relatively low in India because of the meagre organic carbon content of most soils. This is attributed to the soils being subjected to crop cultivation over longer period without adequate application of organic manures which help keep the soils healthy. Besides, even in the case of chemical fertiliser application, the use of plant nutrients is weighed heavily in favour of urea.

Though the ideal ratio of consumption of nitrogen (N), phosphorus (P) and potash (K) should be 4 : 2 : 1, the actual is generally far from it. No doubt, the imbalance in nutrient use has tended to narrow down in past few years thanks to conscious efforts to promote the balanced fertiliser application, but it was still quite skewed at 6.9 : 2.6 : 1 in 2003-04. For 2004-05, this ratio is estimated at a little better 5.5 : 2.2 : 1.

Of the three major forms of fertilisers - nitrogenous (mainly urea), phosphatic and potassic fertilisers - nitrogenous fertiliser urea is the only one which can be produced wholly indigenously, without any import component. But the overall urea production potential is constrained by the supply crunch of feedstocks like natural gas, naphtha, fuel oil or others. The costs of all these have shot up in recent time to a-level that has upset the economics of fertiliser production.

The bulk of the phosphatic fertiliser is produced from the imported raw material like phosphoric acid or other intermediaries. The indigenous availability of rock phosphate is inadequate besides being of indifferent quality. For potassic fertilisers, the country depends almost entirely on imports as it does not have any indigenous source of potash.

In the case of urea production, naphtha dominated as the feedstock of the industry in the 1960s and 1970s. Subsequently, thanks to the availability of the gas from the Bombay High, the preference shifted towards natural gas as the energy consumption in gas based plants was less than that in the naphtha based fuel oil based plants. The cap investment in setting up gas based plants was also relatively low. Thus, nearly 60 per cent of the installed urea production capacity is now natural gas based though naphtha based plants still account for about 30 per cent of the output. But, due to the increased demand of natural gas from other potential consumers and the increase in its cost, liquefied natural gas (LNG) in now being contemplated as the feedstock for urea plants.

What is noteworthy is that the development of indigenous technology for fertiliser production has managed to keep pace with the growth in the fertiliser industry over the years. The available technology now pertains to almost all aspects, including planning and development of process know-how, design engineering, expertise in project management, and execution of the projects. The Indian consultancy organisations, too, have grown in tandem with the fertiliser industry and are now capable of undertaking execution of fertiliser projects right from the stage of conception of commissioning of the plants. However, there still is need for further research and development effort, especially in technological innovations to slash energy requirement of ammonia and urea plants to cut down production costs.

Indeed, the main problem besetting the Indian fertiliser sector is the lack of consistent, long term policies concerning pricing and controls on fertilisers. The inclusion of fertilisers in the list of essential items under the Essential Commodities Act has given the government the powers to fix prices expose distribution controls on fertilisers.

Under the fertiliser pricing policy, first measure that aimed at providing fertilisers to the farmers at affordable price and ensuring adequate returns to the fertiliser producer was the retention price cum subsidy scheme introduced in 1977. Under this, the difference between the statutory notified maximum retail price and the cost of production - as assessed on the basis of 12 per post tax returns - was paid as subsidy to the manufacturers.

This scheme worked quite smoothly for a long time. In fact, it helped the country achieve near total self sufficiency in the production of urea and 85 per cent self sufficiency in di ammonium phosphate (DAP) by the late 1990s. But, certain discrepancies were noticed in its implementation of this scheme which made it imperative to search for an alternative pricing policy. It was felt that this policy had encouraged the factories to show inflated production costs to hike their returns.

This led to the appointment in 1997 of a high powered committee, headed by Professor C.H. Hanumantha Rao, to review the retention price scheme and suggest a possible alternative pricing policy. This committee recommended in its report it April 1998 the replacement of the retention price scheme with a uniform normative referral price for the gas based units and grant of feedstock differential cost reimbursement for the unit based on naphtha, fuels oil or other feedstocks.

However, even before the government could take a decision on the recommendations of this committee, the Expenditure Reforms Commission (ERC) came up with the recommendation of an annual seven per cent increase in the farm gate price of urea and a group based concession scheme for the fertiliser industry. It envisaged classification of the urea units into five groups based on feedstock and vintage of the plant and fixing of a price for each group computed on the basis of weighted average retention price of all the units in that group. It also mooted phased decontrol of distribution of urea.

Though the government decided to implement the recommendations of the ERC, barring the one concerning periodic hike in farm gate prices of fertilisers, but the problem of pricing of urea was indeed far farm resolved. It was soon realised that the main drawback of the group average based retention pricing methods was that it created gainers and losers among the urea producers. As such, it amounted to rewarding inefficiencies and punishing efficiencies. Thus, the search for a proper pricing policy for urea continued, resulting in appointment of more committees. The process is still continuing.

Where phosphatic and potassic fertilisers are concerned, the policy framework has remained as fluid as that for the urea sector. The first major policy intervention in this sector came in August 1992 when, on the recommendation of a joint parliamentary committee, the government decontrolled the phosphatic and potassic fertilisers.

But, responding to the demand for keeping the farm gate prices of these nutrients low, the government soon introduced an ad hoc scheme for providing concession on the actual market price of these fertilisers. The price concession was applicable to both indigenously produced as well as imported fertilisers. Thus, though the decontrol move was aimed primarily at reducing fertiliser subsidy, but this objective could not be fulfilled as the subsidy level remained high due to the price concession.

This policy for the phosphatic and potassic fertilisers has continued since then with some minor modifications. In 2004-05, the government decided to change the methodology for recognising the price of imported phosphoric acid for determining the concession. It opted for normative prices instead of the one negotiated by the industry with the foreign suppliers.

An expert group has now been set up under the chairmanship of the Planning Commission member, Professor Abhijeet Sen, to revisit the pricing policy for the phosphatic fertilisers.

The single super phosphate (SSP) fertiliser, the only phosphatic fertiliser that also contains sulphur which is generally deficient in the Indian soils, is facing a peculiar problem. This fertiliser has been separated from the phosphatic and potassic fertilisers for the fixation and quarterly review of the concession on the sale price. As such, while the effective retail prices of other decontrolled fertilizers are fixed by the Union government, that for SSP are determined by the state governments. The delay in changing the retail prices in response to the changes in the cost of imported raw materials and intermediaries for the production of SSP often result in financial crunch in this segment of the fertiliser industry. This also results in occasional shortage in supplies as witnessed in 2005.

All this apart, which is really significant is the country is dependent, albeit partly, on the imports of all kinds of fertilisers and their raw material to meet its requirement and is likely to remain so in future. Though the demand for urea was almost stagnant in the past several years till 2003-04, obviating the need for imports, but the surge in demand subsequently has again made urea imports imperative. Consequently, about 0.64 million tonnes of urea was imported in 2004-05. Almost an equal quantity of DAP was also imported during that year. The import of potassic fertilisers, notably muriate of potash (MOP), was far higher at about 3.41 million tonnes. The international prices of fertilisers were quite high in 2004-05.

However, if the annual agricultural growth has to be stepped up to four per cent to achieve an annual gross domestic product (GDP) growth of eight per cent and above, the fertilizer consumption would have to be increased. The studies on the response of crops to fertiliser application had indicated that 1 kg of fertiliser causes crop yield to go up by 10 kg (in foodgrains like wheat and paddy). Of late, though this response has tended to decline due to several ecological and technological factors, the increase in yield is still believed to be as high as six to 7 kg per kg of applied fertiliser.

The economics of fertiliser application, however, is quite dynamic as it is a function of the prices of fertilisers and the crops which keep changing. But the time series data collected by the Fertiliser Association of India (FAI) shows that over the years the quantity of foodgrains required to buy a kg of fertilisers has gradually declined due to steady increase in the procurement prices of foodgrains without much rise in fertiliser prices. This indicates that the economics of fertilisers use has become more favourable over a period of time.

The FAI data reveals that 72, about 3.79 kg of paddy, required to buy one kg of nitrogen. But in 2004-05, a kg of nitrogen could be purchased only with 1.91 kg of paddy. Similarly, in the case of wheat, the economics of fertiliser consumption has improved substantially. While in 1970-71,

2.64 kg of wheat was required to buy a kg of nitrogen, in 2004-05 only 1.64 kg wheat was needed for that purpose. Similar is the trend for phosphatic and potassic fertilisers as well.

No doubt, organic farming (without the use of any chemical through fertilisers or pesticides) is gaining in popularity the world over. The demand for the products grown without chemical inputs is also swelling. But the prices of such products are relatively high and they are meant only for a niche market segment. The use of fertilisers cannot indeed be dispensed with, considering the growing requirement of food, fiber and fuel wood which needs to be met through produce of the land.

Importance of zinc as a micronutrient

This nutrient element is an important component of various enzymes that are responsible for driving many metabolic reactions in all crops. Growth and development would stop if specific enzymes were not present in plant tissues. The following pathways have important role to play in photosynthesis and sugar formation, protein synthesis, fertility and seed production and defense mechanism against diseases. In rice, zinc, is essential for several biochemical processes such as : cytochrome and nucleotide synthesis, auxin metabolism, chlorophyll production, enzyme activation, and membrane integrity.

Zinc is required in small amounts but its critical concentration for the growth of plants is very important. The critical limit for zinc has been set at 0.6 ppm in soils. Critical concentrations of zinc in plant tissues are set at 15 ppm for rice, 22 ppm for maize, 20 ppm for wheat and 8 ppm for sorghum to mention a few, a general figure of 20 ppm is generally used for most crops.

Zinc deficiency in rice and wheat

In India, Y.L. Nene was the first scientist to report Zn deficiency in rice at G.B. Pant University of Agriculture and Technology, Pantnagar in 1966. The symptom of Zn in rice were brown spots in the top leaves to begin with followed by almost complete rust browning of leaves. Because of this rust brown colour the disease was named as `Khaira'. Zn being a less mobile nutrient, most of the deficiency symptoms is in young leaves at the growing point. Physiologically, Zn deficiency can adversely affect the leaf chlorophyll content, stomatal conductance and net photosynthesis. Because of this in some cases the Zn deficiency symptoms in some crops such as maize may be confused with Fe deficiency which also produces yellowing or whitening of leaves. Wheat plants show dusty brown spots on upper leaves of stunted plants, shoot growth is more inhibited than root growth, tillering decreases, spikelet sterility increases, midrib becomes chlorotic particularly near the leaf base of younger leaves, leaves lose turgor and turn brown in blotches and streaks appear on lower leaves.

Causes of zinc deficiency

The causes for occurrence of Zn deficiencies are related to the introduction of high yielding varieties, neglect of application of bulky organic manures, imbalanced use of fertilizers and lo Zn uptake and accumulation which depends upon the soil pH, soil organic matter, temperature, light intensity, crop species, etc. Zn deficiency is quite widespread in the Indo-Gangetic plains and other important cereal growing states like Punjab, Uttar Pradesh, etc., which account for almost three fourths of the country's food grain production. The total area under Zn deficiency is about 10 million hectare in India and approximately 85% of rice wheat cropping system takes place in the Indo-Gangetic plain, which has calcareous soils with high pH and thus low Zn availability. Improving production from this cereal belt is therefore vital for sustaining grain production in the country. Zinc occurs in soil as sphaleite, olivine, homblende, augite and biotite; however, availability of Zn from these sources is guided by several factors mentioned above. Zinc deficiency is the most widespread micronutrient disorder in rice. One or more of the following reasons causes zinc deficiencies.

- Small amount of zinc available in the soil
- Planted varieties are susceptible to zinc deficiency (i.e. zinc inefficient cultivars)
- Neutral and calcareous soils containing large amounts of bicarbonates
- Alkaline soils under aerobic conditions where solubility of zinc decreases
- Zinc is precipitated as sparingly soluble zinc hydroxide (ZnO) when pH increases in acidic soils following flooding
- Bicarbonate accumulation takes place due to reducing conditions in calcareous soils with high organic matter or because of large concentration of bicarbonate in irrigation water
- Depressed zinc uptake due to increase in Fe, Ca, Mg, Cu, Mn and P after flooding
- Formation of zinc phosphates following large application of phosphatic fertilizers
- Formation of zinc complexes with organic matter in soils having high pH and high organic matter content
- Precipitation of zinc as zinc sulphide when pH decreases in alkaline soils following flooding
- Excessive liming

Realizing the impact of micronutrient deficiencies on the crop production, Indian Council of Agricultural Research launched ``All India Coordinated Research Project on Micro and Secondary Nutrients and Pollutant Elements in Soils and Plants" way back in 1967. The purpose of this project was to carry out research to: diagnose micronutrient deficiency, delineate deficient areas, and develop suitable amelioration technologies for correcting micronutrient disorders and develop methods for increasing crop production.

Under this project extensive work was carried in 20 states of the country to collect and analyze soil samples to identify areas for zinc and other micronutrients deficiencies. The work showed that 48% soil samples and 44% plant samples were found deficient in zinc.

Soil testing is a widely accepted method to determine the need for zinc in fertilizer programmes. Research has shown that soil testing measures only the quantity of micronutrients present and not their availability. Plant analysis is also very important in deficiency diagnosis. By combining plant analysis with soil analysis more accurate assessment of the micronutrient status could be made. The work of agricultural scientists under this project over the years was successful in creating awareness amongst the farming community, extension workers and the administrators concerned with development about the existence of micro and secondary nutrient deficiencies and their economic importance. The impact of the work done under the project has been more in the northern states has declined. Zinc deficiency is widespread on coarse textured, calcareous or alkaline alluvial soils of Indo-Gangetic plains of North India, fine textured calcareous black soils of Deccan Plateau and highly leached rice growing red and other associated soils.

Zinc Fertilizers

Five different types of zinc compounds, which are used as zinc micronutrient fertilizers are: (i) Inorganic compounds (ii) Synthetic chelates and (ii) Natural organic complexes (iv) Inorganic complexes, and (v) Organic manures.

(i) Inorganic compounds : These include zinc sulphate, zinc oxide, zinc carbonate and zinc chloride. Zinc sulphate is the most commonly used source in the world as well as in India. It has added advantage of having sulfur as well (11% S). It exists both in crystalline form (zinc sulphate heptahydrate) and granular form (zinc sulphate monohydrate) both are water soluble. Monohydrate is less soluble and not as effective as the crystalline form. Zinc sulphate is usually used to supply needed amount of zinc when dry fertilizer materials are used. This material can be either broadcast or incorporated or used in a starter fertilizer. Zinc oxide is insoluble in water and available in either granular or powder form. The granular form is less effective whereas powder form is difficult to use.

(ii) Synthetic chelates : ZnEDTA is the most popular chelate and is regarded as the most effective source of zinc as a micronutrient. It is 2-5 times more effective than zinc sulphate but it is also 5-10 times costlier as well. Chelates are applied to the soils in solid form (usually banded) or in solution form for foliar spray.

(iii) Natural organic complexes : Reacting zinc salts with organic by products from paper and pulp industry usually derives these complexes. These organic compounds include lignosulphonates and polyflavonoids. These complexes are less expensive than chelates but less effective too.

(iv) Inorganic complexes: Amongst inorganic complexes, ammoniated zinc sulphate solution is very popular. This complex is a source of nitrogen (10-15%), zinc (10%) and sulphur (5%). It is often combined with ammonium polyphosphate as a starter fertilizer. This complex being in liquid form is compatible with most of the liquid fertilizers. In all there are 17 sources of zinc available globally, which are used for different crops. Of late zinc compounds blended/incorporated with fertilizers are becoming popular as it saves labour in application.

(v) Organic manures : In addition to the above sources of zinc, organic manures like compost, farmyard manures, green manures or azolla are rich sources of micronutrients. Organic manures play important role in maintaining long term fertility of soils through improvement of physico chemical and biological properties.

Integrated effect of organic manures and micronutrients is far more than micronutrients alone. It is estimated that use of farmyard manures, poultry and piggery manures is capable of reducing zinc requirements by about 50%. If all conceivable forms of organic residues are appropriately recycled and used in agricultural lands it is projected to supply approximately 3.34 million tonnes of micronutrients in soil. Against this vast potential estimation made by Ministry of Agriculture shows that currently only about 90 thousands tonnes are being exploited. Biogas slurry has the potential to be used as a source of micronutrients. Press mud from sugar mills is also a good source of micronutrients. Presently it is being used as a source of either nitrogen or phosphorous. About 55,78,000 tonnes of press mud is available in the country.

Zinc fertilizers available in India

In India there are four zinc compounds, which serve as zinc fertilizers. The fifth source is zinc incorporated with urea and known as zincated urea. These are :

Source of zinc	Zinc content (%)
1. Zince Sulphate Heptahydrate	21
2. Zinc sulphate Monohydrate	33
3. Zinc EDTA	12
4. Zincated Phosphate Suspension (12.9% P_2O_2)	19.4
5. Zincated Urea	2

Zinc sulphate heptahydrate is the most commonly used zinc source in India and it has the advantage of containing sulphur as well. One of the major problems with zinc sulphate is its quality control and a lot of spurious material is sold in the market. The use of other zinc sources is on limited scale and these are made only in small quantities. Some fertilizer manufacturers are making zincated urea on experimental basis.

Conclusion

It is concluded that coating of prilled urea with 1.0% zinc oxide in neem oil emulsion may be sufficient to get higher productivity and increased Zn efficiency over prilled urea alone in aromatic rice wheat system.

Chapter - 33

Balanced Fertilization -- Real Benefits for Agriculture Sustainability

K.N. Tiwari & Rakesh Tiwari

Fertilizers feed the world by feeding the soils and in turn plants, and, if the world is not to go hungry, fertilizers will continue to play the key role in food production. No country in the world has been able to increase agricultural productivity without expanding the use of mineral fertilizers. In India, contribution of fertilizers towards increase in food grain production is estimated to be 50%. Farmers, their agricultural advisors, economists and policy makers all know this very well. What many seem less clear about is that simply adding some nutrients in excessive amount and ignoring the others altogether continuously is not only constraining crop production but the excessive use of any nutrient what the crop has already absorbed to capacity, is proving to be unproductive, expensive, wasteful and damaging to the environment. Simultaneously, utilization of those nutrients which were not replenished through external sources, in particular fertilizers, was mined causing severe nutrient deficiencies and thus constraining the crop productivity. So it is not so much ``fertilizer" that feeds people but the ``balanced use of plant nutrients through external sources mainly fertilizers" that will raise agricultural production and make more food available to a hungry world.

It has been estimated that fertilization accounts for nearly 50 per cent of all crop yield in India. In other parts of the world, where farm land has been abused for centuries or where new land is brought into production and quickly mined of its nutrients, fertilization might contribute as much as 75 per cent of total food production. Proper crop fertilization is essential to prevent massive global starvation. Herein, we should consider all the roles that soil plays in the production of food and fiber for the world's people. It is the medium in which plants grow and the source of most plant nutrients. Soil, water and air bathe plant roots and help keep them and above ground plant parts healthy and growing. The quality of soil in which plants grow is extremely important in determining yield as well as the sustainability of crop production. The key role of balanced used of fertilizers in maintaining soil fertility is well established. The present article deals with the real benefits of balanced fertilization for agriculture sustainability in India.

Balanced Fertilization Improves Soil Health: As science progressed, it was discovered that long term sustainability of crop production was dependent on building and maintaining soil fertility, an important soil quality measurement. Later, it was demonstrated that organic matter levels could be maintained and even increased through balanced fertilization. One of the greatest benefits of crop fertilization, aside from increasing crop yields and improving farmer profit, is its effect n soil organic matter. Harvested crop yields increase as a result of crop fertilization, as does unharvested plant biomass left on the soil surface and crop residues (roots) remaining in the soil. Most of the unharvested surface biomass and underground residues become soil organic matter positively influences structure, tilth, bulk density, water infiltration rates, water holding capacity, and water and air movement within the soil, thus improving soil quality. Organic matter helps to bind soil particles together, reduces soil crusting, increases the stability of soil aggregates, acts as a reservoir for plant nutrients, and reduces soil runoff and erosion losses.

Data from 12 long term experiments (LTE) conducted in India under the Indian Council of Agricultural Research (ICAR) Coordinated Project on Cropping Systems were analyzed to evaluate the effect of different sources of organic matter (farmyard manure (FYM) and green manure (GM)) in combination with inorganic fertilizer, on the productivity of rice wheat systems. The fertilizer treatments showed no significant effect on final yield, although the initial yield was significantly higher with 100% NPK than with FYM. The average rice yield with 100% NPK was still significantly higher than with FYM. In wheat, average yields and yield trends were not significantly different among the fertiliser treatments.

Other long term rotation studies in India have also demonstrated that moderate amounts of fertilizers increase soil organic matter quantity and quality. The positive benefits of fertilization have been directly attributed to the amount of crop residues returned to the soil. In addition to higher grain yields, fertilizer increased straw and root production, the precursors of soil organic matter. However, the realities surrounding short supplies of FYM because of burning of cow dung for fuel and the high labour and transportation costs continue to restrict its extensive and widespread application in agriculture. It is most likely that the most immediate solution to sustaining crop yields in the absence of adequate FYM supply will come from the regular use of site specific application of inorganic fertilizers.

Balanced Beyond NPK

Balanced fertilizer use today implies much more than NPK application. In India, almost 50 per cent of over 200,000 soil samples analyzed have tested low (deficient) in zinc. Soil S deficiencies once considered to be confined to coarse textured soils under oilseeds are now estimated to occur in a wide variety of soils in close to 130 districts and S induced yield increases under field conditions have been recorded in over 40 crops. Likewise, in specific areas, the application of magnesium (Mg) and boron (B) has become necessary for high yields, greater nutrient use efficiency and enhanced profits. These nutrient combination represent the many facets of balanced fertiliser use (Table 1).

Table 1. Balanced nutrient application for a number of soil/crop combinations in India

No.	Situation	Component of Balance
1	Alluvial soils, rice-wheat belt	N, P, K, S, Zn and B
2	Red and lateritic soils	N, P, K, S, B with lime
3	Areas under oilseeds	N, P, K. S, Zn and B
4	Man land area of Karnataka	N, P, K, S and Mg
5	High yielding tea in South	N, P, K, Mg, S and Zn

Source : Tiwari (2000)

Therefore, feeding crops for high yields in India is no longer a simple NPK story. This is no way minimizes the importance of NPK (fertiliser pillars) but emphasizes that the efficiency of NPK and returns from their application can be maximized only when due attention is also paid to other nutrient deficiencies.

Table 2. Effect of balanced (NPK) fertilization on agronomic efficiency of nitrogen (AE_N)

Crop	Control yield (kg ha^{-1})	N applied (kg ha^{-1})	AE_N		Increase in AE_N (%)
			N alone	+ PK	
Rice (wet season)	2,740	40	13.5	27.0	100
Rice (summer)	3,030	40	10.5	81.0	671
Wheat	1,450	40	10.8	20.0	85
Pearl millet	1,050	40	4.7	15.0	219
Maize	1,670	40	19.5	39.0	100
Sorghum	1,270	40	5.3	12.0	126
Sugarcane	47,200	150	78.7	227.7	189

Source : Prasad (1996)

Balanced Fertilization Improves Nutrient Use Efficiency (NUE) : Research has shown that when N is balanced with P, K and other essential plant nutrients, N use efficiency increases. That means more N is used by the crop and less is left in the soil as a potential pollutant. Thus, efficiency of applied N depends not only on the N applied, but also on the availability of other nutrients. When balanced fertilization is practiced, one nutrient often increases the efficiency of the other through synergistic interaction. Data from a large number of multi location experiments conducted under the ICAR's LTE project, and on farmer's fields under the AICARP, clearly bring out the importance of balanced fertilization in increasing NUE (Table 2).

Balanced Fertilization Improves Water Quality and Water Use Efficiency

A common perception among non agriculturists is that fertiliser use damages the environment, specifically water quality. The truth is that balanced fertilization goes hand in hand with high, efficiently produced crop yields, and environmental protection including soil conservation and water quality.

Proper nutrition helps to produce a healthy, fast growing crop having a vigorous root system and establishes a dense canopy to protect the soil surface, resulting in:

- less runoff and erosion;
- increased water infiltration to supply crop needs while boosting yields and slowing water decent to rivers, thus reducing flooding;
- more biomass left after crop harvest to help keep the soil stable and to contribute to organic matter levels.

By developing nutrient management plans and fertilizing according to complete soil tests, farmers help to assure that most of the fertiliser nutrients they apply are taken up by the crop being grown, not left in the soil for possible entry into nature's water system. Nitrogen and P are the only nutrients of concern with regards to potential water problems from fertilization. But, when used in balance with other essential nutrients when needed, such as K, secondary and micronutrients, within systems utilizing best management practices, there is little danger to either surface water or groundwater.

In order to protect water quality, care should be taken to avoid over fertilization. However, significant danger to water quality is also associated with too little fertilization. When crops are produced without proper nutrition, their growth is less robust, and they offer little protection from the potential impacts of wind and water erosion. If the crop can't take up the nutrients it needs because of low soil fertility or improper fertilization, erosion with the potential loss of soil P to surface water - is more common, as is N leaching into groundwater. Needless to say, farmers their produce lower yields per acre they can feed fewer people, their incomes suffer and their soil resource degrades, slipping them deep into a poverty cycle.

Water use efficiency can be increased by as much as 200% and more, simply by supplying essential nutrients in the proper balance. There is increasing competition between urban areas and agriculture for limited surface and ground water supplies. Thus, anything agriculture can do to increase water use efficiency is obviously a good thing for both sectors. People must have clean water to drink, but they must also have food to eat.

Balanced Fertilization : Important for Desired Food Quality and Premium Price

Balanced fertilization improve the quality of the produce and attracts a premium price. Balanced fertilization can bring prosperity to farmers and to rural communities. On a national scale, higher agricultural production can reduce the need for imports, may increase export earnings and, by encouraging a thriving rural economy, help to stem the flow of young people seeking their fortune in major towns and cities. When farmers incomes rise, they spend their extra money on non agricultural items, bringing, on average, more than double the value to the national economy, So balanced use of fertilizers brings benefits to the nation as a whole.

In India, people often say that the food quality has deteriorated due to increasing use of fertilizers. The old taste of the local varieties is being compared with changed taste of the high yielding/hybrids of cereals, pulses, vegetable and fruits and innocently and ignorantly, fertiliser is being considered as the main culprit for this change, People saying this fail to understand that they can't have the same taste of wheat and rice in high yielding varieties as to local varieties, similarly of local varieties of tomato and other vegetable crops in hybrids of these crops. If one plants Dashehari mango, can he get the taste of Deshi Mango. So where is the fault of fertilizers, is it not the effect of high yielding varieties and hybrids which they are growing in their own interest and also in the interest of the nation to have higher yield and profits and also to meet increasing demand of food and other commodities? Balanced fertilization, undoubtedly, makes nutrient use more efficient Leading to increased proteins, vitamins and mineral contents and also the taste. But this should always be compared with the produce of the same variety grown under two different nutrient management practices *i.e.* with imbalanced balanced fertilization.

Balance fertilization Reduces the Risks and Bad Weather: Certain yields are reduced in ``Bad years". That fact has to be recognized. As important as yields are to profitable production, the opportunity for reducing risk in those unpredictable years is even more important to the farms financial health. Good fertilizer management strategies minimize the drastic consequences of `bad' year effects. Adequate potassium helps in two ways: it offsets severe yield reduction caused by high stress conditions; and, it minimizes the cash flow and income problems which result with lower than anticipated yields. Better yet, adequate plant nutrients put everything in place for top yields in the years with excellent growing conditions. The important role of potassium in mitigating abiotic and biotic stress is well established. Unfortunately, in the entire history of Indian agriculture use of potash never exceeded 10% of the total nutrient supply through fertilizers with the result mining of soils, potassium reserves is continuing at an alarming rate. The estimated share of potassium in total nutrient mining in Indian agriculture is about 55%.

Chapter - 34

Land Use and Agrarian Relations

Francis Kuriakose & Deepa Kylasam Iyer

Land is a finite resource and there is conflicting and competing demands on it. For 80% of the world, agriculture land is the primary source of life and livelihood. India holds 2.4% of the world's geographical area (328.73 mha) but supports 17.5% of the world's population. India is home to 18% of the cattle population of the world while owning a mere 0.5% of the total grazing area. Of the total 328 mha (total geographical area). Land use statistics is available for approximately 305 mha (93%) of the total land. 228 million ha (69%) of its geographical area falls within dryland that encompasses arid, semi-arid dry and sub humid land as per Thornthewaite classification.

India is blessed with a wide range of soil pattern, each particular to the locale. The alluvial soil (78 mha) that covers the great Indo-Gangetic Plains, the valleys of the rivers Narmada and Tapti (Madhya Pradesh), the Cauvery Basin (Tamil Nadu) supports cereals, oil, pulses, potato and sugar cane. The Black Cotton soil (51.8 mha) found in Maharashtra, Gujarat, Madhya Pradesh, Uttar Pradesh, Karnataka, Rajasthan and Andhra Pradesh supports cereals, cotton, citrus fruits, pulses, oil seeds and vegetables. The red soil of South India and Madhya Pradesh. West Bengal and Bihar supports rice, millets, tobacco and vegetables. The laterite soil (12.6 mha) and desert soil (37 mha) are not found suitable for agriculture.

Water is a resource precious and scarce in India. The variability of precipitation spatially and in quantity can be inferred by the fact that rainfall has been recorded as low as 100 mm in West Rajasthan and 9000 mm in Meghalaya in North Eastern India. India receives 4000 cubic kilometre of precipitation in the country in its 35 meterological sub divisions. Of this amount, only 50% is put to benefit due to topographical and other constraints. The fact that water is crucial to agriculture in a country that has 68% of its net cultivated area as rain fed, can hardly be exaggerated. Of the total cultivated area of 142 mha, 97 mha is rainfed. The full irrigation potential of the country has been revised to 139.5 mha out of which 58.5 mha is watered by major and minor irrigation schemes,

15 mha by minor irrigation schemes and 40 mha by groundwater exploitation. India's irrigation potential increased from 22.6 mha (1951) to 90 mha (1995-96) but water usage efficiency is a meagre 30-40%. That is why more than 50% of the total cultivated area is still rainfed. The state of soil and water that mainly determine land and its utility in agriculture is of prime importance to maintain sustainable development. We need to define and examine land use pattern with an emphasis on a viable land use policy taking the above factors into consideration.

Land Use Pattern

Land Use Pattern is determined by physical, economical and institutional framework, i.e. the action and interaction of the physical characteristics of land, the economic factors like capital and labour, location of land with respect to factors of infrastructure like transport and institutional framework that determines the interrelation between all the factors involved. In other words, land use pattern is a complex phenomenon determined by the dynamic equilibrium of factors of agrarian relations, economic development, infrastructure and policy making. It is the synthesis of physical, chemical and biological process on one hand and human process on the other.

The pattern of land use in India can be determined by looking at the post independence scenario. Till 1949-50, land area was divided into a five fold classification. This was inadequate to meet the agricultural demands as there was lack of uniformity in definition and scope of classification. Hence it was difficult to compare and utilise the classification to improve the existing land pattern. To break up the existing tracts of land into smaller constituencies for better utility and monitoring. The Technical Committee on Co-ordination of Agricultural Statistics (Ministry of Food and Agriculture) recommended a nine-fold use of land in the country. There was the area under agriculture that was the mainstay of farmland. Three fourth of this area was shared by the states of Bihar, Gujarat, Madhya Pradesh, Karnataka and Maharashtra with Maharashtra topping the chart with the highest percent of the net sown area. The area under non agricultural use comprised the land under water, land used for the construction of buildings, roads, railways and barren agricultural land. The area under forest was 76.52 mha (State Forest Department, 1999). It was classified as Reserve, Protected and Unclassed. Using Remote Sensing, Technology, it was ascertained that the actual forest cover was only 63.73 mha. The ownership of forest land was left to the Government of India and community clans wherever applicable. The per capita availability of forest land was 0.08 hectares whereas the optimum area of land required for meeting the basic needs was 0.47 hectares. This immense pressure on forest cover led to the search of potential areas for expansion of forest cover in culturable land tracts. 13.94 mha of the total land form wetland, fallow land and land put to other uses. Forests form an important part of land use. Land allocation for forestry include forest land and land allotted for agro forestry, farm woodlots, wind belts, shelter belts, avenue trees, urban forests, homestead forests and sacred groves. The state of Natural forest in India can be deciphered from Table 1.

Table 1. State of Natural Forests in India

Area of Natural Forest	51.3 mha
Total growing stock in Natural Forests	2431.30 million cu.m
Total biomass in Natural Forests	4805.7 million tonnes

Source : NFAP, MOEF, Government of India, 1999

Forests in India show the greatest variation and range depending rainfall topography and climatic factors. Forests are both a resource and a habitat and of the 16 detailed forest types given, 38.2% is topical deciduous forests and 30.2% is moist deciduous. The benefits of natural forests include soil protection, fertility water flora and fauna conservation, microclimate, genetic resource conservation, use of genetic breeding and bio technology, integrated watershed management and regeneration of eco systems.

11 mha of the total land comes under permanent pastures and grazing lands. Rajasthan, Uttar Pradesh, Madhya Pradesh. Andhra Pradesh and Orissa cover 75% of the grazing land in India. The forests of India support 40% energy needs of the country out of which 80% needs are in the rural region and 30% fodder needs of cattle remain significant. The live stock statistics of India given in the table is relevant in this context. It is evident that as land remains constant, the increasing livestock population and their needs could be met only with judicious planning and sustainable use of land.

Table 2. Livestock population in India

Year	Total livestock population in (000)	Cattle (in 000)
1977	369,645	180,140
1982	419,742	192,453
1987	445,286	199,645
1992	470,860	204,584

Source : Agricultural Statistics at a glance, 2001, Directorate of Economics and Statistics, Ministry of Agriculture, Government of India. Note that livestock includes cattle, buffalo, sheep, goat, horse, pig, donkey, mule, camel, yak and mithun.

Area under common Property Resource (CPR) includes the land that caters to the basic needs and services of the vulnerable sections of the rural poor. This includes village forestry, grazing and watershed drainage to help the farmers in crisis. CPRs have property rights in the land allocated, wasteland is the ecological characteristics coined to initiate developmental programmes for the recovery of degraded lands irrespective of property rights. Velayutham (2000) has shown that the area under CPR has diminished during the period 1950-1997. Grazing pressure, land degradation resulting from a burgeoning cattle population that increased from a livestock population of 292 million to 462 million during the period resulted in the gross erosion of CPR changing them into wastelands.

Case for Land Use Policy

The way land is used as a means for life and livelihood is not just dependent on the defect users, it is exposed to a wider realm and is decided by all the factors directly and indirectly involved. One of the main problems that is faced today is the depletion of the quality of land and land degradation. Approximately 5-7 million hectares of usable land is lost every year through land degradation. The relative influence of land degradation is 39% in Asia. This translates to half a billion people in the developing world with no irrigation facilities, 400 million living on soil unsuitable for agriculture, 200 million on slope dominated regions and 130 million in fragile forest ecosystem. 73% of the earth faces severe and significant problems in agricultural investment while trying to sustain a rising population. A recent pioneering study by three UN agencies including FAO, UNDP and UNEP estimate the severity and cost of land degradation in South Asia to be 2% of the Gross Domestic Product of the region and 7% of the agricultural output. The statistics given below reaffirm the finding.

Table 3. Extent of Land degradation in India (area)

Source of Erosion	Area in mha
Water	103.90
Wind	13.10
Physical Agents	12.23
Chemical Agents	10.30
Other Agents	7.20

Source : National Bureau of Soil Survey and Land Use Planning.

The rising trend in land use degradation can be attributed to the following reasons

1. Deforestation

2. Inadequate land use
3. Unsustainable farming and grazing practices
4. Demographic pressure
5. Lack of adequate technology implementation
6. Markets and legal instruments
7. Climate fluctuation

Demographic Pressure

Demographic pressure is one of the foremost reasons of land degradation as increasing population puts more pressure on arable land, grazing, forestry, wild life, tourism and development. Not surprisingly, population demands for food, fuel and employment is going to double in the next five decades. This will involve expansion of fragile marginal lands for utility in developing countries as poverty is endemic and institutional capacity for land management is weak. Urbanisation and industrialisation outstrips land capacity. There are serious concerns about land, environmental degradation, decreased productivity and growth rate in the developing world. The population of 1.3 billion living on fragile land is set to double. the vulnerable segment of the population notably the rural poor with moderate assets, land, tradition social capital, human capital and indigenous knowledge are not developed by the institutions. These invisible millions living in disperse settlements in an informal economy are not picked up by the development juggernaut. They lay neglected along with the environmental distress signals.

Land degradation as a result of External features

The net value of land is the sum of two factors the present value of the revenue stream and the present value of the terminal value of land. There are a number of factors that diminishes the value of land. Intensive farming practices are the foremost among these. Green revolution in India brought in petrochemical technology, pest intensive agricultural method, cross breeding and single species forest plantations which were mindlessly adopted from other parts of the world. Over application nitrates has led to groundwater contamination, soil degradation and an imbalance in micro nutrients. The extension of area under irrigation has jumped from 19% to 38% in terms of net sown area in four decades. This has led to water logging and salinity. National Remote Sensing Agency and Forest Survey of India has brought out the fact that 60% of the total area under cultivation is degraded. More than one source of irrigation has increased the salinity and alkalinity of soil. Low precipitation coupled with unscientific use of water and drainage facilities take a toll on water resources. Improper cropping patterns and intensive farming practices degrade the quality and value of land.

The consequences of large scale and degradation are two-folded.

The on site costs The technological break through that the Green Revolution offered led us to produce short duration high yielding crops. Intensive land use, increased area under irrigation, prolific use of chemicals to raise the efficiency of production also brought in on site costs like soil erosion, alkalinity, salinity, micro nutrient deficiency, water logging, depletion and contamination of ground water.

The off site costs The off site costs include river and dam siltation, damage to roadways and sewers, siltation of harbours and channels, loss of reservoir storage, disruption of stream ecology, damage to public health and increased frequency of flooding.

Policy Intervention

The rationale for policy intervention should be based on two factors

i) The significance of off site costs as a result of land degradation

ii) The costs of on site degradation even when it is not apparent in the immediate context

This requires a foresight and vision for long term sustainable development through policies, action and awareness brought out through education, training and extension programmes. The objective of the policy intervention should be the following

i) Restore efficiency to meet the growing consumption needs

ii) Suitable mechanism for scientific management, conservation and development of land resource

iii) Expansion of forest cover to restore ecological balance

iv) Conjunctive use of surface and ground water

v) Preservation of agricultural land

The Integrated Approach

For effective and efficient use of land we need eminently practical plans for land use management. this is included in the integrated approach. To reduce the conflicts and to make trade offs link social and economic development with environmental protection, sustainable development is the key. The essence of integrated approach is the sectoral planning management. There are a number of issues to consider while adopting approaches and policies. For land use pattern through sectoral approach, we need to plan linkages, formulate economically viable project for each sector and use technology. This would include making Land Use Atlases, system database on land utilisation land records at district, state and national levels. Better legal, political and administrational will is also the key.

We need strict laws for land use conversion, survey of land based on climate, water and soil particulars to improve investment and training orientation, publicity and awareness based on local needs. Effective reclamation is needed to check degeneration. This can be done through effective watershed management, reduction of regional imbalances and diversification of land use. Preventive measures on adverse effects from industrial wastes and effluent and development of agro-based industries are also keys to developing an integrated approach.

To monitor the better use of land, Remote sensing satellite technology like Geographical Information system and Global Positioning System can be used. One of the Problems frequently encountered while measuring the loss of land value is the difficulty in measurement itself as there are so many variables involved. Empirical or process based models have to be so complex to take into consideration the effects of all the variables. One of the methods is to estimate long term average annual soil loss from arable land using Universal Soil Loss Equation (USLE) or its revised form (RUSLE). There are various mathematical simulation models based on physical process involved in soil detachment, transportation and deposition. Use of Iso-erosion rate map (Singh *et al.*, 1992) is an example. Soil erodability factor can also be measured. Loss of soil value due to land degradation is needed to understand the environmental costs of agriculture. Production approach that assesses the impact, preventive cost approach that focuses on conservation and defensive expenditure and replacement cost approach that relies on the cost of restoration are the different ways to measure this. There are various econometrics models that can include and evaluate the inputs for alteration and cropping pattern. In India, soil and land survey conducted by Department of Agriculture and co operation developed land degeneration mapping in the eighth five year plan through District Information system where soil information system of 30 districts in diverse agroclimatic zones were formulated. Similarly, the Department of Land Resources, Ministry of Rural Development has brought out the Wasteland Atlas of India 2000 after studying different types of degraded wastelands in the country.

Reclamation of wasteland is one of the most important aspects of sustainable land use. Agrarian practices can be modified for reclaiming wasteland. For example, application of gypsum consecutively for three years with reduced application in the second and third year will reduce salinity. Integrated watershed management is a preventive method in which soil and water is conserved and cropping pattern is altered to improve land use. Percolation of water into subsoil, reduction of surface water run off, elimination of soil erosion and increase water availability are the chief aims of such sustainable management practices. For attaining these objectives, check dams along gullies are constructed, bench terracing, contour bunding, land levelling, planting grass along the contours, good vegetal cover on the watershed are deployed. Difference can be brought through Governmental Intervention and policy making. The Soil and Water Conservation Division, Ministry of Agriculture plans to manage 86 mha under 30 projects through Integrated Water Management. 30,000 hectares of shifting and semi stable land dunes have been treated with shelter belts and strip cropping as a conservation measure (TERI Report, 1997).

The National Land Use and Wasteland Development Council (1985) was set up with the objective of formulating a National Policy and Perspective Plan for Conservation and Management of Land Strategy. It is time to set right some policies unsuitable for sustainable development. For example, the governmental policy of heavily subsidising electricity for tube well irrigation and chemicals led to poor land quality and eventual abandoning of land. Similarly, the New Economic Policy that encouraged relaxation on land acquired by Non Resident Indians, conversion of agricultural land into non agricultural land, ceiling of agricultural land holdings eventually led to distorted market value due to speculation. The encouragement given to export oriented agriculture and concessions given to agro processing industry adversely affected Indian agriculture by increasing the investment costs. Rational Policies to face regional imbalances should be brought in. The commitments of Tropical Forestry Action Plan, World Food Programme, UNCED led Forest Principles and the Government of India's National Conservation Plan should be adhered to. Rational Pricing Policy combined with resource efficiency in agro-processing industry is the need of the hour.

Economic incentives for soil conservations practices, conjunctive use of chemicals with biological inputs, classification of Land use statistics and studying the land use impact of agriculture will help at the macro level. Use of remote sensing technology to study different dimensions of the problem is mandatory Legislation is in place for conservation of bio diversity and forestry but not to protect soil relations. Such gaps in law should be filled in with appropriate legal protection. New technology and crop management practices should emphasise the integrated system approach. Meaningful farm research practices will address the concept of linking agriculture with environment. The aim of agriculture should be sustainable crop production with enhanced production envisioned for the long term. Diversification of agriculture should be encouraged. Farming oilseeds and pulses in place of cereals and horticulture wherever applicable demand less water and encourage crop rotation. This permits an understanding of agro climatic conditions, favourable topographic conditions, efficient land use, conservation of soil and maximum use of land resources. Integration of farm forestry with agro forestry will reduce the tremendous pressure on land. Growing a combination of species like agri-silviculture, farm and grove system will make management approach complementary, improve biomass production, regeneration of land resources and increased generation of employment and income.

Thus integrated and sustainable land use comprises prioritisation of critical land sensitivity, understanding land use and forest response, integrated strategy for forest and pest management, diversification of agriculture, crop combination, use of people's indigenous knowledge to attain food and nutritional security, increased productivity and address the environmental concerns. This is the way forward towards an evergreen revolution.

Chapter - 35

Land Acquisition in India Need for A Paradigm Shift

Dr. L. Rathakrishnan & K. Ravi Kumar

Land is the base for economic development and poverty alleviation of a country. In recent years, land acquisition for private sector projects and Public Private Partnership (PPP) projects like Singur, Nandigram, Yamuna Expressway, POSCO, etc created a lot of noise. Few lakhs crores rupees of investment is hanging in balance in the country from both domestic and Foreign Direct investment (FDI) sources because of failure of government to provide land for the projects and also failure of the land market to provide sufficient land for development. Is land acquisition process in India seriously flawed? Is the Right to Fair Compensation Resettlement Rehabilitation and Transparency in Land Acquisition Bill 2012 (RFCRRTLA Bill 2012) solution to this problem? Are land institutions of India not market friendly in the post 1991 economic reform era? These questions, which provide the basis for this paper, are examined through field observation and field experience of the authors.

Doctrine of Eminent Domain

The power of the sovereign state of acquire or expropriate private property for public use/purpose is driven from doctrine of Eminent Domain. The origin of the term ``Eminent Domain" can be traced to the legal treatise written by the Dutch Jurist Hugo Grotius in 1625 and described as follows :

The property of subjects is under the eminent domain of the state, so that the state or he who acts for it may use and even alienate and destroy such property, not only in the case of extreme necessity, in which even private persons have a right over the property of others but for ends of public utility, to which ends those who founded civil society must be supposed to have intended that private end should give way. But it is to be added that when this done the state in bound to make good the loss to those who lose their property (Neil, 2009).

Almost all sovereign states in the world have law for land acquisition or expropriation. Pakistan and Bangladesh are using the same land Acquisition Act 1894 (LA Act 1894). Even through all sovereign state are acquiring or expropriating private properties, why land acquisition becomes a hindrance for economic development in India? The fundamental conceptual difference is defining the purpose of land acquisition : public use Vs public purpose. Most of the western countries acquire land for public use like roads, public purpose. Most of the western countries acquire land for public use like roads, public safety, health, etc and not for the project in which private profit motive is involved even though project has public purpose. On the other hand in UK common law system, land is acquired for public purpose, which is followed throughout all Commonwealth Nations including India.

In Indian Jurisprudence also, when LA Act 1894 was enacted public purpose included in the definition was roads, canals and social purpose of state run schools and hospitals. By an amendment in 1933, railway companies were included in public purpose. But the amendments introduced in 1984 in the LA Act 1894 by amending section of the original act to insert the words ``or for a Company" after ``any public purpose". This opened the floodgates to acquisition of land by the state for private and public sector companies and again this is embellished in the proposed bill. If we put ban on land acquisition for private projects and PPP project with present land system in India, we strongly believe, the economic development of India will be seriously affected because of inherent problem in our land institutions.

Land System of India

The problem of land acquisition in India can be better appreciated by understanding the land system of India. Modern day land system of India has its base from the land revenue system introduced by Akbar's revenues minister Todar Mal. The Salient features of Todar Mal's systems were measurement of land, classification of land and fixation of rates (Appu, 1996). After the decline of Mughal dynasty, East India Company and the British Raj were established extractive land institutions on the Todar Mal principle called *Zamindhari* system, *Mahalwari* system and *Raiyatwari* system to extract maximum land revenue from peasants, which was the major source of revenue. In the *Zamindhari* area, British had not hold elaborate administrative arrangement and lowest functionary level was sub divisional level and no proper land records maintained either by British administration or by *Zamindars*. Only land record maintained was land record created after each survey and settlement operation and again by revisional settlement. Because of this reason, elsewhere *Zamindari* area does not have proper land records and weak administration. In *Mahalwari* areas, the land revenues were fixed for each or group of villages in which one family or person who was responsible to collect and pay land revenue.

The Raiyatwari system covered the erstwhile Madras (except North Madras) and Bombay Presidencies and Part of the central provinces and Barer. The *Raiyatwari* system was based on the assessment of land revenue on sight fields or holdings, surveyed, numbered and marked out on the ground (Appu 1996). Because of elaborate arrangement for revenue collection and administrative step created during British Raj, these areas of India is having better land records than rest of India even today. Another wisdom of British was creation of primitive land institution in excluded and partially excluded areas to separate tribal and others deprived people of these areas with plain and Hindu population by perpetuating divide and rule policy. This primitive land institutions created by British was responsible for creation of scheduled V and VI areas and poverty and deprivation of these regions.

During the first four five year plan periods, India introduced radical land reform on socialism land reform model to increase agricultural production and to provide social justice without any role for market forces. During this land reform period, there was no respect for private property rights and no land institutions was created for allocation land resources for industrialisation and urbanization through market forces. Till date, land system of India is suited for subsistence agriculture using manual labour and does not have major provision for industrialisation, urbanization and mining activities. By introducing *Zamindari* abolition law and ceiling law on agriculture land and also on urban land into the India's land system, Indian land holding become too small and restriction on transfer and lease, which further reduced the size of holding. In the name of distribution of government land and redistribution of surplus land, we distributed waste land, barren land and dry land for agriculture which could have been kept as construction land. This led to non availability of large plots of land in thousands of acres for industrialisation and urbanisation.

Since 1980s, the land reform in India was abandoned as lost cause. Only visible activity in land resources is computerisation of land records, that too also happening only in already developed states. Through tenancy or tenure system reform, the land market of India become more of socialist model and become less suitable for market economy. For example, under Santhal Pargana Tenancy Act 1949, no land is transferable, leasable, mortgageble in six districts of Jharkhand. Under Bombay Tenancy and Agricultural land Act 1963, only farmers can own agriculture land in Maharashtra. Land system of India is not friendly for industrialization, urbanization, mining etc. Thus, land acquisition through government is the only option for large private sector projects and PPP projects. In short, land acquisition procedure cannot be seen in isolation from the land system of India. By orienting the old age land system designed during British Raj and socialism period toward market economy, we can allocate land resource for industry, mining, infrastructure projects and urbanisation without creating much noise in our democracy. Can we design a land system for India which can allocate just three per cent of its geographical area for Industrialisation which is amount to 2.43 crores acres of land?

Multiple Land Acquisition Laws and Multiple Authority of Land Acquisition

On the basis of Doctrine of Eminent Domain many land acquisition laws were enacted in India. The Central and State Acts are :

(1) The land Acquisition (Mines) Act 1885

(2) The Indian Tramways Act 1886

(3) The Land Acquisition Act 1894

(4) The works of Defence Act 1903

(5) The Damodar Valley Corporation Act 1948

(6) The Resettlement of Displaced Persons (Land Acquisition) Act 1948

(7) The Requisitioning and Acquisition of Immovable property Act 1952

(8) The National Highways Act 1956

(9) The Madhya Pradesh Municipal Corporation Act 1956

(10) The Coal Bearing Areas Acquisition and Development Act 1957

(11) The Ancient Monuments and Archaeological Sites and Remains Act 1958

(12) The Atomic Energy Act 1962

(13) The Petroleum and Minerals, Pipelines (Acquisition of Right of user in Land) Act 1962

(14) The Metro Railways (construction of works) Act 1978

(15) The Railways Act 1989

(16) The Electricity Act 2003 read along with the Indian Telegraph Act 1885

(17) The Special Economic Zones Act 2005 and

(18) The Cantonments Act 2006

Beyond this knowledge, there are some state laws for acquisition of land for state highways and also these acts have undergone number of amendments. Multiple acts in India lead to multiple authority for land acquisition. All legislation had its own procedure for acquisition and calculation of compensation and this led to confusion in the ground zero and also multiple land records of land acquisition. Multiple land acquisition in one district under different acts at the same time not only creates confusion among land losers and requesting agency and also revenue departmental officials and staff in the district.

Some of the acts listed above are not in tune with market economy and out dated. When the Coal Bearing Areas acquisition and Development Act 1957 enacted, India was a Socialist country and factors of production are mostly nationalised including coal mining. The Coal India Ltd was given authority to acquire land and rehabilitate the affected persons through appropriate compensation. After disinvestment in the post 1991 economic reform, The Coal India Ltd., is a listed company in the stock exchange; it has one of the highest market capitalisation of the country. The Rehabilitation and Resettlement (R &R) policy of the Coal India Ltd., and its subsidiaries are decided at their board meeting. The motto of a listed company in the stock exchange in any market economy can be maximisation of production at minimum cost. It can't be expected that the welfare activity to the projected affected people is one of the main function of the company. We do not understand the rational of retaining the power of eminent domain with a listed company which is a prerogative of a sovereign state. Again, The power Grid Corporation of India Ltd has authority to erect and pass over anybody land without paying any compensation to the land except damage for crop under electric wire (The Electricity Act 2003 read along with the Indian Telegraph Act 1885). Once the high transmission tower is erected, land owner can't grow trees below the transmission lines and can't construct building. Again, each transmission tower is erected on the land, which occupies around two decimals of land but the land owner is not eligible for any compensation for land and he is eligible only for damage to standing crops of that season only.

Just Compensation for Land

Calculating just compensation for land in land acquisition is a very difficult process throughout the world because there should be a balance between land losers demand and willingness of requesting body to pay. under Indian land acquisition process, land price is calculated without any negotiations between land losers and requesting body and it is done through mechanical calculation. One of the important highlight of the proposed RFCRRTLA Bill 2012 is that the land owner in the rural areas will get four times the market value and the land owners in the urban area will get two times the market value. When we see these promises in the contest of land system in India, it is very difficult to realize practically. Just assume, the highlights of compensation of four time and two times of market value is true, as the case may be, then, this compensated land owners in rural areas can purchase four times the land of similar nature in the same locality and also land owners in a urban area can purchase two times the land of similar nature in the some locality. But this assumption is false because of the following field reality in the land system of India :

(1) Clause 26 of the RFCRRTLA Bill 2012 authorises the District Collector to determine the market value. Value mentioned in clause 26(1) (a) is not market value but what is known in common parlance as the circle value. This circle value can by no means be called as the market value and it is always much lesser than the market value.

(2) As soon as notification is issued, land registration/sale in that area are stopped. As per the proposed bill, the land acquisition process has to be completed within two years from the date of notification excluding the days wasted in court proceedings. Again, the District Collector will calculate this price average of 50 per cent of highest sales deed of last three years preceding this notification. Assume that there is no court cases, the price arrived by the District Collector will be at least more than 3.5 years older sales deed values. In other words, what this District Collector calculated is 3.5 years old historical sales deed value but the demand of the land owners are the present market value of land.

(3) In order to avoid stamp duty, income tax payable and also to park black money, the value of land/property mentioned in the sales deed is always much below the actual price paid. Even if the land looser able to purchase a land/property using the compensation money, the purchaser has to pay stamp duty, registration fee, land broker charges in the unorganised land market, speed money at the registration office and other miscellaneous expenses related with document registration. After this document registration, the land loser has to approach the revenue authority for mutation of land in order to create patta in his name. For this mutation, he has to pay the mutation fee, speed money to revenue authorities and other miscellanies expenses related to this mutation process.

(4) Land market in India is un organised and speculative market. The land price in the locality increases as soon as a new project is announced. Very few projects in India, which led to fall in land price. Due to long gap between the announcement of project and actual receiving of compensation amount, the land losers will find it difficult to purchase land in the locality by that time land price might have increased many fold.

(5) All land acquisition for urbanisation involves re classification of land from agricultural category into commercial using the discretionary power given to government which led to many fold increase in land price without doing any development work on the land. This price increase led

to allegation of quick money and ministers, builder and bureaucracy nexus and farmers got cheated in terms of paying less.

Taking all above five factors into consideration, we doubt, the land losers can be able to find a property of equal nature in the similar locality or nearby. Again when we look at the land institutions in the scheduled V and VI areas of constitution of India and other tribal areas, there is no proper land market. This area includes Jharkhand, Chhattisgarh, Orissa, Part of Madhya Pradesh, six districts of Andhra Pradesh, tribal belt of East Gujarat, and the whole of North Eastern India, which are poorest region in the country. In general, lands in scheduled V areas are inheritable, only partially transferable with lot of condition. Lands in scheduled VI areas are inheritable, not transferable and land is owned by Clans/Community. Land laws in other tribal area like Manipur, Nagaland, is also similar. Again, the RFCRRTLA Bill 2012 says, land value of scheduled V and VI areas will be decided by state government. In principle, any price fixation administratively and arbitratively by a government is not prudent in market economy and these process of land price fixation does not reflect the real value of the land. Having negatively criticised the fixation of compensation, we strongly believe that the compensation proposed in RFCRRTLA Bill 2012 is much better than the LA Act 1894 and other acts in the country. However the misleading word ``market value" in the RFCRRTLA Bill 2012 should be replaced by ``circle rate", sales deed price" and ``just compensation" at the appropriate places.

Lack of Developed Land Market

Land Institutions of India do not support developed land market for allocation of large plots of land for industry, urbanisation and even for large scale commercial farming. Elements like ceiling on agricultural land and urban land, restriction on lease in and lease out of land, restriction and even ban on transfer of land, tribal tenure/tenancy system, restriction on transfer of land by land reform beneficiaries of ceiling and surplus land, ban on transfer of allottees of government waste land, changing the nature/classification of land are acting as hindrance for development of land market in market driven economy. Industrialist and Builders are dependent on government for allocation of large plot of land for development purpose. In Scheduled V and VI areas, there is no land market at all and these regions are poorest region of the country. There is no provision for allocation of land for other sectors of economy through market forces like industry, urbanisation, lack of movement of land resources from inefficient user to efficient users and credit market incompatibility in these areas.

Land for Land Demand

Land for land is a difficult demand in land acquisition in India. Normally with present land system in India, conceding to this demand will create a vicious cycle of land acquisition and there will be no end to it. RFCRRTLA Bill 2012 tried to address this issue by providing one acres of land in command area for each irrigation project affected families and 20 per cent of developed land from urbanisation project in cost basis. When we go deep into irrigation projects, in general more land is acquired in catchment area, which is normally located in hilly terrain areas for water storage and construction of dam, and less land is acquired in command area for canals and distributaries. Providing one acre of land in command area by again acquiring land means, the land loser in hilly terrain has to be shifted to plain area where land is allotted. Is this proposal will be acceptable to hilly terrain people? It will again start the cycle of land acquisition in common area. in case of urbanisation project, 20 per cent of developed land will be given to land losers on cost basis, then who will pocket the profit of the rest 80 per cent of developed land in the urbanisation project, which is having less gestation period project and less cost involved in development of land. When we again analyses this issue and being working in this field, the demand for land for land arise due to two major reasons, namely:

(i) Most of the land for land demand arises from scheduled V and VI areas of costitution of India and other tribal areas. Land in these areas is mostly inheritable and it has little or no transferability. Most of the land looser, land once lost is lost forever, neither they themselves nor their future generation can purchase land from the land market, because there is no developed land market or land market at all. For example in Jharkhand The Santhal Pargana Tenancy Act 1949 (SPT Act) covers six district of Jharkhand namely, sahibganj, Pakur, Godda, Dumka, Deoghar and Jamtara. Section (20)(1) of SPT Act says ``no transfer by a raiyat of his right in his holding or any portion here of, by sale, gift, mortgage, will, lease, or any other contract or agreement, expressed or impiled, shall be valid unless the right to transfer has been recorded in the record of rights, and then only to this extent to which such right is so record" (Prasad, 2007). Law like this makes impossible for a land loser to purchase another piece of land using the compensation amount. Everyone knows the story that court in the Uttar Pradesh ruled that Amitabh Baachan is not a farer and hence he is ineligibles to purchase agricultural land. Here, there is mutual consent of both parties for consideration, why should state intervene? This distortion in land market is making land acquisition difficult and strengthening demand land for land.

(ii) As discussed in just compensation for land, the amount compensation paid in land acquisition is not sufficient to purchase same amount of similar nature land in similar location, in many cases.

False Food Security Alarm

Again, whenever any land is acquired or government decision is taken for declassification of any land from agriculture to commercial use, there is point that agricultural land is being acquired or declassified; it will lead to food insecurity. Problem of food security in India is not due to unavailability of agricultural land, but due to low productivity, lack of storage and transportation facility, lack of access of poor person to food grains and lack of incentive for farmers to cultivate. As industry and urbanisation needs large amount of land in thousands of acres in single stretch, land acquisition and declassification agricultural land is inevitable. On the other hand, agricultural land with irrigation facility is costly; a well meaning business man will not prefer an agricultural land when he has option to purchase a waste land or dry land.

Rehabilitation and Resettlement

Rehabilitation and resettlement is a pet topic for civil society activists, public interest litigants and mass media. As a nation, we have to confess that we are doing a raw deal to our land losers in land acquisition in the name of economic development and public purpose. RFCRRTLA Bill 2012 included components of R & R in the land acquisition legislation. However, there is a legal discussion, whether parliament has jurisdiction to legislate on transfer and alienation of agricultural land, which is a state subject under item 18 of state list in the seventh schedule of constitution. It is very difficult for us to understand the rational for inoculation of R & R in land acquisition legislation. First, India's diversity of land laws is much more than the diversity of the country and needs of the project affected people varies from project to project and region to region. Second, law on any system/subject brings in rigidity to the system. These issues can be better handled by guidelines and policy, which is more flexible.

We have to understand that the states are governed by democratically elected governments and responsible to their electorate directly. They are closer to the pulse of the people and have greater familiarity with the ground level situation. They can better safeguard the interest of the communities whose land is being acquired on the one hand and the requirement of the project conceived in the national interest on the other hand. When we compare the Jharkhand Voluntary land Acquisition Rule 2010 enacted under section 11(2) of LA Act 1894 along with the Jharkhand Rehabilitation and Resettlement policy 2008 on the one hand and recently introduced RFCRRTLA Bill 2012 on the other hand, former will be more suitable for local ground condition and flexible to the demand of local community and the requirements of the specific projects. In the post economic reform period, there is a fierce competition between states to attract investment into their state. Wisdom of state government will be in better position to balance between incentive for industrialist to attract investment and fulfilling the aspiration of electorate taking into consideration of ground reality.

RFCRRTLA Bill 2012 also included the private company which purchases land directly from market without government support if the area of land is more than 100 acres in rural area and 50 acres in the urban area within the ambit of R & R package. This provision not only violated. The Indian Contract Act 1872 and the Transfer of properly Act but also failed to understand the competition of industry in the globalisation era. In this globalisation era, new industries with R & R package not only have to competitive with other similar industry in India without R & R condition and also similar industry throughout the world. This provision takes back us into the socialist era and it will create a new type of dispute called project affected people disputes for which we do not have specialised body for dispute resolution. And in addition to this, the RFCRRTLA Bill 2012 included non land losers under R & R ambit, which will make the identification of beneficiary difficult and local unrest will be order of the day.

Transparency and Accountability

It is quite often said in land acquisition discussion that agriculture land or farmers lands are being looted at cheep rate and exploitation of poor land owners. In the proposed RFCRRTLA Bill 2012, there is a proposal to establish new institutions like National LA & RR Dispute Settlement Authority, National Monitoring Committee, State LA & RR Dispute Settlement Authority, Chief Secretary Committee, State Commissioner for RR, District Collector Committee, Administrator for RR, RR Committee and new documentary requirements like Social Impact Assessment Report, Expert Group Report, Collectors Report about non-availability of waste land for industry location. In the process of establishing these institutions we forgot about single window system of clearance and hassle free clearance for industrial projects in the name of transparency and public participation. We believe, this mandatory provision will lead to more corruption and delay at all levels. Mischief mongers will become very active in land acquisition process and it will lead to increase in number of judicial intervention on blimpish grounds. The whole system will lead to delay in payment of compensation to land owners and economic development will slow down. In India, we are having a large number of transparency enforcing institutions like CBI, CVC, State Vigilance Dept., Information Commission, Enforcement Directorate, Committees of legislative bodies, Departmental Proceedings if Civil servant is erroneous, High Courts and Supreme Court, etc. Have we lost faith in these institutions? This issue reminds us a unwritten law (flaw) in the Indian bureaucracy, "Corruption and delay are proportional to number of dharogas (Inspectors) appointed but never failed to appoint dharogas, otherwise you will be blamed for lack of vigilance".

Giving complete recipe for an inclusive land institution in a short article is very difficult, however, the following summarisation of foregoing discussion on land acquisition and related issue for allocation of land resources for industrialisation, mining, infrastructure projects and urbanisation will provide a bird's eye view.

(i) Combine all central and state land acquisition legislation into one land acquisition legislation with power to state government to bring amendments according to their diversity in land laws and local conditions. Public purpose/use should be well defined. R & R should be delinked from LA Act.

(ii) Strengthening the provision of section 11 (2) of LA Act 1894 which provides for consent award. Some redundant sections of LA Act 1894 should be removed to reduce delay in payment of compensation to land owners.

(iii) Long term lease of land for 15 to 30 years should be provided in land acquisition procedure for purposes like mining or temporary purpose in which land can be return to the land owner. If land is not returned after expiry of lease period, again compensation has to be paid at current market rate.

(iv) Purchase of land directly from land owners by requesting agency for development projects. Making land owners as partners or share holder of the project wherever possible. Sale/lease of government land without going to cabinet approval as practiced in some states. Transfer of community land with consent of Gram Sabha/Gram Panchayat with adequate compensation and also replacement of community property wherever possible, reclassification of land usage by local authorities with permission of local revenue authorities may be permitted.

(v) Transfer of tribal and non tribal land in Scheduled V and VI areas for public use like education, health, large scale employment generation activities, urbanization, infrastructure development, etc through market forces should be allowed. This public use transfer is necessary to reduce economic deprivation and marginalization of tribal themselves.

Last but not the least, government should free land institutions of India from outdated socialism model land legislations and, industrialists and builders should try to create inclusive land institutions rather than purchasing land through forceful land acquisition.

Chapter 36

Shrinking of Rivers and the Global Water Crisis : Issues and Solutions

Manzoor, K.P. & Dhurjati Mukharjee

A recent study of 900 rivers in the world has found that the Ganga is one of the world's rapidly shrinking rivers. One of the country's most culturally and economically important rivers, the Ganga is among 45 in the study that showed a statistically significant reduction in discharge to the ocean. This group includes the Colombia, Mississippi, Niger, Parana, Congo and a few others.

According to the study titled, Changes in Continental Freshwater Discharge, conducted by the National Centre for Atmospheric Research in Colorado, the Ganga in 2004 had 20 per cent less water than it did 56 year ago. In the coming decades, it is likely to shrink even faster and could disappear in another 50 years. The waning of the 2500 km. Ganga has obviously huge ecological and economic ramifications to India. It will reduce the country's supply of drinking water and water for irrigation. The region will lose a crucial vehicle for channeling sewage into the sea.

The Ganga is losing water for two reasons - the glaciers that feed it are in retreat which means they are losing mass and rainfall in the region has been decreasing over the years. In fact, most climate models predict a weaker monsoon over South Asia as carbon dioxide induced warming continues. It is a well known fact that glaciers all over the world are in retreat because of global warning. Moreover rainfall over north India has gradually fallen over the years. The causes may be attributed to the El Nino effect and atmosphere above the Indian Ocean becoming warmer as also the weakening of the south-west monsoon.

Temperature fluctuations have become the order of the day and may be related to global warming, as the IPCC reports have pointed out. The increase in floods, droughts and other natural calamities specially in the tropical countries, is also linked to the El Nino effect.

Apart from the Ganga, the Jamuna has been under threat. A report prepared by the National Environmental Engineering Research Institute (NEERI) largely in the context of the Delhi stretch observed that ``the river too has died its natural death without fresh water from upstream".

As per official data in the 11th Plan document, the Jamuna's BOD (biochemical oxygen demand) level recorded in Delhi's Nizamuddin Bridge is 31, the figure for Agra canal is 28 while at Mathura, Agra City and Etawah this is around 15 (summer average, March-June, 2006).

These figures reveal that despite massive amounts spent under the Jamuna Action plan and also the Ganga Action Plan (Phase I and II) which also covered the major tributaries of the two rivers. There is massive pollution in both the rivers, primarily because of industrial effluents and also sewage and the objectives of these action plans have virtually failed. It is estimated that around Rs. 900-1000 crores or even more have been spent on the Ganga Action Plan over the last two decades.

Meanwhile seven IITs have plunged into a government effort to clean the Ganga, promising to recommend a slew of river management and technology strategies to improve its ecological health. The consortium will develop a Ganga River Basin Management Plan under an agreement signed recently. The plan will take into account the future growth of population and the increasing demand for land, energy and water as well as requirements of sewage treatment plants in cities along the river. It is also expected to ensure that the biological water flow remained unaffected throughout the year, according to Prof. Vinod Tare of IIT, Kanpur who is coordinating the Rs. 15 crore project.

Meanwhile, the Centre promised the Supreme Court recently that the Ganga will be pure and free of pollutants by 2020. In its affidavit before the Court, the National River conservation Directorate under the Ministry of Environment & Forests, said: ``An assistance of $ 1 billion has been indicated in the first phase by the World Bank. A project preparation facility advanced of $ 2.96 million has been sanctioned by the WB".

It may be mentioned here that in India, as also in many other countries pollution of rivers has been a big problem. The developing world, particularly India and China, needs to learn from Europe's experience of reviving and maintaining rivers. In our country, the Supreme Court judgements along these lines have been quite significant but not much effective action has been taken in this regard. The projects that have been taken up are far from satisfactory and not efficiently monitored. As per 2006 official audit of the Ganga Action Plan, only 39-40 per cent of its sewage treatment target has been accomplished and the plan is behind schedule by over 13 years. The same holds true of the Yamuna Action Plan where progress has also been not quite unsatisfactory.

In the case of Yamuna, apart from the problem of sewage entering the river, the large scale extraction of water in the upstream of Delhi for drinking and irrigation purposes, leading to negligible flow in the river after Wazirabad, as per reports of the Ministry of Environment & Forests has indeed been a big problem. This problem has also been witnessed in Kolkata (of the Hooghly river, an offshoot of the Ganga) after the water sharing agreement formula was signed between India and Bangladesh.

Some of the states are facing severe water crisis, both in the urban and rural areas, while Assam, Bihar face floods almost every year or one in two years. Meanwhile the 11th Plan aimed at expanding irrigation by 2.5 million hectares a year, and, at meetings of the National Development Council (NDC), most states have voiced the need for more allocations for increasing their irrigated area. In such a scenario, there is need for judicious management of water and ensuring its optimum use throughout the country which possibly the new plan of the Ganga proposes to do.

Another aspect of the problem is the drying up of rivers. In the case of West Bengal, ecologists have found that as many as 9 rivers such as Saraswati, Adi Ganga, Kana Damodar Maja Damodar, Ganur, Behula, Anjana, Bhoirab and Bidyadhari have complete gone dry. The condition of Jalangi, Ichchamati, Churni and Yamuna is extremely poor as they are about to dry up. It is understood that considering the gravity of the situation, the West Bengal Pollution Control Board (WBPCB) has embarked upon a drive and appointed a special committee to assess the status of many rivers navigating across the state. The study, headed by Kalyan Rudra, a river scientist, is expected to be completed in the next two years and will ascertain areas such as sediment load, toxicity, ecological flow, velocity and changing course.

It is quite obvious that demands of water would increase considerably more so because of the rapid pace of industrialization and urbanization in the coming years. It is thus necessary that the rivers, at least the major ones, have to be protected and all matters pertaining to water sharing, water pollution and water management have to be seriously examined by the Central authorities, if necessary in consultation with the respective state governments. It is in this context that the question of river interlinking may need to be reconsidered in a judicious manner by experts, keeping into consideration the geological, environmental, economic and practical aspects.

It is also necessary that the Himalayan rivers may be allowed to meander for which vast stretches of free river banks are needed. They fulfill the crucial role of conserving flood water and rain water in the porous land. This ecological role is particularly important for cities like Delhi that face water shortage. Experts believe that concrete structures should not be built on the river bank land and `flood plains' should be opposed in view of the fact that substantial parts of these plains have already been lost to the urban sprawl and what remains need to be saved.

With the rivers drying up and the rivers deprived of the minimum requirement of fresh water, the consequences would be disastrous unless specific steps are taken at this juncture. The per capita water availability is declining and India is expected to fall in the list of water stressed countries by the year 2014-15 along with the added problem of contamination of groundwater, resulting in a jump in water borne diseases. A water management and river conservation plan needs to be drawn up at the earliest to recharge our water resources and save them from extinction.

A clean flowing river could be of immense benefit to the country. Apart from its tourism potential, it could also create lot of employment. Asian rivers have been untouched by the lëisure economy except for pilgrimage tourism in India. Which the polluter pay principle should be enforced, rivers and its banks would also have to be kept environmentally clean.

We would like to believe there is an infinite supply of water on the planet. But the assumption is tragically false. In fact available freshwater represents less than half of one per cent of the world's total water stock. The rest is seawater, or inaccessible in ice caps, ground water and soil. And supply of this is finite. Most disturbingly, we are diverting, polluting and depleting that finite source of freshwater at an astonishing rate. United Nations argue that 31 countries are facing the water stress and scarcity and over one billion people lack adequate access to clean drinking water. By the year 2025, as much as two thirds of the world's population estimated to have expanded by an additional 2.6 billion people will be living in conditions of serious water shortage and one third will be living in conditions of absolute water scarcity.

Mismanagement of Water Resources

The water crisis is a serious human issue that exists when supply of water is less than demand. Environmental economists argue that the demand for water is increasing due to the subsequent growth of population in all over the world. When population grows there is no large source of water to satisfy the wants of all human kind, because the water becomes scarce. Demand for water among the urban people is very high compared to urban places for jobs and business opportunities are the most important reasons for growing water demand in urban areas of many countries. An Amsterdam based ecological management foundation argues that the rainfall is our only renewable source of fresh water. It creates a constant global supply of 40,000 to 50,000 cubic km water per year. But the world population increases roughly by 85 million per year.

Therefore, the availability of fresh water per head is decreasing rapidly. The growing population at geometrical rate and subsequent fall in water supply disturbs the demand and supply curve of water. Here is the importance of producing the additional quantity of water for the existence of human beings in next decade. The ground water researchers find that our fresh water source has been declining in many countries since last so many years. The scientific studies conducted in major rivers of Asia and pacific shows the over all decline in the water amount flowing to world's ocean. The climate changes strongly influence the reduction of water amount and quality because the large level of increase in carbon emission by the effects of global warming pollutes the drinking water resources dangerously. Mountainous forested watersheds are the most important fresh water yield area in the world. But we are loosing the supply of quality water due to the poor land management and the deforestation. Therefore, the sustainable and equitable water management is necessary for saving the world from human and economic tragedy.

Liberal attitude of governments towards water use leads to decrease in water resources in major industrialist countries. This liberal attitude of governments are exploited by the big corporate for misusing water resources for their industrial purpose. The big plants misuse thousand gallons of clean water to create electricity for working their machines and tools. Thus, the industrialists are building their merchant power plant in rural areas where the plenty of natural water can acquire. Experts opine that industrialists use 84 million gallons of water (15% of total water use) a day in southwest Louisiana.

Solution for Water Crisis :

(1) Water harvesting : This is a traditional technique first used by the Greeks and island people to save the water and obviously, it can be the first solution for water scarcity. Harvesting water means harvesting river. People are identifying the truth that the supply of water comes from the sky, which is rain. But in some countries are not getting adequate rain to meet their water demand. Therefore water harvesting being a great solution for those people. These people can find the solution for their water scarcity by the caught, storage of rain water. Water harvesting raises the supply of water. Consequently, we can manage the supply demand curves of the water. Government should promote the rain water harvesters. In India, Delhi government announces the award for the `best rain harvesters' which provides 2 lakh rupees for group and one lakh rupee for individual.

(2) Develop sea water desalination technologies : Desalination of sea water is another big solution for water scarcity. Singapore and Middle East countries wholly depends this technology for increasing their water supply. Many technologies are using for the desalination of sea water in different nations. More research and development on sea desalination technology will help us to reduce the energy consumption because the major portion of water resources extremely using for the creation of energy. Unfortunately, water desalination technologies are untouchable to poor nations due to its expensive nature.

(3) Organize the training workshop on water use and recycle : the water work shops help the society to be more vigilant on water preservation. The work shops on water economics, water policy and water ethics promotes the strong awareness on water scarcity and water use.

(4) Water regulation and strict water policies : `safe drinking water' policy is necessary for the smooth distribution of water and intervenes of private parties on water distribution creates water conflicts in the society. Strong environmental law can control over pumping and ground water exploitation.

Conclusion

The United Nations reports that two third of world population will face serious problems from the shortage of drinking water by 2025. Thus shortage in water supply in coming years will have a great impact on our food security, because we use major portion of water (70% of global use) for agricultural purpose. The impact of water crisis is very dangerous, because the unlimited use and pollution of water resources creates an adverse effect on our bio diversity. A strong environmental laws necessary to save the world from water pollution and water scarcity. The government can play an important role in detaining the water exploitation by the strict water policy and strong water regulation. But, instead of protecting our water resources, some industrialist nations privatize the water supply that creates water conflicts in the society. Public controls 97% of water distribution in the poor countries and private investments in water supply on these countries can have a negative impact on the living status of poor people.

Chapter - 37

Inter Linking of Rivers

Dr. S.B. Verma, Dr. Arvind Kumar Shrivastawa & K. Perumalammal

The schemes of inter linking of rivers of the country is under consideration with Government, the Ministry of Water Resources (MOWR) (erstwhile Ministry of Irrigation) had formulated a National Perspective Plan (NPP) for water Resources Development in 1980 envisaging inter basin transfer of water from surplus basins to deficit basins/areas which comprises two components, namely, Himalayan Rivers Development Component and Peninsular Rivers Development Component. The National Water Development Agency (NWDA) was set up under the MOWR in 1982 for carrying out various technical studies to establish the feasibility of the proposals of NPP and to give concrete shape to them. Based on various studies conducted, NWDA has identified 30 links (16 under Peninsular Component & 14 under Himalayan Component) for preparation of Feasibility Reports (FRs). Out of these, FRs of 14 links under Peninsular Component and 2 links (Indian Portion) under Himalayan Component have been completed.

Five links under Peninsular Component namely (i) Ken Betwa, (ii) Parbati-Kalisindh Chambal, (iii) Damanganga Pinjal, (iv) Par Tapi Narmada & (v) Godavari (Polavaram) Krishna (Vijayawada) were identified as priority links for building consensus among the concerned states for taking up their Detailed Project Reports (DPRs). DPR of one priority link namely, Ken-Betwa was completed and sent to the Governments of Madhya Pradesh and Uttar Pradesh for comments. The Government of Madhya Pradesh suggested an alternative proposal based on which two phases of the project are envisaged. The DPR of the phase-I has been prepared and sent to the Government of Madhya Pradesh and Uttar Pradesh in May 2010 for comments. Further, NWDA has taken up the DPRs of two more priority links after concurrence of the concerned states, namely Par Tapi Narmada and Damanganga Pinjal which are planned to be completed by December, 2011. Another priority link namely, Godavari (Polavaram) Krishna (Vijayawada) link is part of the Polavaram project of the Andhra Pradesh. Planning Commission has given investment clearance to the Polavaram Project and the Government of Andhra Pradesh has taken up the above project including link component as per their proposals.

The names of the links and the names of rivers likely to be interlinked under NPP and their present status is given below :

List of Water Transfer Links Identified Under NPP and Their Status

Peninsular Rivers Development Component

1.	Mahanadi (Manibhadra) - Godavari (Dowlaiswaram) link	-FR completed
2.	Godavari (Polavaram) - Krishna (Vijayawada) link*	-Taken up by the state as per heir own proposal
3.	Godavari (Inchampalli) Krishna (Pulichintala) link	- FR completed
4.	Godavari (Inchampalli) - Krishna (Nagarjunasagar) link	- FR completed
5.	Krishna (Nagarjunasagar) - Pennar (Somasila) link	- FR completed
6.	Krishna (Srisailam) - Pennar link	- FR completed
7..	Krishna (Almatti)-Pennar link	- FR completed
8.	Pennar (Somasila) - Cauvery (Grand Anicut) link	- FR completed
9.	Cauvery (Kattalai) - Vaigai - Gundar link	- FR completed
10.	Parbati-Kalisindh- Chambal link*	- FR completed
11.	Damanganga - Pinjal link*	- FR completed & DPR taken up
12.	Par - Tapi- Narmada link*	- FR completed & DPR taken up
13.	Ken - Betwa link*	-DPR of phase - I Completed
14.	Pamba- Achankovil - Vaippar link	- FR completed
15.	Netravati - Hemavati link	-PFR completed
16.	Bedti-Varda link	-FR taken up

Himalayan Rivers Development Component

1.	Kosi-Mechi link	-Entirely lies in Nepal
2.	Kosi-Ghaghra link	-S&I works taken up
3.	Gandak Ganga link	-S&I works taken up
4.	Ghaghra Yamuna link	-FR completed (for Indian portion)

5.	Sarda Yamuna link	-FR completed (for Indian portion)
6.	Yamuna Rajasthan link	-S&I work completed
7..	Rajasthan Sabarmati link	-S&I work completed
8.	Chunar Sone Barrage link	-S&I work completed
9.	Sone Dam - Southern Tributaries of Ganga link	-S&I works taken up
10.	Manas-Sankosh-Tista-Ganga (M-S-T-G) link	-S&I works taken up
11.	Jogighopa-Tista-Farakka (Alternate to M-S-T-G) link	-S&I works taken up
12.	Farakka Sunderbans link	-S&I work completed
13.	Ganga Damodar Subernarekha link	-S&I work completed
14.	Subernarekha Mahanadi link	-S&I work completed

* Priority links

PFR-Pre-Feasibility Report; FR - Feasibility Report; DPR-Detailed Project Report
S & I-Survey & Investigation in Indian Portion

The link proposals firmed up by NWDA under NPP envisage additional irrigation benefits of 25 million hectares of irrigation from surface waters, 10 million hectare by increased use of ground waters and generation of 34000 MW of poor apart from the benefits of flood moderation navigation, water supply, fisheries, salinity, pollution control etc.

However this is subject to the consensus among the States involved in each project.

By transferring water from a surplus basin to a water deficient basin, problems relating to irrigation and potable water can be mitigated significantly.

Agriculture has continued to be the main source of livelihood for the majority of the population and forms as backbone of the Indian economy. At present it has engaged over 60 per cent of population in India. Agriculture contributes a major share to the national income. 2001 census estimates show that 72.22 per cent total population was living in the villages, the rest in urban areas. Over 74 per cent of the population inhabiting over 95 per cent of the geographical area which makes rural India, cannot be taken lightly in any strategy that the country may adopt for into socio-economic development.

Irrigation

Water is indispensable to agricultural production. In areas where rainfall is plentiful and well distributed over the year, there is no problem of water. But rainfall in certain areas in very scanty as well as incertain.

This is so in Deccan and central India. Punjab and Rajasthan, in this areas, artificial irrigation is absolutely essential for without it cultivation is almost impossible. There are certain food and cash crops such as rice and sugarcane, which require abundant, regular and continuous supply of water.

During the 50 years since Independence, the Government had spent about Rs. 231,400 crores (at 1996-97 prices) on major, medium and minor irrigation works. As a result, the country's irrigation potential has increased from 23 million hectares in the pre-plan period (i.e.) 1950-51) to 89 million hectares at the end of 1996-97. India has the largest irrigated area among all the countries in the world.

Sources of irrigation Since 1950-51, considerable importance was attached to the provision of the canal irrigation. Canal irrigated area had increased from 8.3 million hectares to 18 million hectares during 1950-51 and 1999-2000. Even then, its relative importance has come down from 40 percent to 31.5 percent.

Well irrigated area has increased from 6 million hectares to 34 million hectares during last 50 years well irrigation in 1999-2000 accounted for nearly 59 percent of the local irrigation area as compared to only 29 per cent in 1950-51.

In India, irrigation words are classified into major medium and minor irrigation works. Since 1951, the major irrigation projects were defined as those costing more than Rs. 5 crores Medium irrigation as those costing between Rs. 25 lakhs, and minor irrigation works costing less than Rs. 25 lakhs.

Irrigation potential

When India started planned economic development in 1951, the irrigation potential from major and medium irrigation was about 10 million hectares, to taking 23 million hectares. By 1999-2000, the total irrigation potential created had increased to 94 million hectares.

India is one of the few countries with abundant land and water resources. The average precipitation over the country is about 400 million hectare metre. Annual water resources in the various river basins of the country is estimated at 187 million hectare metre.

Need for Interlinking rivers

The distribution of water resources of India over the land is uneven. This is because of the seasonal and regional distribution of rainfall. Most of the rain occurs during the monsoon months, the precipitation during the remaining part of the year being insignificant, large areas in Western, Central and South India have a very low rainfall while in the northern and eastern regions, heavy monsoon raths cause extensive floods and large volumes of water flows into the sea.

The uncertainty of rainfall marked by continuity for a long time, dry spells and fluctuation in seasonal and annual amount is a serious problem for the country water grids have been conceived for remedying this imbalance to a certain extent by transferring water from surplus regions to deficit regions by interlinking the various river basins so that trans basin transfer of water becomes possible.

India now ranks among the more important dam building nations in the world. Dams were regarded as ``the temples of modern India" by Jawaharlal Nehru. Till now ``irrigation and hydropower have been the major objectives of water resources development in India.

Aimed at linkage of rivers : The National perspective plan prepared in 1980 comprises Himalayan river development plan Peninsular river development plan including diverting a part of the water of the west flowing rivers in Kerala and Karnataka to the east for irrigating the drought prone areas of Karnataka, Tamil Nadu and Andhra Pradesh.

The Garland Canal Scheme : 1. Himalayan Canal 2. The Central and Southern Garland Canal

The National Water Grid : The National water grid was formulated by K.L. Rao and sponsored to some extent by the central water and power commission during the 1970. The scheme includes several water links from one river basin to another. 1. The Brahmaputra - Gangal link 2. The Narmada river - link 3. The Chambal river, Rajasthan - Gujarat link 4. The Ganga - Cauvery link, passing enroute through the basins of the Narmada, the Tapti, the Godavari, the Krishna and the Pannar. 5. Canal from Narmada to Western Rajasthan. 6. Canal from chambal to central Rajasthan. 7. Canal from the Mahanadi to serve coastal area of Orissa and Andhra Pradesh.

Conclusion

Predominantly agricultural states like Bihar, Uttar Pradesh, Orissa and Andhra Pradesh, have generally unprogressive traditional agriculture. Agriculture over the greater parts of the country, still works as a shock absorber for seasonal frictional and technological unemployment generated in the non-agricultural sectors. The agricultural sector of a country is particularly influenced by physical factors like the topography, type of soils climate, and rainfall. Though these projects involve high investment the results will be fruitful and India may turn as supreme power in the world. It is said ``Indian economy gambles with monsoon. This can be changed when Ganga and Cauvery are linked.

Table 1. Area irrigated by Sources in India

Sources of Irrigation	1950-51		1999-2000	
	Area*	Percentage	Area*	Percentage
Canals	8.3	40	18.0	31.5
Wells and tube wells	6.0	29	33.6	58.7
Tanks	3.6	17	2.7	4.7
Others Sources	3.0	14	2.9	5.1
Total	20.9	100	57.2	100.0

* Million hectares; Source : CSO, Statistical Abstract 2002.

Table 2. Some of Multi Power River Valley Project in India

Project	Across the River	Benefited State
1. Damodar	Damodar	West Bengal, Bihar
2. Hireakud	Mahanadhi	Orissa
3. Bhakna Nangal	Sutlej	Punjab, Haryana
4. Tungabadra	Tungabadra	Andhra Pradesh Karnataka
5. Kosi	Kosi	Bihar
6. Chambal	Chambal	Madhya Pradesh
7. Mayurakshi	Mayurakshi	West Bengal
8. Nagarjuna Sagar	Krishna	Andhra Pradesh

REFERENCES

1. V.K.R.V. Rao, Dr. Panse Memorial Lecture on, New challenged before Indian Agriculture.
2. B.D. Dhawan, Irrigation in India's Agricultural Development.
3. Indian Economy, Ruddar Datt & K.P.M. Sundharam.

Chapter - 38

Interlinking of Rivers in India : Problems and Prospects

S. Manicham

Interlinking of Rivers (ILR) is Government of India's proposal to link 37 rivers through 30 links, dozens of large dams and thousands of miles of canals, making it the largest water project in the world. It aims to transfer water from water surplus to water deficit areas and thus proposes to provide a permanent solution to the paradox of floods and drought. Of the 30 links proposed, 14 are in the Himalayan and 16 in the Peninsular component.

How, when and by whom - A short history of ILR

1972 : Ganga Cauvery link proposed by Union minister Dr. K.L. Rao. 1974 : Garland Canal, Proposal by Captain Dinshaw J Dastur, a pilot. Both plans were rejected due to technical infeasibility and huge costs and 1980 : Ministry of Water Resources frames the National Perspective.

Plan (NPP) envisaging inter basin transfer

1982 : The National Water Development Agency (NWDP) set up to carry out pre feasibility studies. These form the basis of the ILR plan.

1999 : A national commission (NCIWRDP) set up to review NWDA reports concluded that it saw no imperative necessity for massive water transfers in the peninsular component and that the Himalayan Component would require more detailed study. Aug. 15, 2002 President Abdul Kalam mentioned the need for river linking in his independence day speech, based on which senior advocate Ranjit Kumar filed a PIL in Supreme Court. Oct 2002 - Supreme Court recommends that the government formulate a plan to link the major Indian rivers by the year 2012. Dec. 2002 - Govt. appointed a Task Force (TF) on interlinking of rivers (ILR) led by Mr. Suresh Prabhu. The deadline was revised to 2016.

Uneven water availability

a. India has highly uneven water availability in space and time b. The country receives rain fall for only 3-4 months c. The Brahmaputra

Barak Ganga basin accounts for 60% of surface water resources d. This region is also rich in ground water e. Western and southern India experience severe deficit in both surface and ground water. f. 60% of the country experiences water deficit, while parts of the country suffer from floods.

Himalayan Rivers Development

1. Construction of storage reservoirs on the principal tributaries of the Ganga and the Brahmaputra in India, Nepal and Bhutan, 2. Along with interlinking canal systems to transfer surplus flows of the eastern tributaries of the Ganga to the West. 3. Apart from linking of the main Brahmaputra and its tributaries with the Ganga and Ganga with Mahanadi.

Benefits

The Himalayan component would provide additional irrigation of about 22 million hectare and generation of about 30 million KW of hydropower, besides providing substantial flood control in the Ganga & Brahmaputra basins. It would also provide the necessary discharge for augmentation of flows at Farakka required *inter alia* to flush the Calcutta port and the inland navigation facilities across the country.

Peninsular Rivers Development

This component is divided into four major parts.

1. Interlinking of Mahanadi Godavari Krishna Cauvery rivers and building storages at potential sites in these basins. This is the major interlinking of the river systems where surpluses from the Mahanadi and the Godavari are intended to be transferred to the needy areas in the South.

2. Interlinking of west flowing rivers, north of Bombay and south of Tapi. This scheme envisages construction of as much optimal storage as possible on these streams and interlinking them to make available appreciable quantum of water for transfer to areas where additional water is needed. The scheme provides for taking water supply canal to the metropolitan areas of Bombay; it also provides irrigation to the coastal areas in Maharashtra.

3. Interlinking of Ken-Chambal Rivers. The scheme provides for a water grid for Madhya Pradesh and Uttar Pradesh and interlinking canal backed by as much storage as possible.

4. Diversion of other west flowing rivers. Heavy rainfall on the western side of the Western Ghats runs down numerous streams, which empty, into the Arabian Sea.

Benefits

Construction of an interlinking canal system backed up by adequate storages could be planned to meet all requirements of Kerala as also for transfer of some waters towards east to meet the needs of drought affected areas. The peninsular Component is expected to provide additional irrigation of about 13 million hectare and is expected to generate about 4 million KW of power.

Interlinking or networking of rivers entails construction of dams and canals and other connected hydraulic engineering works for mass transfer of water across the river basins. Basically, the scheme is to convey flood water in the Ganga and Brahmaputra river basins to the arid and semi-arid areas of Rajasthan and Madhya Pradesh, and to the peninsular rivers of south India. There are essentially three methods to achieve the same. 1. Canal option - to construct lengthy canals, 2. Tunnel option - to convey water under mountains, and 3. Pumping option - to pump water over mountains.

The enormous drain of water into the seas, the paradioxical and perennial shortage of water for irrigation and drinking, and the floods in many parts of the country have prompted the idea of networking the rivers. The former President Dr. Abdul Kalam has said that the plan must be accorded top priority, it is hoped it will kick start the economy and mitigate the problem of unemployment. This according to him will convert the country into a developed nation. The project is also certain to integrate the rural and urban economies and bridge the gap in the great rural urban divide.

Dr. Kalam had adumbrated certain requirements so that the grandiose plan is successfully implemented. They are as follows :

a. The need to develop greater tolerance, compassion, hard work, dedication, and an ability to feel and realize the problems of others and the readiness to help. Avoiding narrow political ambition and greed, leaders must foster inter and intra communal harmony. The country as a whole must realize the economic need for such a project that would stimulate growth.

b. The second priority is political. Water must be moved from the State List and included in the Concurrent List, with over weaning Central control. The need for bringing water under Central control has been amply demonstrated by the non implementation of several river water sharing awards between the States.

c. An immediate dialogue with Pakistan and Bangladesh to seek their approvals for the networking is an essential priority as per Dr. Kalam. The project will be not be successful without linking the Ganga. the indus and the Brahmaputra. The Ganga, the Yamuna and the Brahmaputra combine in Bangladesh before entering the Bay of Bengal.

Similarly, the indus and its tributaries - the Ravi, the Beas, the Sutlej, the Jhelum and the Chenab merge in Pakistan before entering the Arabian Sea.

d. The next priority would be to look at and review the land acquisition laws. This river networking project would require a lot of land across the country and also would need access rights from several million landowners.

e. The fifth priority is to design an acceptable management structure to plan this project, and implement and monitor it. Once completed this network would last several generations and change the face of this country. This would also cost an enormous amount of money.

Expected benefits of interlinking of rivers : Surface water irrigation : 25 million Ha Ground water irrigation : 10 million Ha Hydro-power generation : 34 million KW Improved agriculture : it will help in ensuring food security Flood and drought control Alternative means of transport : river transport is a cheap and non polluting Higher GDP growth : creation of more employment opportunities will approximately lead to a 4% growth in the GDP. Lead to national unity and national security.

Disadvantages of this networking project have been enumerated below and later there are some details elaborating the same. 1. No inclusion of people's participation 2. Lack of consensus among citizens 3. Criss-cross construction of dams and canal systems, which will cause displacement of people 4. Submergence of land, forests and reserves 5. Negative impact on flora and fauna. 6. Acquisition of large tracts of land 7. If control is transferred to the center then decisions might be taken under political pressure.

Arguments against interlinking of rivers

1. Legal angles and election tangles: At present, there are serious disputes between various states of the Indian Union concerning sharing of river water. The disputes occur on account of the Chief Executive of any State having to take decisions and make claims in the interest of the people of his/her State since after all, that is the purpose for which he/she is elected. A Central Law to dictate water sharing between all the states from the network has the potential to precipitate new problems. This is because that there is no guarantee for change in the very political climate that causes interstate disputes in the first place, despite the present of river sharing agreements and authorities. Furthermore, if control is transferred to the center then decisions might be taken under political pressure.

2. Financing : The effect of huge borrowing an enormous amount of money (estimated today at Rs. 5.6 lakh crores as conveyed by government of India to the Supreme Court, but it would surely increase) needs to be reconsidered. This especially when India is almost in a debt trap with rising debt servicing almost equaling loans received from financial institutions like World Bank or Asian Development Bank. It is also necessary to consider whether we will be in a financial and physical position to maintain the huge assets when created (dams, canals, tunnels, captive electric power generation plants, etc.) in order for the system to continue to function and give the benefits for which it is designed.

If we cannot maintain the network, the capital assets created will deteriorate and be lost and the benefits of the project and incomes from it will not be available, though the loan liability would remain. This will inevitably lead to take over of assets by the creditor Banks to consolidate the entry of foreign interests into India. The political aspect of forcible project implementation is increasing disaffection among displaced people who already number tens of millions since independence.

3. Flood period : The basic idea of networking rivers is to convey unwanted floodwaters from one place to another where it is deficient and needed. But this idea does not consider that the period when it is surplus in the donor area (July to Oct. in the Ganga-Brahmaputra basins) is not the time when it is needed most in the recipient area (January to May in the peninsular rivers). In such a situation, it will be necessary to construct enormous holding reservoirs that will add to financial, social and environmental costs.

4. Desertification : Flooding per se is not undesirable because it results in deposition of alluvium particularly in the delta areas of rivers to maintain the fertility of the land by compensating loss of topsoil due to natural erosion. Any system that prevents or severely reduces natural flooding (by diversion of floodwater) will cause land fertility to gradually reduce over the years, thus desertifying the land. The greatest loss that land can suffer is desertification by loss of topsoil. The land that will be so lost to cultivation is the most fertile delta land, and therefore the impact of this on total food production needs to be factored into the discussion. History tells us that entire civilizations have vanished due to desertification.

5. River pollution: Annual floods flush industrial and municipal pollution in the Ganga down to the ocean. Reducing the flow in the Ganga by diversion will increase the concentration of pollution in the river. A live example is the Yamuna, from which Haryana and Delhi draw so much water that it barely flows after Delhi and the water quality at Delhi is so poor as to be positively poisonous. It is relevant to note that the expensive project to clean the Ganga has not succeeded even with annual flooding. This is not to argue that pollution of river water is inherent and may never be checked at source, but that this factor is yet another that needs to be included in the legitimacy check for the project.

6. Security : India has a national electric power grid that functions with difficulty because supply does not meet demand. However, it is kept functional because electric power can be switched from one circuit to another in the grid. Further it is not easy to deliberately interfere physically with the flow of very high voltage (upto 132 kV) electricity on overhead conductors atop huge pylons. But a national water grid is entirely different because water does not flow instantaneously like electricity, it cannot be switched like electric power, and it can very easily be tampered with enroute to divert, pump out or interrupt flow.

A canal breached deliberately or due to natural circumstances combined with poor maintenance would spell disaster for the areas around the breach. Water is basic for human survival unlike electric power, and motivation for interference is that much more. Maintenance of a network of canal, dams, etc., will have to be done under central supervision. Flow can be prevented or caused by the simple expedient of taking control of sluice gates as demonstrated by farmers during the recent Cauvery water problem. Thus security of the network will be an enormous load on security forces of Central and State Governments. In contrast, decentralized systems can be maintained, repaired and protected by those who benefit from them and live nearby.

7. Land acquisition : One cannot consider the acquisition of 8000 sq km of land when acquisition of land even in acres is a vexed issue, which has taken years. Even if fresh legislation makes it possible within a short period, its implementation will cause untold misery and injustice to the displaced people in obtaining compensation due to systemic corruption. Besides, land for resettlement is mostly not available. Thus, we must scrutinize closely and guard against our tendencies to address the political challenges of progressive policy and lawmaking for resolution of conflicts over natural resources with technology heavy solutions.

Conclusion

The proposed river network is a mega project comprising of a system of interlinked projects and has to be therefore, subjected to multidisciplinary scrutiny. The people involved in the decision making about networking of rivers do not look into the holistic view of the situation but only examine it by associating it with their knowledge and expertise in special fields. Democratic action and enlightened self interest by all citizens of India is the need of the four. The ILR should be implemented in an eco friendly manner and that the benefits will surely outweigh the costs.

References

1. Radha Singh, ``Interlinking of Rivers, ``Economic and Political Weekly, Vol. 38, No. 18 (May 10-16, 2003), pp. 1885-1886.
2. Ali, I., Interlinking of Indian Rivers. *Curr. Sci.,* 2004, 86, 498-499.
3. Jain, S.K., Agarwal, P.K. and Singh, V.P., *Hydrology and Water Resources of India,* Springer, The Netherlands, 2007.

Chapter - 39

Saving Water in Rice Cultivation : Technological Options

Amod K. Thakur & Rajeeb K. Mohanty

Rice, a global food grain is grown in 114 countries with 90% of world's rice is produced and consumed in Asia. The demand for rice increases with population, which is expected to rise by a further 38% within 30 years, according to the United Nations.

To save water in rice cultivation, farmers can use following technological options during land preparation, before main crop establishment, and actual crop growth period.

Land Preparation

Land preparation lays the foundation for the whole cropping season and it is important to plan and adopt proper water management during land preparation itself. It includes field channels, land leveling, tillage and puddling, and bund preparation.

Field channels

In India, most of the rice fields get irrigation water from plot to plot (flooding method of irrigation) due to absence of field channels and the amount of water flowing in and out of rice field cannot be controlled and field specific water management is not possible. Mostly farmers may not be able to drain their fields before harvest because water keeps flowing in from other fields. The water that continuously flows through rice fields also removes nutrients from the field. Constructing separate channels to convey water to and from each field (or to a small group of fields) greatly improves the individual control of water and is the recommended practice in any type of irrigation system.

Land leveling

A well leveled field is a prerequisite for proper water management and good crop. Unleveled fields results in uneven crop emergence and uneven early growth, uneven fertilizer distribution, and extra weed problems.

Tillage and Puddling

Rice fields with few rat holes or cracks rapidly lose water by seepage and percolation during soaking of field. Shallow soil tillage before land soaking close the cracks and amount of water used in wet land preparation can be reduced by 350 mm.

Thorough puddling results in a good compaction of soil, reduces permeability and percolation rates throughout the crop growing period. The efficacy of puddling in reducing percolation depends on soil properties. Puddling is very effective in clay soil (that form cracks during fallow period) and ineffective in coarse soils (do not have enough fine clay particles). Puddlings is not necessary in heavy clay soils and in such soils, direct dry seedling (without puddling but tilled) with minimum percolation losses can be practiced.

Soil compaction using heavy machinery is another option to decrease soil permeability in sandy soil with loamy sub soils having at least 5% clay. Government support can only make this technology feasible on a large scale as most of the farmers cannot afford to compact their soils by their own.

Bund preparation and its maintenance

Good bunds are a prerequisite to reduce seepage and underbund loses of water. Bunds should be well compacted and any cracks or rat holes should be plastered with mud at the beginning of the crop season. To avoid overbund flow during heavy rainfall, bund height should be at least 20 cm. Researchers have used plastic sheets in bunds in field experiments to reduce seepage losses which are probably financially not feasible to farmers.

Raised beds

One of the recently proposed innovations to deal with water scarcity in the rice wheat system in the Indo Gangetic Plain is the use of raised beds. In the system of raised beds, rice is grown on beds that are separated by furrows through which irrigation water is coursed. Though dimensions may vary, beds are usually around 35 cm wide, separated by furrows that are 30 cm wide and 25 cm deep. Rice can be transplanted or direct seeded on the beds. So far, the raised bed system has mostly been tested with current lowland rice varieties, and yield gains can be expected when suitable aerobic varieties are developed/used. Tractor-pulled equipment has been developed that shaped the beds and drills seed (sometimes together with fertilizers) in one operation.

Among the suggested benefits of raised beds are improved water use and nutrient use efficiency, improved water management, higher yields, and when the operations are mechanized reduced labor requirements and improved seedling and weeding practices. In the Indo Gangetic Plain, farmers are experimenting with raised beds for rice and other crops with different degrees of success. More information on raised beds can be obtained from the Rice Wheat Consortium (www.rwc.cgiar.org/index.asp).

Before main crop establishment

Minimizing the time between land soaking for wet land preparation and transplanting reduces the period when no crop is present and when outflows of water from the field do not contribute to production. In large scale irrigation systems with plot to plot irrigation, farmers raise seedlings in part of their main field and whole main field is soaked when the seedbed is prepared and remains flooded during the entire duration of the seedbed. Under such condition, time between land soaking and transplanting can be minimizing by the installation of field channels, the adoption of common seedbeds, or the adoption of direct wet or dry seedling. With field channels, water can be delivered to the individual seedbeds separately and the main field does not need to be flooded. Common seedbeds, either communal or privately managed, can be located strategically close to irrigation canals and be irrigated as one block. With direct seeding, the crop starts growing and using water from the moment of establishment onward.

Crop growth period

1. Irrigation methods

a. Saturated Soil culture

In saturated soil culture (SSC), the soil is kept as close to saturation as possible, which decreases the seepage and percolation flows. SSC in practice means that a shallow irrigation is given to obtain about 1 cm of ponded water depth a day or so after the disappearance of ponded water. By following SSC, water input decreased on average by 23% from the continuously flooded check. Raised beds can be an effective way to keep the soil around saturation. Rice plants are grown on beds and the water in the furrows is kept close to the surface of the beds.

Alternate wetting and drying

In alternate wetting and drying (AWD), irrigation water is applied to obtain flooded conditions after passing certain number of days of disappearance of ponded water. The number of days of nonflooded soil in

AWD before irrigation is applied can vary from 1 day to more than 10 days. AWD treatments results in yield reductions varying from 0% to 70% compared to flooded fields with increased water productivity. Total (irrigation and rainfall) water inputs may be decreased by 15-30% without a significant impact on yield. The benefits of AWD are improved rooting system, reduced lodging (because of a better root system), periodic soil aeration, and better control of some diseases such as golden snail. On the other hand, rats find it easier to attack the crop during dry soil periods.

Crop establishment methods

a. Dry seeding

Under direct dry seeding method of rice cultivation, dry seeds are broadcasted on plowed and well leveled soil (without puddling). The soil should be moist but not saturated from sowing to till emergence, or else the seeds may rot in the soil. After sowing, a flush irrigation is applied if there is no rainfall to wet the soil. The soil needs to be saturated the soil when plants have developed three leaves, and gradually depth of ponded water is to be increased with increasing plant height.

b. Wet seeding

In wet seeding method, farmers generally broadcast pre germinated seeds directly on the puddled and levelled field, which are free from standing water. The soil should be kept at saturation from sowing to 10 days after emergence, and then the depth of ponded water should gradually increase with increasing plant height. This method saves water up to 25% by minimizing water loss through reducing the time between land soaking and transplanting. This method is also suitable for areas where labour availability is major constraint during peak transplanting time.

Aerobic rice

A fundamentally different approach to reduce water requirements from rice fields is to grow the crop like an upland crop, such as wheat or maize. Unlike lowland rice, upland crops are grown in non puddled, non saturated (i.e. ``aerobic") soil without ponded water. The amount of irrigation water should match evaporation from the soil and transpiration by the crop (plus any application inefficiency losses). The potential water reductions at the field level when rice can be grown as an upland crop are large, especially on soils with high seepage and percolation rates. Besides seepage and percolation losses declining, evaporation decreases since there is no ponded water layer, and the large amount of water used for wet land preparation is eliminated altogether. Upland rice varieties are drought tolerant, but have a low yield potential and tend to lodge under high levels of external inputs such as fertilizer and supplemental irrigation. Alternatively, high yielding lowland rice varieties grown under aerobic soil conditions, but with supplemental irrigation, have been shown to save water, but at a severe yield penalty. Achieving high yields under irrigated but aerobic soil conditions requires new varieties of ``aerobic rice" that combine the drought tolerant characteristics of upland varieties with the high yielding characteristics of lowland varieties.

Conservation agriculture

With aerobic rice, technologies of conservation agriculture, such as mulching and zero - or minimum tillage as practiced in upland crops, become available to rice farmers as well. Various methods of mulching (e.g., using dry soil, straw, and plastic sheets) are being experimented in non-flooded rice systems in China and have been shown to reduce evaporation as well as percolation losses while maintaining high yields.

Chapter - 40

Drainage for Sustainable Agriculture

R.M. Singh & D.K. Singh

The natural drainage systems are severely affected by the development processes and thus increased in waterlogged and salt affected areas. In major and medium irrigation projects due to inadequate designs coupled with poor management practices has raised groundwater table and in turn sizeable command areas are being affected both by water logging and soil salinization, Water logging and salt problem have been experienced in irrigation projects all over the country. Both adversely affect the growth and the yield of the crops. An area with water table within 2 m from the land surface is called as water logged area. It is potential to water logging if water table is between 2-3 m from the land surface. It has been estimated that around 16.71 million hectare land is affected by salt and waterlogging in India. The drainage is the remedy to these problems for sustainable agriculture.

Drainage is defined as the natural or artificial removal of surplus grounded surface water and dissolved salt from the land in order to enhance agriculture Production. In the case of natural drainage the excess waters flows from the fields to lakes, swamps, streams and rivers. However, in an artificial system surplus ground or surface water is removed by means of sub surface or surface conduits. Improved drainage create a healthier environment for plant growth, It conserve soil and water; and provide drier field conditions for ease in farm operations for the crop production. Agricultural drainage is must to realize the full benefit of the irrigation.

Certain drainage criteria must be used to determine drainage need. Drainage is broadly divided into two types i.e., Surface and subsurface drainage. The excess surface or sub surface water of an agricultural field can be removed by applying the appropriate drainage method. Drainage system may consists of field drains system and main drainage system. Water of field is designed through field drainage and sent to main drainage for moving towards outlet. Field drainage system may be divided into surface drainage by gravity flow and subsurface drainage by gravity or pumped flow. The main drainage may be divided into deep collectors consisted of pipe or ditches for subsurface drainage and shallow collectors consisted of channels or ditches for surface drainage. The collector drains flows to disposal drains and to outlets of the drainage system to some stream or depressions.

Drainage coefficient is defined as the amount of water that runs off from a given area and is to be removed in 24 hours. While designing surface drainage system, a low value of the drainage coefficient will lead to partial improvement in drainage though the cost of design may be relatively low, whereas a high value would increase the cost substantially without any additional gain in the removal of surface congestion. Estimation of 24 hr rainfall depth that might occur with a probability level generally of 20% or a return period of 5 years should be considered for agricultural drainage.

Field drains for a surface drainage system have a different shape from field drains for subsurface drainage. Those for surface drainage have to allow farm equipment to cross them and should be easy to maintain with manual labour or ordinary mowers. Surface runoff reaches the field drains by flow through row furrows or by sheet flow. In the transition zone between drain and field, flow velocities should not induce erosion. Field drains are thus shallow and have flat side slopes. Simple field drains are V-shaped. Their dimensions are determined by the construction equipment, maintenance needs, and their cross ability by farm equipment. Side slopes should not be steeper. Nevertheless, long field drains under conditions of high rainfall intensities, especially where field runoff from both sides accumulates in the drain, may require a transport capacity greater than that of a simple V-shaped channel. Without increasing the drain depth too much, its capacity can be enlarged by constructing a flat bottom, thereby creating a shallow trapezoidal shape.

All field drains should be graded towards the collector drain with grades between 0.1 and 0.3%. Open collector drains collect water from field drains and transport it to the main drainage system. In contrast to the field drain, the cross-section of collector drains should be designed to meet the required discharge capacity. Besides the discharge capacity, the design should take into consideration that, in some cases, surface runoff from adjacent fields also flows directly into the field drains, which then require a gentler side slope. When designing the system, maintenance requirements must be considered.

Attention must also be given to the transition between the field drains and the collector drains, because differences in depth might cause erosion at those places. For low discharges, pipes are a suitable means of protecting the transition. For higher discharges, open drop structures are recommended. A free board of 25% of designed depth is kept. Permissible values for average velocity of flow to avoid scouring may be adopted.

Subsurface drainage improvement is designed to control the water table level through a series of drainage pipes that arc installed below the soil surface. The subsurface drainage network generally outlets to an open ditch or stream. Subsurface drainage requires some minor maintenance of the outlets and outlet ditches. For the same amount of treated area, subsurface drainage improvements generally are more expensive to construct than surface drainage improvements.

Subsurface drainage may be achieved by tubewell drainage, open drains or subsurface drains (pipe drains or mole drains). Tubewell drainage and mole drainage are applied only in very specific conditions. Subsurface (groundwater) drainage for water table and soil salinity in agricultural land can be done by horizontal and vertical drainage systems. Horizontal drainage systems use open ditches (trenches) or buried pipe drains. Parallel, herringbone, targeted and double main system layout could be adopted for subsurface drainage system. The spacing of drains could be evaluated using Hooghoudt or Ernst or Child method. Drainage system requires several materials to be used such as tiles, pipes, and envelope materials for its better functioning. Envelope materials such as gravel envelop and filter including geotextile filters can be used as per requirement. Various coefficients developed may be used for proper selection of the materials to realize effective drainage for sustainable agriculture.

Introduction

The natural geo-physiographical and agro-ecological situations of India are one of the major factors in causing surface water logging and development of salt affected areas in. The natural drainage systems are severely affected by the development processes and thus increased in waterlogged and salt affected areas. The other major factor is the development of man-made major and medium irrigation systems, where huge quantity of water is being transported into new geo- hydrological arid and semi-arid regions. The lack of working experiences in these regions caused inadequate designs coupled with poor management practices has raised groundwater table and in turn sizeable command areas are being affected both by water logging and soil salinization. National Commission on Agriculture, Govt. of India (NCA 1976) defined an area as waterlogged when the water table causes saturation of crop root zone soil, resulting to restriction to air circulation, decline in oxygen and increase in carbon dioxide levels.

The Working Group on Problem Identification in Irrigated area, constituted by the Ministry of Water Resources, Govt. of India (MOWR 1991) adopted the following norms for identification of waterlogged areas:

(i) **Waterlogged area** : Areas with water table within 2 m from the land surface'

(ii) Potential area for waterlogging : Areas with water table between 2-3 m from the land surface

(iii) Safe area : Areas with water table below 3 m from the land surface.

The physical effects of waterlogging are lack of aeration in the crop root zone, difficulty in soil workability and deterioration of soil structure. Its chemical effect is soil salinisation. Both adversely affect the growth and the yield of the crops. The extent of drop damage depends upon the magnitude, duration and frequency of the waterlogged condition and the degree of soil salinity. Salt problem is a major cause of decreasing agricultural production in many of the irrigation project areas. Salinity may be a major problem in many non-irrigated areas where cropping is based on limited rainfall. The various agencies evaluated the status of water-logging and soil salinization problems in these areas. However, the officially accepted one is the estimates of Working Group, (1991).

Table 1. Water logged and Salt affected areas in million hectares

Source	Irrigated Command Area				Country as a whole			
	Water logged	Saline	Alkali	Total	Water logged	Saline	Alkali	Total
Working Group of MoWR	2.46	3.06	0.24	5.76	-	-	-	-
MoA, GoI	-	-	-	-	8.53	5.50	3.58	17.61

Water logging and salt problem have been experienced in irrigation projects all over the country. The examples are Chambal Command areas in Rajasthan and M.P., Indira Gandhi Canal Project in western Rajasthan, Kosi and Gandak Project Commands in Bihar, the Tungabhadra Project area in Karnataka, the Nagarjunasagar Project area in Andhra Pradesh and the Kakrapar Project area in Gujarat. Construction of drainage canals, field drains and avoiding wastage of canal supplies have been adopted as remedial measures. However, lack of maintenance, operational constraints of large irrigation projects, construction of highways, railway embankments and other obstructions, without providing for adequate drainage facility are still the major factors for water logging (Singh *et al.*, 2011). In the Chambal Command area soils became water logged with a few years of introduction of irrigation. In many coastal areas excessive groundwater exploitation has caused seawater intrusion, worsening the salinity problem. Extent of waterlogged and salt affected areas for some states in India has been presented in Table 2.

There are extensive low lying areas in the rice growing coastal belts of eastern and south eastern regions of India where poor drainage seriously affects crop production in the monsoon season. The agricultural drainage is the remedy to these problems for sustainable agriculture. Reclamation of water logged/saline affected land by scientific and cost-effective methods should form a part of command area development programme. The drainage system should form an integral part of any irrigation project right from the planning stage. Some examples are the Sardar Sarovar Project in Gujarat the Narmada Canal Project in Rajasthan, Madhya Pradesh, the Indira Sagar Project in Madhya Pradesh, the Subarnarekha Barrage project in West Bengal, the Arjun Sahayak Pariyojana in UP, the Bodwad Parisar Sinchan Yojana in Maharashtra and many others (Singh et al. 2011). Sustainable agriculture could be achieved if components of sustainable development, environment, society and economy remain in balance (Ott 2003 and Adams, 2006). Different types if agricultural drainage systems to manage water logging and soil salinity have been discussed in this chapter.

Table 2. Geographical, waterlogged and salt affected areas of some states in India

State	Geographical area, million ha hectares	Water logged area, million ha hectares	Salt affected area, million ha hectares
Andhra Pradesh	27.44	0.339	0.813
Bihar	17.40	0.363	0.400
Gujarat	19.60	0.484	0.455
Haryana	4.22	0.275	0.455
Karnataka	19.20	0.036	0.404
Kerala	3.89	0.012	0.026
Madhya Pradesh	44.20	0.057	0.242
Maharashtra	30.75	0.111	0.534
Orissa	15:54	0.196	0.400
Punjab	5.04	0.199	0.520
Rajasthan	28.79	0.348	1.122
Tamil Nadu	12.96	0.128	0.340
Uttar Pradesh & Uttaranchal	29.40	1.980	1.295
Total	258.43	4.528	7.006

(Source : Ghosh 1991 and Tyagi 1999)

Drainage

Drainage is defined as the natural or artificial removal of surplus ground and surface water and dissolved salt from the land in order to enhance agriculture production. In the case of natural drainage the excess waters flows from the fields to lakes, swamps, streams and rivers. However, in an artificial system surplus ground or surface water is removed by means of sub surface or surface conduits *(Source: FAO Glossary of Land and Water Terms).* Improved drainage create a healthier environment for plant growth, It conserve soil and water; and provide drier field conditions for ease in farm operations for the crop production. Adequate drainage is required to improve soil health and soil water plant interaction for enhanced water productivity to ensure sustain ability (Singh et al., 2009). Agricultural drainage is must to realize the full benefit of irrigation.

1.1. Drainage criteria

Certain drainage criteria must be used to determine drainage need. A groundwater balance of the drainage area is the most accurate tool to calculate the volume of the water to be drained. Besides agricultural drainage criteria, technical drainage criteria (relating to minimization of the cost of installation and operation of system while maintaining the agricultural criteria), environmental criteria (relating to the minimization of the environmental damage), and economic drainage criteria (relating to the maximization of the net benefits i.e. the difference between benefits and costs, and damages) should also be taken into account.

1.2. Types of drainage system

Drainage is broadly divided into two types i.e., Surface and subsurface drainage. The excess surface or sub surface water of an agricultural field can be improved by applying the appropriate drainage method. Drainage system may consists of field drains system (internal) and main drainage system (external). Water of field is drained through field drainage and sent to main drainage for moving towards outlet. Field drainage system may be divided into surface drainage by gravity flow and subsurface drainage by gravity or pumped flow both divided into regular system and checked system (Figure 1).

2. Surface Drainage System

Surface drainage uses the potential energy due to land elevation to provide a hydraulic gradient for the movement of water. Surface drainage improvements are designed to minimize crop damage resulting from water ponding on the soil surface following a rainfall event, and to control runoff without causing erosion. Surface drainage can affect the water table by reducing the volume of water entering the soil profile. This type of improvement includes: land leveling and smoothing; the construction of surface water inlets to subsurface drains; and the construction of shallow ditches and grass waterways, which empty into open ditches and streams. These have disadvantage of requiring annual maintenance; and extensive and expensive earthmoving activities, and land grading might expose less fertile and less productive subsoils. Also, open ditches may interfere with moving farm equipment across a field.

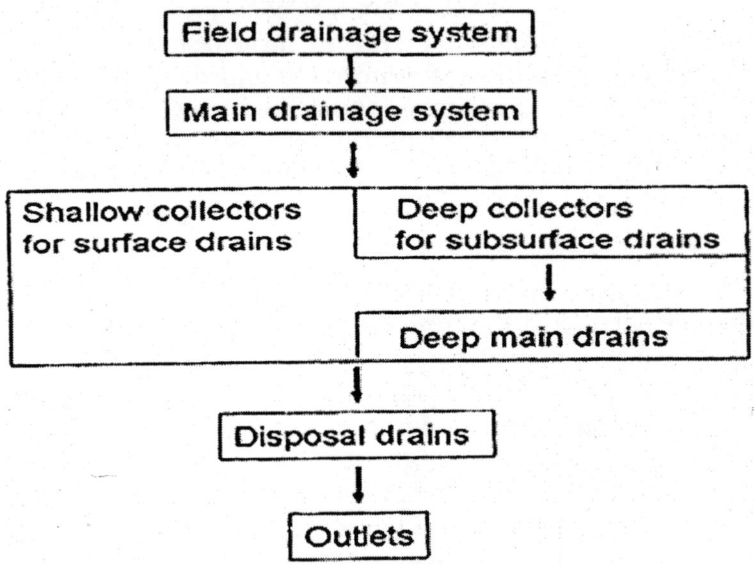

Figure 1 : Agricultural drainage system.

The main drainage may be into deep collectors consisted of pipe or ditches for subsurface drainage and shallow collectors consisted of channels or ditches for surface drainage. The collector drains flows to disposal drains and to outlets of the drainage system to some stream or depressions.

2.1 Causes of surface drainage

Low-lying flat areas, heavy soils with low permeability and lands in humid tropical or sub-tropical regions with high intensity storms, are subject to surface inundation and therefore, require surface drainage. Surface drainage problem in many cases might not be of local origin as in the case of low lands where problem is caused due to the flow of surface runoff from uplands. Overflow from rivers or natural channels sometime contribute to the drainage problem of an area. The following reasons could be ascribed to the problem:

(i) Flat land surfaces causing hindrance in the natural runoff from the upper catchment area. The problem is severe in heavy textured soils and in humid climate,

(ii) Inadequate capacity of the drainage channels particularly during critical periods could cause surface stagnation. It is one of the main reasons of surface stagnation in many parts of the Indo-Gangetic plains. During intense storms the main drainage channels are full to the brim thereby reducing the capacity of the lateral and collector channels causing inundation upstream.

(iii) Inadequate outlet conditions partly due to developmental works, which obstruct the flow and partly due to choking of the outlet of the natural drainage system,

(iv) Non-availability of outlet due to backwater flow particularly in coastal regions,

(v) Waterlogging in agricultural lands is called as critical, if water table fluctuates between 0-2 m below ground surface. It is treated as semi-critical, if water table fluctuates between 2-3 m below round surface.

2.2. Factors affecting drainage

Generally, climate, soil, depths of water table and crop affect the drainage requirement. Climatic conditions in a particular region decide the degree and frequency of surface water stagnation. Thus, besides the seasonal and annual rainfall, frequency analysis procedures are used to determine various return period rainfall events, number of storms and the dry periods to plan an effective drainage and reuse strategies.

Kind of the soils and topography determines the degree of surface congestion. For the same event, degree of stagnation would be more in a heavy than a light textured soil. A flat terrain would be more prone to congestion than a rolling topography. Water absorption into the soil profile would depend upon the depth to water table. An area with relatively high water table would be subjected to relatively greater depth of water stagnation and for periods larger than the area with deep water table. Drainage is influenced to a great extent on the kind of crops grown and their tolerance to surface stagnation. The crop tolerance to water stagnation varies from 1-7 days depending upon which group of crops is grown.

2.3. Open-surface drainage

Open-surface drainage is defined as the diversion or orderly removal of drainage water by means of improved natural or constructed channels, supplemented when necessary by the shaping and grading of land surfaces of such channels.

2.4. Parts of surface drainage system:

The surface drainage system consists of three parts, (1) collection system, (2) conveyance or disposal system, and (3) outlet (Figure 2).

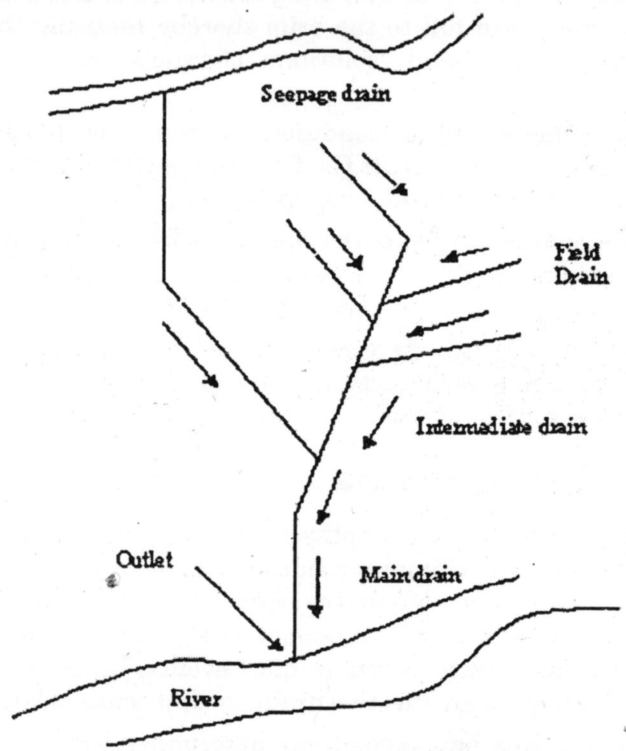

Figure 2: A typical drainage system.

Water to be drained from the individual field is collected through collection system (consists of field drains) arid moves through disposal system (consists of intermediate and main drains) to outlet.

2.5. Design criteria for field drainage

The very purpose of a good surface drainage system is to prevent the harmful effects of waterlogging on crops. Selection of an appropriate drainage coefficient is the key to design a successful surface drainage system. Drainage coefficient is defined as the amount of water that runs off from a given area and is to be removed in 24 hours. While designing surface drainage system, a low value of the drainage coefficient will lead to partial improvement in drainage though the cost of design may be relatively low, whereas a high value would increase the cost substantially without any additional gain in the removal of surface congestion. The waterlogging tolerance of a crop should be considered while estimating the drainage coefficient for a surface drainage project. Although there are several methods for estimation of drainage coefficients, a simple method involves following steps for its estimation:

1. Estimation of 24 hr rainfall depth that might occur with a probability level generally of 20% or a return period of 5 years should be considered for agricultural drainage
2. Evaluate the basic infiltration rate of the soil and determine the expected potential evapotranspiration.
3. Estimate the crop tolerance to surface congestion at sensitive growth stages in days. Multiply the infiltration rate and expected potential evapotranspiration with crop tolerance period.
4. Subtract the value calculated in step (3) from estimated probable rainfall value.

The resulting value when divided by crop tolerance period (days) will give the drainage coefficient in depth of water/day. These steps could be described mathematically in the form of an equation as:

$$Q= [R-n(E + I)\}/n \qquad ...(1)$$

Here, q is the drainage rate in mm/day, R is the rainfall in mm, E and I are potential evapotranspiration and infiltration rate in mm/day and n is number of days. When n is greater than one, it may be useful to increase the duration of rainfall in step 1 also from 1-day maximum to n-day maximum. The drainage coefficient thus calculated will be closer to the actual field values. For rice crop, the drainage coefficient would be less as at the end of drainage period some depth of water is allowed to stand in the fields. This depth is subtracted before dividing by number of days, n.

2.6. Design of surface drains

Field drains for a surface drainage system have a different shape from field drains for subsurface drainage. Those for surface drainage have to allow farm equipment to cross them and should be easy to maintain with manual labour or ordinary mowers. Surface runoff reaches the field drains by flow through row furrows or by sheet flow. In the transition zone between drain and field, flow velocities should not induce erosion. Field drains are thus shallow and have flat side slopes. Simple field drains are V-shaped. Their dimensions are determined by the construction equipment, maintenance needs, and their cross ability by farm equipment. Side slopes should not be steeper. Nevertheless, long field drains under conditions of high rainfall intensities, especially where field runoff from both sides accumulates in the drain, may require a transport capacity greater than that of a simple V-shaped channel. Without increasing the drain depth too much, its capacity can be enlarged by constructing a flat bottom, thereby creating a shallow trapezoidal shape.

All field drains should be graded towards the collector drain with grades between 0.1 and 0.3%. Open collector drains collect water from field drains and transport it to the main drainage system. In contrast to the field drain, the cross-section of collector drains should be designed to meet the required discharge capacity. Besides the discharge capacity, the design should take into consideration that, in some cases, surface runoff from adjacent fields also flows directly into the field drains, which then require a gentler side slope. When designing the system, maintenance requirements must be considered. Attention must also be given to the transition between the field drains and the collector drains, because differences in depth might cause erosion at those places. For low discharges, pipes are a suitable means of protecting the transition. For higher discharges, open drop structures are recommended. Permissible values for average velocity of flow to avoid scouring may be adopted from Table 3.

Table 3. Maximum allowable velocities in channels for different soil textures.

SI. No.	Soil texture	Max. Allowable velocity (m/sec)
1.	Very light silty sand	0.30
2.	Light loose sand	0.50
3.	Coarse sand	0.75
4.	Sandy and sandy loan	0.75
5.	Silty loan	0.90
6.	Firm clay loan	1.00
7.	Stiff clay or stiff gravelly soil	1.50
8.	Coarse gravel	1.50
9.	Shale, hardpan, soft rock etc.	1.80
10.	Hard cemented conglomerates	2.50

Slightly higher velocities are allowed if water contains colloidal silt. If the land slopes are steeper to create scouring velocity, the same has to be reduced by a gentle slope of the drain through provision of suitable drops/falls in the channels.

2.6.1. Side slopes

The Side slopes of the drains in general are recommended as given below Firm soil 1.0:1 (horizontal: vertical) Loam soil .1:5:1 Sandy soil 2:5:1

However, it is desirable to design the side slope of a channel from consideration of angle of repose of the soil.

2.6.2. Channel grade

The channel should be as uniform as possible. The grade should be as steep as possible provided the maximum allowable flow velocities are not exceeded. Design grades should be from 1- 0.3 % and should never be less than 0.05%.

2.6.3. Channel depth and width

The depth of channel should be sufficient to carry the design discharge. In general the depths of main and sub main should be kept between 2-3 m. The bed width depends on peak design discharge at different points.

2.6.4. Channel bottom width

The bottom width can be computed for a given discharge after the channel grade, grade depth and side slopes are selected. The bottom width for most efficient cross section (hydraulic radius is one half the depth) The minimum bottom width should be 1.2 m except in small lateral. The cross section is designed to meet the requirement of the capacity, velocity, and side slope, bottom width. The parabola has the smallest wetted perimeter and it is well suited for concrete channels. A trapezoidal section with minimum recommended bottom width of 0.60 m is recommended for earthen channels. The following factors may be considered for designing: A deeper ditch gives a higher velocity than a shallow one and also may provide future opportunity for pipe drainage it will remain effective for a longer period sediment bars may cause less obstruction, it requires less waterways than a shallow one. It may uncover unstable layers of soil. Shallow drain may be more practical to maintain by pasturing or by mowing flat side slopes, the depth should be. related to a good outlet condition ns, Design velocity should be selected so as to maintain the ditch cross s section with time. In channels that flow intermittently, some scouring may be desirable at high flows to counteract sediment deposition that occurs at low flows.

2.6.5. Berms and spoil banks

Berms are required to provide for work areas and facilitate spoil bank spreading, Prevent excavated materials falling back into the ditch. The berm width for drains should not be less than the depth of cutting. The minimum berm widths are given in the following Table 4. The berm width should be increased in unstable soil where it is feared that the drain will enlarge in the section. The spoil is spread until the height is reduced to an economical figure usually not more than lm.

Table 4. Minimum berm widths

Depth of drain, m	Minimum berm width, m
0.6-1.2	1.2
1.2-1.8	1.8
1.8-2.4	3.0
>2.4	4.5

2.6.6. Free board

The free board is additional depth above the design water level used to provide a safety factor for the design storm. A free board of 25% of designed depth is kept.

2.7 Location, spacing and alignment

Drain ditches should be located in a way to provide the most effective drainage and to cause the least interference with irrigation system and farm operations. These serve as outlet for surface runoff frond rainfall, as outlet for excess irrigation water, or as disposal ditches for pipe drains. They may be located parallel to canal embankments to collect seepage water. For controlling water table they may be installed parallel and at regular intervals with the same depth and spacing as pipe drains. Ditches should be normally close to the low point depression. Crossing with irrigation watercourses should be avoided. Grade control and crossing structures should be minimized.

2.8. Patterns

Two main types of surface drainage patterns are random and parallel. Each includes lateral ditches that permit water to flow from drainage system to a suitable outlet. The chosen pattern depends upon the soil type and topography of the land.

2.8.1 Random

The random ditch pattern is practiced to slowly permeable soils having depression areas that are too large to be eliminated by land smoothing or grading (Figure 3). Soil from the ditches can be used to fill minor low spots in the field. Field ditches should extend through most of the depressions for complete drainage, and they should follow the natural slope of the land.

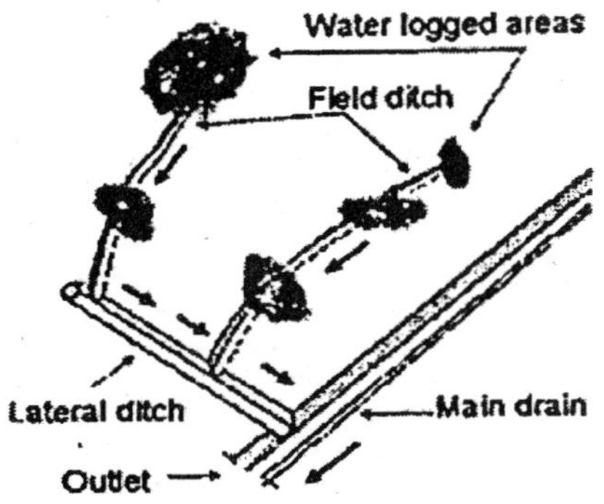

Figure 3 : Random drain system.

Field ditches connect the major low spots and remove excess surface water from them. They are generally shallow enough to permit frequent crossing by farm machinery

2.8.2 Parallel

The parallel ditch pattern is suitable for flatter, poorly drained soils that have numerous shallow depressions (Figure 4). Dead furrows are neither desirable nor necessary. Although the ditches must be parallel, they need not be equi-distant. The spacing between them depends upon the permissible length of row drainage for the soil type and upon the amount of earth and the distance it must be moved to provide complete row drainage. The maximum length of the grade draining to a ditch should be 200 m. The success of a parallel pattern depends largely upon proper spacing of the parallel ditches and the smoothing or grading between them. During the grading operation, fill all depressions and remove all barriers. Excavated material from ditches can also be used as fill for establishing grades. 2.9. Shaping the surface.

2.9.1. Grading

Land grading (also termed precision land forming) is the reshaping of surface of land with tractors and scrapers to planned grades. Its purpose is to provide excellent surface drainage although the amount of grading will depend upon the soil and costs. To do a good job of land grading, you need a detailed engineering survey and construction layout. To assure adequate surface drainage, eliminate all reverse surface grades that form depressions. The recommended surface grades range from 0.1 to 0.5 per cent and may be uniform or variable. The cross slopes normally should not exceed 0.5 percent. Minimum grade limits should include a construction tolerance that will permit the elimination of all depressions either in original construction or in post-construction touch up.

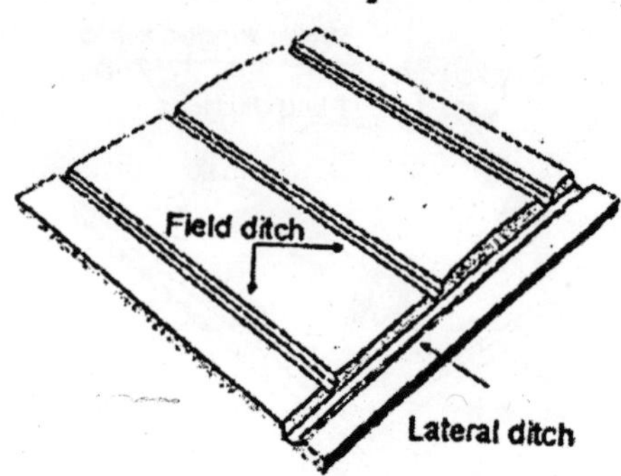

Figure 4: Parallel drain system.

In fields that can be cultivated up and own slope, parallel field ditches are installed across the slope to break the field into shorter units Of length and make it less susceptible to erosion. The field should be farmed in the direction of the greatest slope.

Reverse grades can be eliminated with relative ease in a field that has "minimum grades of 0.2 per cent. Unusual precision in construction is required to eliminate reverse surface grades in fields that have 0.1 per cent and flatter grades. Land grading is hampered by trash and vegetation. This material should be destroyed or removed before construction and kept under control while the work is being done. The fields should be chiseled before construction if there are hard pans. The field surface should be firm when it is surveyed so that rod readings taken at stakes will reflect true elevation. Do not grade fields when they are wet because working wet soil impairs physical condition of soil.

2.9.2. Smoothing

Land smoothing removes irregularities on the land surface and should be done after land grading and may be useful in other situations. Special equipment such as a land plane or land leveler should be used (Figure 5).

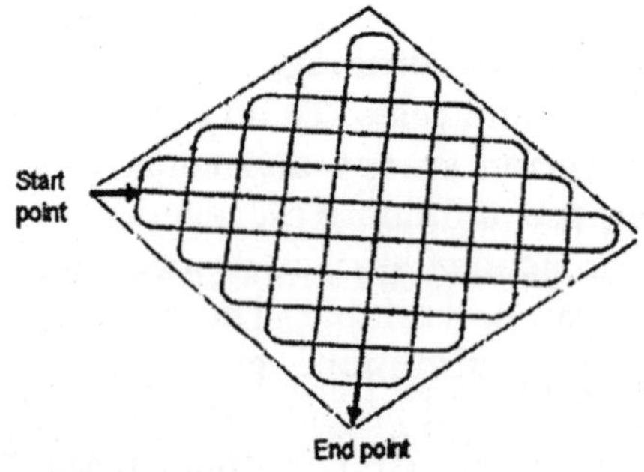

Figure 5 : Smoothing operation.

The purpose of land smoothing is to improve surface drainage. The smoothing operation may ordinarily be directed in the field without detailed surveys or plans, although grid surveys may be needed for some critical parts of the field. A smoothing operation consists of a minimum of three passes with a land leveler.

Make the first two passes on opposite diagonals as noted in Figure 5 and the last pass in the direction of cultivation. Either before or after the final land smoothing operation chisels fields to loosen the cut surfaces and to blend the fill material with the underlying soil. The finished surface should be free from minor depressions so that runoff will flow unobstructed to field or lateral ditches.

2.10. Good surface drainage practice

Following are the points to be remembered for good drainage:

(i) Gravity is the primary vehicle for carrying runoff away. There must be a continuous minimum fall in the ground level to assure drainage, and a minimum slope of 1% to 5% for grass swales. 1% minimum slope for smooth interior pipe is a general guideline for pipe conveying run off water to a discharge point

(ii) Large amounts of water should not cross a sidewalk to reach the street storm drain. Use drains or install piping to cross walks or other pedestrian walkways to prevent hazards.. Consult an engineer or architect for minimum slope in critical applications,

(iii) Break up one large drain to several smaller drains to:

   (a) Prevent erosion on steep landscapes by intercepting water before it accumulates too much volume and velocity.

   (b) Provide a safety factor. If a drain inlet clogs, other surface drains may pick up water

   (c) Improve aesthetics. Several smaller drains will be less obvious than one large drain.

   (d) Spacing smaller drain inlets will give surface runoff a better chance of reaching the drain. Water will have farther to travel to reach one large drain inlet.

(iv) Erosion is a big problem in drainage - slopes must be carefully calculated to ensure continuous flow, yet not steep enough to erode,

(v) Slow moving water will create a bog, while water moving too fast will cause erosion, form gullies and weaken foundations. Design a drainage system that will eliminate both extremes.

(vi) Design paved areas so they are graded almost level – avoid wildly sloping paved areas or dramatic changes in slope,

(vii) Runoff water must never be directed purposefully from one property onto another property. It is acceptable for water that flows naturally from one property to the other to continue, but you must never increase this flow artificially through grading and piping,

(viii) Check local code requirements and their applications

(ix) When designing a system, work from the discharge point towards the highest elevations,

(x) Design a secondary drain route to allow for overflow conditions during severe rainfall or in case the primary drain system fails,

(xi) Many systems require a grate or "clean out" fitting every 50 to 100 feet or at alignment changes of 45 degrees or greater to clean out the pipeline. Clean outs are normally constructed at grade,

(xii) Keep it simple. Over-design in storm water systems is expensive.

3. Subsurface Drainage

The objective of subsurface drainage is to drain excess water and salt from the plant root zone of the soil profile by artificially lowering the level of water table (Figure 6) Subsurface drainage improvement is designed to control the water table level through a series of drainage pipes (or tubing) that are installed below the soil surface (Figure 7). The subsurface drainage network generally outlets to an open ditch or stream. Subsurface drainage improvement requires some minor maintenance of the outlets and outlet ditches. For the same amount of treated acreage, subsurface drainage improvements generally are more expensive to construct than surface drainage improvements. The main objective of drainage is to remove excess water quickly and safely to reduce the potential for crop damage.

3.1. Types of subsurface drainage systems

Subsurface drainage aims at controlling the water *table and a control that may be* achieved by tubewell drainage, open drains or subsurface draims (pipe drains or mole drains). Tubewell drainage and mole drainage are applied only in very specific in conditions. Subsurface (groundwater) drainage for water table and soil *salinity in* agricultural land can be done by horizontal and vertical drainage systems. Horizontal drainage systems use open ditches (trenches) or buried pipe drains.

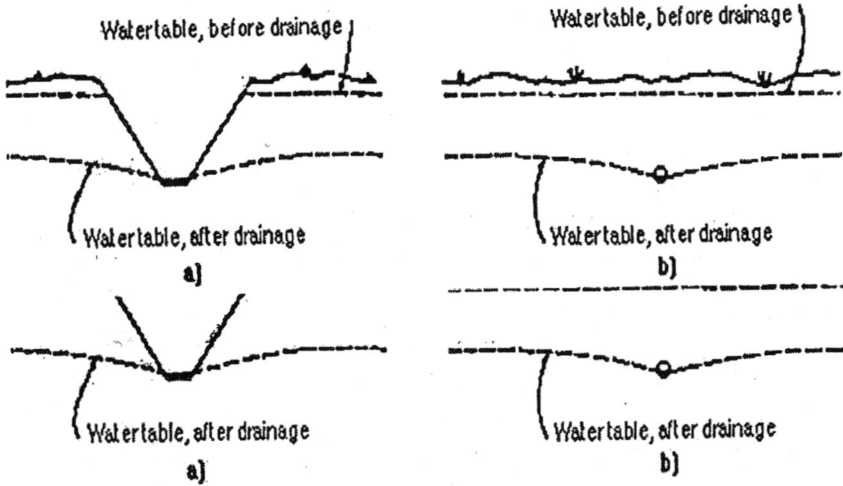

Figure 6 : Level of water table before and lowered level after drainage improvement: (a) surface drainage ditch; (b)subsurface drainage pipe. (USDA-ERS, 1987)

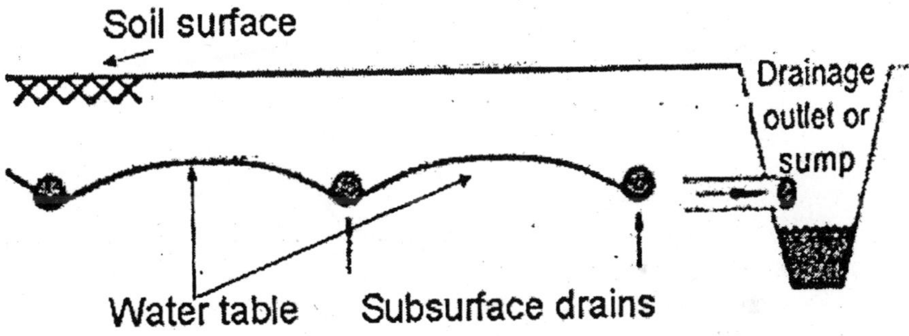

Figure 7: A controlled level of water table profile under subsurface drainage system.

Draining excess water from the soil profile where plant roots grow help aerates the soil and reduces the potential for damage to the roots of growing crops. It produce soil conditions more favorable for farming operations.

3.1.1 Open drains

Open drains have the advantage that they can receive overland flow directly, but the disadvantages often outweigh the advantages. The main disadvantages are the loss of land, interference with the irrigation system, the splitting-up of the land into small parcels, which hampers mechanized farming operations, and a maintenance burden.

3.1.2. Tile drainage

Tile drainage is an agriculture practice that removes excess water from subsurface soil. Whereas irrigation is the practice of adding additional water when the soil is naturally too dry, drainage brings soil moisture levels down for optimal crop growth. While surface water can be drained via pumping and/or open ditches, tile drainage is often the best recourse for subsurface water. Too much subsurface water can be counterproductive to agriculture by preventing root development, and inhibiting the growth of crops. Too much water also can limit access to the land, particularly by farm machinery.

3.1.3. Mole drainage

Heavy soils of low hydraulic conductivity (less than 0.01 m/day) often require very closely spaced drainage systems for satisfactory water control. With conventional pipe drains, the cost of such systems is usually uneconomic and hence alternative techniques are required. Surface drainage is one possibility; the other is mole drainage (Figure 8).

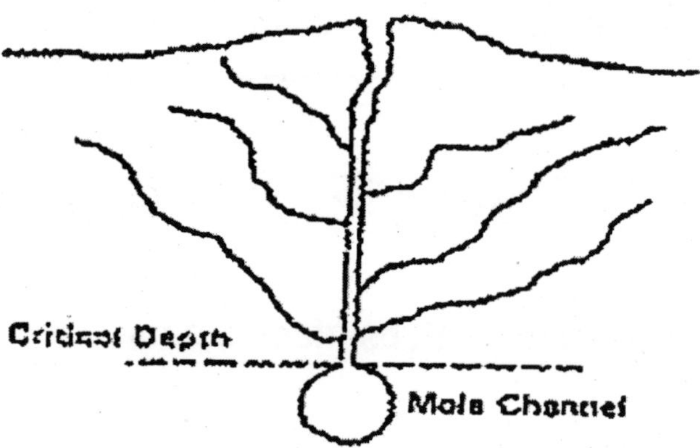

Figure 8 : Mole drains are unlined circular soil channels which function like pipe drains.

Advantage of mole drainage is their low cost and hence they can be installed economically at very close spacing. Their disadvantage is their restricted life but, providing benefit/cost ratios are favourable, a short life may be acceptable. The success of a mole drainage system is dependent upon satisfactory water entry into the mole channel and the mole channel stability. Mole drains are formed with a mole plough (Figure 9), which comprises a cylindrical foot attached to a narrow leg, followed by a slightly larger diameter cylindrical expander. The mole plough is attached to the drawbar of a tractor and the mole channel is installed at depths between 0.4 and 0.7 m. Common lengths of run vary from 20 to 100 m.

Figure 9 : A mole plough. The foot and expander form the drainage channel and the leg generates a slot with associated soil fissures which extends from the surface down into the channel.

3.1.4. Vertical drainage

Vertical drainage systems use pumped wells, either open dug wells or tube wells. Agricultural land is drained by pumped wells to improve the soils by controlling water table levels and soil salinity. Tubewell drainage refers to the technique of controlling the water table and salinity in agricultural areas by pumping, from a series of wells, an amount of groundwater equal to the drainage requirement. The success of tubewell drainage depends on many factors, including the hydrological conditions of the area, the physical properties of the aquifer to be pumped and those of the overlying fine-textured layers.

3.2. Subsurface drainage coefficient

Drainage coefficient is the volume of water per unit area to be removed in 24 h. The drainage coefficient is important in subsurface drainage design and a dependable drainage coefficient is difficult to obtain. It depends on land use, rainfall, runoff, infiltration and evapotranspiration. Incorporation of all these factors in a single physical parameter is difficult to measure effectively.

In arid regions, drainage coefficients are calculated on the basis of irrigation management and leaching requirements for salinity control. Since excess salts in the root zone are critical to the reclamation process than improvement in the aeration. In humid areas, the drainage coefficient is taken to be the depth of rainfall removed in 24 h. Drainage coefficients can be calculated from a soil water balance method. Selection and use of suitable drainage coefficients have always been problems in the design of subsurface drainage system. Use of a low value will reduce the effectiveness of the drainage system whereas use of a high value will raise its cost.

Drainage coefficient is the most important parameter that decides the lateral drain spacing, size of the laterals and collectors and capacity of the pump to dispose off the drainage effluent. The drain spacing is less in cases where drainage coefficient is more as compared to a case where drainage coefficient is less. As such, the cost of the system depends largely upon this parameter. Therefore, the need to select an appropriate value for this parameter has always been emphasized. The drainage coefficients for some of the sites in India have been observed to be in the rane of l-5mm. In the case of subsurface drainage design based on non-steady state conditions, the drainage criterion is based on time to lower the water table from a predecided original to the final level. Usually the original water table is considered at the soil surface while the final level is taken as 30cm below the soil surface. As per the recommendation the capacity of the drainage system should be sufficient to lower the water table by 30 cm in 2 days time. In general, design rates area likely to be in the following DC ranges given in Table 6.

Table 6: Range of drainage coefficient suitable for various conditions.

DC, mm/d	Suitable conditions
<1.5	Soils low infiltration rate
1.5-3.0	Most soils, with higher rate for more permeable soils and where cropping intensity is high
3-4.5	Extreme conditions of climate, crop and salinity managements, an under poor irrigation practices.
>4.5	Special conditions, e.g. rice irrigation on light textured soils

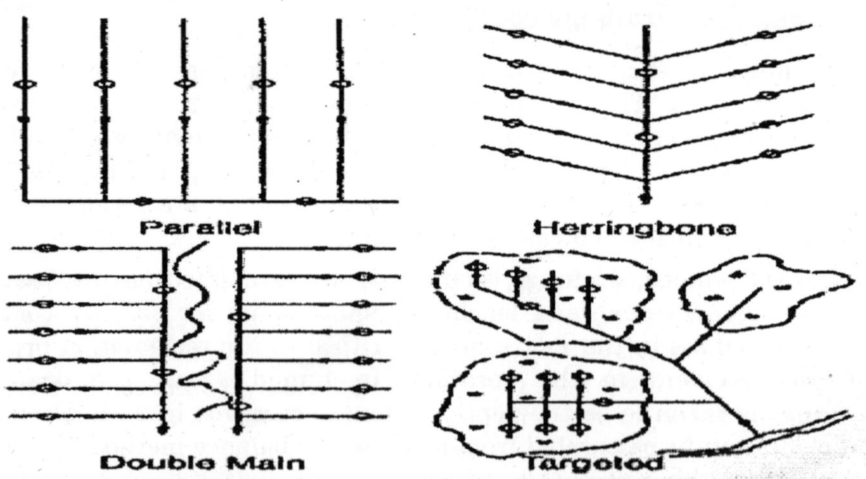

Figure 10 : Various drainage system layout. The objectives of layout include removing water from an isolated area, improving drainage in an entire field, intercepting a hillside seep, and so on.

3.3. Drainage system layout

Although there may be many possible layout for a given field specific drainage objectives should be evaluated to find the best layout (Figure 10). System layout and drainage needs should be based anticipating future needs where possible. Additions to a system will be much easier to make if the established mains are already large enough and located appropriately.

3.4. Spacing between drains

When there is a linear flow of ground water, spacing between drains is given by the relationship,

$$S^2\,4K = \frac{(H^2 - h^2 + 2dH - 2dh)}{Q} \qquad 2$$

where S = Spacing between drains

K = hydraulic conductivity

Q=rate of discharge,

H=Maximum height of water table above bottom of drains and

H = height of water in the drain.

d = depth of impervious layer below the drain.

3.5. Measurement of in situ hydraulic conductivity

Measurement of hydraulic conductivity on disturbed soil samples is not a very reliable approach. Similarly, determining these values on undisturbed samples drawn with a core sampler is also not a flawless method. Considerable errors are caused due to presence of root holes. Methods have, therefore, been developed to measure hydraulic conductivity in situ. Three simple methods in this respect are those of hooghoudt, Ernst and Childs. These are described below.

3.5.1. Hooghoudt's method

The methods is suitable for homogeneous soil. In this method a hole is made in the soil below a water table with the help of post hole digger. The water level in the bore is allowed to come in equilibrium with the water table in the soil and the water is pumped out from the bore. The rate of rise of water in the bore is recorded and the hydraulic conductivity is calculated with the following formula

$$\frac{k = 2.3as}{(2d + a)\Delta t} Log10 \frac{Y_0}{Y_l} \qquad (3)$$

when auger hole is drilled upto the impermeable layer (3)

Where,

a = hole radius

Δt = time in seconds,

yo - initial water level in cm,

yi = final water level in cm,

K = hydraulic conductivity in meter/second,

d = distance from bottom of the hole to water table and

S = constant of proportionality which is a distance from bottom of the hole to the impermeable layer and is equal to ad/0.19.

3.5.2 Ernst's method for layered soils

In this method two auger holes Bi and B2 of different depths are made in the soil. The bottom of the first hole is kept 10 cm above the second layer. The second hole should be deep enough in the lower layer. The entity (d-h) should be greater than 15.

The conductivity in the first hole, KI is determined as usual. The mean conductivity K is determined for two layers by recording rate of rise in the second hole. Hydraulic conductivity in the 2nd layer, K_2 is then calculated by the following relationship.

$$KIdI+K_2(d_2-dI) = Kd_2 \text{ ....} \tag{4}$$

Where,

K_1 and K_2 = hydraulic conductivity of 1st and 2nd layer, respectively.

K = mean hydraulic conductivity of two layers

D_1 and d_2 == distance between water table and bottom of the lsl and 2nd bore, respectively. The results are satisfactory if K_2 is greater than K_1

3.5.3 Child method with two auger holes

Two auger holes, W_1 and W_2 of equal diameter and depth, are made below the water table. Water from W, is pumped out and poured into w_2 to create a difference in hydraulic head Δ H. Hydraulic conductivity K is then calculated as follows-

$$\frac{K=Q}{nL\Delta H} Cosh^{-1}\frac{b}{2a}$$

Where

Q = rate of pumping,

Δ H = hydraulic head difference between the two holes,

L = depth of the hole below the water table,

a = radius of the hole,

b = distance between the two holes,

4. Drainage Materials

The subsurface field drainage systems consist of horizontal or slightly sloping channels made in the soil; they can be open ditches, buried pipe drains, or mole drains;, also, a series of wells. Drainage system requires several materials to be used such as tiles, pipes, and envelope materials for its better functioning. Envelope materials such as gravel envelop and filter including geotextile filters can be used as per requirement. Various coefficients developed may be used for proper selection of the materials.

4.1. Tiles materials

Concrete, burnt clay and PVC are the materials used for tiles. Concrete tiles are easy to manufacture and can be transported without breaks. Concrete tiles can also be locally made using simple moulds made of mild sheets. Concrete 1: 2:4 proportions should be used and tiles should be properly cured. Clay tiles should be well burnt and without any defects. Good clay tiles have distinct ring when tapped with metal objects. These have advantage of being cheaper than concrete tiles and also can be used in acid and alkali soils. However, they are difficult to transport without breakage and will not be able to with stand as much load as concrete tiles can take.

Plastics pipes are increasingly being popular for tile drainage. The common material is the polyvinyl chloride and polyethylene (PVC). PVC pipes are somewhat more resistant to outside pressure. Plastic pipes come in both smooth and corrugated types. The smooth pipes are rigid and come in standard length. Corrugated pipes are flexible and come in coils. They can be conveniently be used with tile lying machines. They have greater hydraulic resistance than smooth pipes and hence larger size pipe is required to drain the same amount of water. Circular hole in the corrugation allows the water enter into tile drains. Around 20 holes per 30 cm length are provided. The Use of envelope material increases the effective diameter and hence fewer perforations per unit length will be required to get the desired flow. Pipes require less plastic material per unit length and have greater resistance to outside pressure.

4.2. High density polyethylene

Polyethylene is an extremely versatile piping material and has some properties that make ideal for use in underground drainage systems. Relatively lightweight, polyethylene allows for easier and less costly transportation and installation costs. It is not easily susceptible to cracking during pipe handling and installation activities. A corrugated

HDPE pipe is resistant to abrasion, corrosion, chemical scouring and is structurally strong with the ability to support large loads. This characteristic offers advantages for use as underground drainage systems.

Corrugated perforated HDPE pipes are generally used for subsurface drainage. Corrugated both inside and outside, perforated pipe is ideal for subsurface drainage used in everything from draining garden to major highways. Lightweight and durable, can handle chemically abrasive environments, is easy to install and not easily susceptible to cracking during pipe handling and installation. Flexibility and perforation are its most important features. Due to its flexibility, perforated pipe is better suited for trenches or where the ground is uneven, and easy to install. These have slots around the entire circumference or half-perforated, i.e., have holes on only one side of the pipe, non-perforated, and perforated covered in filter sock. Perforations allow subsurface water to be collected and transported to desired locations for discharge. The flexibility and lightweight of perforated pipe becomes an advantage for use in harsh conditions.

4.3. Drain envelopes and filters

Drain envelopes and filters are two different techniques used to solve different problems. Drain envelopes are permeable materials, such as gravel place around the drains for the purposes of improving flow conditions. Filters for drains are permeable materials, such as geotextiles, placed around the drains for the purpose of preventing fine-grained materials in the surrounding soil from being carried into the drain by groundwater. Drain envelope is a material placed around a drain pipe to provide either hydraulic function, which facilitates flow into the drain, *or* barrier function, which prevents certain sized soil particles from entering the drain. It is termed as sock also. Drain envelopes are not filters. Filters become clogged over time, but, drain envelopes do not.

Many types of envelope material such as thick gravel and organic fiber to thin geotextiles are in use. The useful life of a synthetic drain envelope is long, if not left in the sun for a long time and exposed to much ultraviolet radiation. Fine-textured soils with a clay content of 25 to 30 per cent are generally considered stable, so they don't need drain envelopes. A geotextile sock is recommended for coarse-textured soils free of silt and clay. These soils are considered unstable even if undisturbed, so that particles may wash into pipes. The need for an envelope in intermediate soils (clay contents less than 25 to 30 per cent) is best left to a professional contractor or soil and water engineer because soil movement is more difficult to predict.

In general gravel, coarse and very coarse sand, silty clay loam, sandy clay and peat soil do not require any filter or envelope materials. Gravelly fine sand, medium sand, fine sand, clay sand, sandy loam, loam and silt require filter materials in low to high degree of urgency. However, silty clay and clay require envelope materials.

4.4. Envelope material

In soils where there is a chance of heavy movement of soil particles into the tile drains and consequently clogging them, filter materials are placed around the tiles. The filter or envelope materials prevent the entry of the relatively coarser particles as well. They act as bedding material and thus allow the tiles to take greater loads. They also increase the effective diameter of the tiles and allow quantities of water to flow into the tiles. A variety of materials have been used as envelope materials in tile. These consist of gravel coarse sand organic materials like straw, coir matting etc. However, gravel has been the most widely used material because of its; efficiency and also of permanent nature.

4.4.1. Gravel envelope

The function of a drain envelope is to improve permeability in the surrounding of the drain. For this reason, the envelope material should have a hydraulic conductivity 7 times higher than the base material. Since envelopes are not designed for their filtration capacity, they do not need to be well graded. The gravel envelope should Prevent movement of soil particles into the tile drains. The thickness should be the same as the sand and gravel filter (i.e. 100 mm around the pipe). All the envelope material should be smaller than 38 mm, $D90 < 19$ mm and the $D10 > 0.250$ mm.

The gravel envelope should be designed taking into consideration the nature of the soil around the tile drains. The gravel should be clean; free from organic matter and bentonite clays. The United States Bureau of Reclamation recommends the following criteria for the design of gravel envelopes. For uniform soils, the ratio of D_{50} of envelope material and D_{85} of soil should range from 5 to 10. However, the ratio ranged from 12 to 58 for graded soils.

4.4.2. Filters

Filters can be either geotextile or well graded gravel and sand. Filters are necessary only where there is something in soil that needs to be filtered out, namely fine sand. If sand is not present, filters are not necessary. Not only do filters add to the cost of drainage but they also constitute an additional barrier to inflow of water and can, therefore, reduce the effectiveness of the drain. Few soils present the danger of sand particles clogging the drainage system. Most soils contain sufficient amounts of clay or organic matter to form relatively stable aggregates of individual soil particles. Filters in these soils are of no benefit and may have reduced drain performance. Instead of filters, a porous envelope is appropriate to ensure good flow conditions at all times at the interface between drain and soil. Drain rock, pea gravel and similar materials meet these requirements and are used extensively in some drainage applications. The first step in design of a proper filter system, either geotextile or sand and gravel, is to perform a particle size analysis of soil at drain depth in field. Usually soils with more than 30% clay content do not require a filter.

4.4.2.1. Geo textile filters

Suitable filters may be used to restrict entry of fine particles of silt and sand from entering the drains in soils with poor cohesion. A properly designed filter stabilizes the soil around the drain and allows free entry of water. There are two basic types of geotextile filters, knitted and non-woven. Knitted geotextiles are usually made of polyester or polypropylene filaments that are knitted or woven together. The most common type has a thickness of 1 mm, a weight of 150 g/m^2 and an Apparent Opening Size (AOS) of 300 microns. For applications requiring more filtration capacity, a sock knitted with velour or pile on one side, that is thicker (< 2 mm), heavier (250 g/m^2) and has an AOS of about 100 microns is needed.

Non-woven geotextiles are made from several layers of randomly distributed fibers that are rolled pressed and usually interconnected by needle punching. It has been found that fabrics about 2 mm thick are very good for silty soils. The general guide for designing a geotextile filter is that the ratio of O_{95} Fabric material and D_{50} Soil should be $=< 2.5$. Where O_{95} is the apparent opening size (AOS) of the geotextile filter D_{85} is the size of which 85% of the particles are finer.

4.4.2.2. Gravel and sand filters

In arid areas sand and gravel filters are used to some extent instead of geotextile filters. Drains usually run deeper and the sand and gravel filters also act as an envelope to improve bedding and permeability characteristics. Filter materials should be well graded. If more than one gradation is used, the layers should be from coarsest to finest material, starting at the pipe. A minimum thickness of 100 mm is recommended for each layer of the filter.

Limits for the filter material are such that ratio of D_{50} Filter material and D_{50} Base material ranges between 12 to 58 and that of D_{15} Filter material and D_{15} Base material from 12 to 40. D_{50} (lhe size of which 50% of particles are passing through the screen) of the base material times 12 and 58 will yield the lower limit and upper limit for D_{50} filter. Provided the filter has no more than 5% finer than 0.074 mm and is relatively well graded.

The chosen filter material should be checked for stability by adopting the criterion that the ratio of D_{15} Filter material and D_{85} Base material should be < 5. The D_{85} size of the filter material with respect to the opening of the drainpipe should be verified using ratio of D_{85} Filter material and Maximum drainpipe opening as > 2.

A well graded filters material is required for good performance. A filter material is considered well graded when all particle sizes from the largest to the smallest are present in a balanced way. As given below, the coefficient of uniformity can be used to verify how well graded the material is.

$$C_u = D_{60} \text{ Filter material} / D_{10} \text{ Filter material} \quad (6)$$

where C_u = Coefficient of uniformity

$$C_c = (D_{30})^2/(D_{10})(D_{60}) \quad (7)$$

where, C_c = coefficient of curvature.

In general, for well graded filter materials maximum size of aggregates should be 38 mm. However, D_{90} upto 19 mm and D_{10} upto 0.25 mm. C_u >4 for sand and >6 for gravel may be recommended. However, C_c might range in between I and 3 for well graded filter materials.

5. Conclusions

An area with water table within 2 m from the land surface is called as water logged area. It is potential to water logging if water table is between 2-3 m from the land surface. It has been estimated that around 16.71 million hectare land is affected by salt and waterlogging in India. The drainage is the remedy to these problems for sustainable agriculture. Drainage is defined as the natural or artificial removal of surplus ground and surface water and dissolved salt from the land in order to enhance agriculture production. In the case of natural drainage the excess waters flows from the fields to lakes, swamps, streams and rivers. However, in an artificial system surplus ground or surface water is removed by means of sub surface or surface conduits.

Improved drainage create a healthier environment for plant growth, It conserve soil and water; and provide drier field conditions for ease in farm operations for the crop production. Agricultural drainage is must to realize the full benefit of irrigation. Certain drainage criteria must be used to determine drainage need. Drainage is broadly divided into two types i.e., Surface and subsurface drainage. The excess surface or sub surface water of an agricultural field can be removed by applying the appropriate drainage method.

The drainage system may consists of field drains system and main drainage system. Water of field is drained through field drainage and sent to main drainage for moving towards outlet. Field drainage system may be divided into surface drainage by gravity flow and subsurface drainage by gravity or pumped flow. The main drainage may be divided into deep collectors consisted of pipe or ditches for subsurface drainage and shallow collectors consisted of channels or ditches for surface drainage. The collector drains flows to disposal drains and to outlets of the drainage system to some stream or depressions.

Drainage coefficient is defined as the amount of water that runs off from a given area and is to be removed in 24 hours. While designing surface drainage system, a low value of the drainage coefficient will lead to partial improvement in drainage though the cost of design may be relatively low, whereas a high value would increase the cost substantially without any additional gain in the removal of surface congestion. Estimation of 24 hr rainfall depth that might occur with a probability level generally of 20% or a return period of 5 years should be considered for agricultural drainage.

Field drains for a surface drainage system have a different shape from field drains for subsurface drainage. Those for surface drainage have to allow farm equipment to cross them and should be easy to maintain with manual labour or ordinary mowers. Field drains are shallow and have flat side slopes. Simple field drains are V-shaped. A free board of 25% of designed depth is kept. Permissible values for average velocity of flow to avoid scouring may be adopted Subsurface drainage improvement is designed to control the water table level through a series of drainage pipes that are installed below the soil surface. The subsurface drainage network generally outlets to an open ditch or stream. Subsurface drainage requires some minor maintenance of the outlets and outlet ditches. For the same amount of treated area, subsurface drainage improvements generally are more expensive to construct than surface drainage improvements Subsurface drainage may be achieved by tubewell drainage, open drains or subsurface drains (pipe drains or mole drains). Tubewell drainage and mole drainage arc applied only in very specific conditions. Subsurface (groundwater) drainage for water table and soil salinity in agricultural land can be done by horizontal and vertical drainage systems. Horizontal drainage systems use open ditches (trenches) or buried pipe drains. Parallel, herringbone, targeted and double main system layout could be adopted for subsurface drainage system. The spacing od drains could be evaluated using Hooghoudt or Ernst or Child method. Drainage system requires several materials to be used such as tiles, pipes, and envelope materials for its better functioning. Envelope materials such as gravel envelop and filter including geotextile filters can be used as per requirement. Various coefficients developed may be used for proper selection of the materials to realize effective drainage for sustainable agriculture.

References

Adams, W.M. (2006). The Future of Sustainability: Re-thinking Environment and Development in the Twenty-first Century. Report of the 1UCN Renowned Thinkers Meeting, 29-31 January' 2006

FAO Glossary of Land and Water Terms. (2001). FAO Glossary of Land and Water Terms http.www.fao.org/landandwater/glossary // wglos.jsp? keyword I 1 = & subject = % 25 & term_e=Drainage+& search = Display. (FAO/IPTRID, drainage and sustain-ability, issues paper No. 3 October 2001).

Gosh, S.P. (1991). Agro-Climatic Zone specific Research-Indian perspective under NATP. ICAR, Krishi Bhawan, New Delhi.:539.

MOWR (1991) Report of the Working Group on Problem

Identification in Irrigated Areas with Suggested Remedial Measures, Ministry of Water Resources, Govt. of India. New Delhi.

NCA (1976). Report of the National Commission on agriculture. Ministry of Agriculture. C(Department of Agriculture), Govt. of India, New Delhi.

National Water Policy. (2002). Ministry of Water Resources. Government of India, New Delhi

Ott, K (2003)). The cases for strong sustainability. In: Greifs Wald's Environmental Ethics. Ott, K. and P. Thapa (eds.). Greifswald: Steinbccker Verlag Ulrich Rose

Singh, D.K., Chandola, V.K. and Singh, R.M., (2009). Sustainable management of irrigation water, An approach to address impact of climate change in agriculture. Flair Book Publication, New Delhi.

Singh, R, Singh, R.M., Kishore, R., and rao, K.V.R. (2011). Agricultural drainage technologies-status and Indian experience. Proceedings of National Seminar on Sustainable Management of Water Resources. Banaras Hindu University, Varanasi, 14-15 *January, 2011:7:28*

Tyagi (1999). Management of salt affected Soil. In : 50 Years of Natural Resource Management Research.(Ed. G.B. Singh and B.R. Sharma.). Division of Natural Resource Management. ICAR, Krishi Bhawan, New Delhi.

United Nations. (1987). Report of the World Commission on Environment and Development. General Assembly Resolution 42/187, 11 December 1987.

Chapter - 41

A New Beginning on Water in the 12th Plan

Mihir Shah

INDIA FACES a major crisis of water as we move into the 21st century. This crisis threatens the basic right to drinking water of our citizens; it also puts the livelihoods of millions at risk. The demands of a rapidly industrialising economy and urbanizing society come at a time when the potential for augmenting supply is limited, water tables are falling and water quality issues have increasingly come to the fore.

Limits to Large Dams

Recent scholarship points to definite limits to the role new large dam projects can play in providing economically viable additional water storage (Ackerman, 2011). The ambitious scheme for interlinking of rivers also presents major problems. The comprehensive proposal to link Himalayan with the Peninsular rivers for inter-basin transfer of water was estimated to cost around Rs. 5,60,000 crores in 2001. Land submergence and R&R packages would be additional to this cost. There are no firm estimates available for running costs of the scheme, such as the cost of power required to lift water. There is also the problem that because of our dependence on the monsoons, the periods when rivers have "surplus" water are generally synchronous across the subcontinent. A major problem in planning inter-basin transfers is how to take into account the reasonable needs of the basin states, which will grow over time. Further, given the topography of India and the way links are envisaged, they might totally bypass the core dryland areas of Central and Western India, which are located on elevations of 300+ metres above MSL. It is also feared that linking rivers could affect the natural supply of nutrients through curtailing flooding of the downstream areas. Along the east coast of India, all the major peninsular rivers have extensive deltas. Damming the rivers for linking will cut down the sediment supply and cause coastal and delta erosion, destroying the fragile coastal eco-systems.

It has also been pointed out that the scheme could affect the monsoon system significantly (Rajamani et al, 2006). The presence of a low salinity layer of water with low density is a reason for maintenance of high sea-surface temperatures (greater than 28 degrees C) in the Bay of Bengal, creating low pressure areas and intensification of monsoon activity. Rainfall over much of the sub-continent is controlled by this layer of low saline water. A disruption in this layer could have serious long-term consequences for climate and rainfall in the subcontinent, endangering the livelihoods of a vast population.

The Crisis of Groundwater

The relative ease and convenience of its decentralised access has meant that groundwater is the backbone of India's agriculture and drinking water security. Groundwater is a Common-Pool Resource (CPR), used by millions of farmers across the country. Over the last four decades, around 84 per cent of the total addition to the net irrigated area has come from groundwater. **India is, by far, the largest and fastest growing consumer of groundwater in the world. But groundwater is being exploited beyond sustainable levels and with an estimated 30 million groundwater structures in play,** India may be hurtling towards a serious crisis of groundwater over-extraction and quality deterioration.

Nearly 60 per cent of all districts in India have problems related to either the quantity or the quality of groundwater or both. According to the Central Ground Water Board's latest assessment (CGWB, 2009), at the all India level, the stage of groundwater development is now 61 per cent. In Punjab, Haryana, Rajasthan and Delhi, this level has crossed 100 per cent, closely followed by Tamil Nadu (80 per cent) and UP (71 per cent).

Need for a Paradigm Shift

Given this apparent emergence of limits to further develop of water resources in large parts of the country, the 12th Plan faced a challenge of how to move forward. It was clear that business-as-usual would not do. New ideas needed to be desperately put into place for which the best scholars and practitioners had to come together. Thus, a new architecture of plan formulation was designed. The Working Groups for the 12lh Plan in the water sector were, for the first time in the history of the Planning Commission, all chaired by renowned experts from outside government. Over the course of several months in 2011-12, a new path was charted out, giving rise to a ten-fold paradigm shift in water resource management in India. This paper outlines the main features of this change.

Ten Elements of the Paradigm Shift

1. Large Irrigation Reform

Given the emerging limits to further development in the Major and Medium Irrigation (MMI) sector, the 12.[h] Plan proposes a move away from a narrowly engineering-construction-centric approach to a more multi-disciplinary, participatory management perspective, with central emphasis on command area development and a sustained effort at improving water use efficiency, which continues to languish at a very low level. Given that nearly 80 per cent of our water resources are consumed by irrigation, an increase in water use efficiency of irrigation projects by the 12th Plan goal of 20 per cent will have a major impact on the overall availability of water not only for agriculture but also for other sectors of the economy.

The key bottleneck so far has been that capacities of irrigation departments in many states to deliver quality services have failed to keep up with the growing MMI investments. While States compete for capital investments in new MMI projects, they do little to manage them efficiently.

This is closely linked to the fact that in many states the Irrigation Service Fee (ISF) to be collected from farmers has been abolished or is as low as 2.-8 per cent of dues. In this way, the accountability loop between farmers and irrigation departments is broken. Wherever ISF gets regularly collected, irrigation staff shows greater accountability and responsiveness to farmers. There is greater contact between the two, there is greater oversight of water distribution and farmers expect at least a minimal level of service if an ISF is demanded of them.

A substantial National Irrigation Management Fund (NIMF) is, therefore, being created to incentivise states to make the required paradigm shift. The NIMF will be a non-lapsable fund that reimburses to state irrigation departments, a matching contribution of their ISF collection from farmers on a 1:1 ratio. In order to generate competition among MMI staff across commands, States would allocate the central grant to MMI systems in proportion to their respective ISF collection. To encourage Participatory Irrigation Management (PIM), the NIMF will provide a bonus on that portion of each State's ISF collection which has been collected through Water User Associations (WUAs). And this will be on the condition that WUAs and their federations are allowed to retain definite proportions /of the ISF, which would not only enable them to undertake repair and maintenance of distribution systems, but also increase their stakes in water management.

Similarly, to encourage volumetric water deliveries, NIMF will provide an additional bonus on that portion of a State's ISF collection which accrues through volumetric water supply to WUAs at the outlet level. The clear understanding is that empowering WUAs is the key to making the process of pricing of water and ISF collection more transparent and participatory. These proposals are based on experience on the ground over the last few years in Andhra Pradesh, Gujarat, Maharashtra and Karnataka.

Our huge investments in irrigation have yielded much less than what they should have, mainly because Command Area Development (CAD) has been consistently neglected and divorced from building of irrigation capacities. The 12th Plan stipulates that all irrigation project proposals (major, medium or small) will, henceforth include CAD works from the very beginning as an integral part of the project.

2. Participatory Aquifer Management

Since groundwater accounts for nearly two-thirds of India's irrigation and 80 per cent of domestic water needs, the 12th Plan advocates the adoption of a participatory approach to sustainable management of groundwater based on aquifer mapping that takes into account the common pool resource (CPR) nature of groundwater.

It is this understanding that underpins the comprehensive programme for the mapping of India's aquifers initiated during the 12th Plan.

This requires an understanding based on the following aspects:

- Relationship between surface hydrologic units (watersheds and river basins) and hydro geological units, i.e. aquifers;
- The broad lithological setup constituting the aquifer with some idea about the geometry of the aquifer - extent and thickness;
- Identification of Groundwater recharge areas, resulting in protection and augmentation strategies;
- Groundwater balance and crop-water budgeting at the scale of a village or watershed;
- Groundwater assessment at the level of each individual aquifer in terms of groundwater storage and transmission characteristics, including the aquifer storage capacity;
- Regulatory options at community level, including drilling depth (or whether to drill tube wells or bore wells at all), distances between wells (especially with regard to drinking water sources), cropping pattern that ensures sustainability of the resource (aquifer) and not just the source (well/ tubewell), comprehensive plan for participatory groundwater management based on aquifer understanding, bearing in mind principles of equitable distribution of groundwater across all stakeholders. Each of these will be the central focus of the National Aquifer Management Programme being launched in the 12th Plan.

3. Breaking the Groundwater-Energy Nexus

The current regime of power subsidies for agriculture has had a major role to play in deteriorating water tables in most parts of India. These very same power subsidies fuelled the Green Revolution but given the emerging stresses on groundwater, an imaginative way needs to be found, which breaks the groundwater-energy nexus, without hurting farmer interests. The single most effective solution found by States has been the physical segregation of power feeders to provide 24x7 electricity to rural habitations and non-farm users, with separate feeders giving 3-phase predictable supply to agriculture, which is rationed in terms of total time, at a flat tariff. This provides requisite power to schools, hospitals and the non-farm economy, while allowing rationed supply of power to agriculture, which can be at off-peak hours. For example, the Government of Gujarat invested US $1250 million during 2003-2006 to separate 800,000 tubewells from other rural connections and imposed an 8 hour/day power ration but of high quality and full voltage. This was combined with a massive watershed development programme for groundwater recharge. The net result has been: [a] halving of the power subsidy; [b] stabilized groundwater draft and [c] improved power supply in the rural economy. Combined with other measures such as High Voltage Distribution System (HVDS), specially designed transformers and energy-efficient pumpsets, this could be a better way of delivering power subsidies that cuts energy losses and stabilises the water table at the same time. Major investments are proposed in this direction in the 12th Plan.

4. Watershed Restoration and Groundwater Recharge

Even while emphasizing the need to improve the efficiency and sustainability of our irrigation systems, the 12^{111} Plan is fully cognizant of the fact that the demands of national food security necessitates a major breakthrough in the productivity of our rainfed areas. A primary requirement for this is a massive programme for watershed restoration and groundwater recharge. The 12th Plan proposes to move in this direction by transforming MGNREGA into our largest watershed programme, giving renewed energy to the reformed Integrated Watershed Management Programme (IWMP) launched in the 11th Plan and launching a completely revamped programme on Repair, Renovation and Restoration (RRR) of Water Bodies.

5. A New Approach to Rural Drinking Water and Sanitation

The fact that the same aquifer is being tapped for both irrigation and drinking water, without any co-ordinated management of the resource, has greatly aggravated availability of drinking water. Indeed, we are close to entering a "vicious infinite regress" (Wittgenstein, 1953, sec. 239) scenario, where an attempt to solve a problem re-introduces the same problem in the proposed solution. If one continues along the same lines, the initial problem will recur infinitely and will never be solved. This regress appears as a natural corollary of "hydroschizophrenia" (Llamas and Martinez-Santos, 2005; Jarvis et al, 2005), which entails taking a schizophrenic view of an indivisible resource like water, failing to recognize the unity and integrity of the hydrological cycle. Thus, tubewells drilled for irrigation are more and more drying up the aquifers being used for drinking water.

Lack of convergence with sanitation, on the other hand, compromises water quality, even as it makes provision of improved sanitation difficult. Water quality has also been affected chemically due to geogenic leaching (arsenic and fluoride).

This understanding of the flaws in the drinking water programme has prompted the adoption of a new approach based on the principle of subsidiarity that seeks solutions to these problems as close to the ground as possible. Decisions on location, implementation, sustainability, O&M and management of water supply schemes will be devolved to local drinking water and sanitation committees with an umbrella role for Gram Panchayats (GPs) for effective implementation. A Management Devolution Index (MDI) will track and incentivise more substantive devolution of functions, funds and functionaries to the GPs. The problem of vicious infinite regress can only be tackled through the sustainable and participatory aquifer management approach described earlier, so that the left hand of drinking water knows what the right hand of irrigation is doing.

Convergence between drinking water supply and sanitation will be strengthened by taking up villages covered with piped water supply to get Open Defecation Free (ODF) status on priority and vice versa. Waste water treatment and recycling will be an integral part of every water supply plan or project. Management of liquid and solid waste will be promoted together with recycling and reuse of grey water for agriculture and groundwater recharge and pollution control. This will be done on priority in Nirmal Gram Puraskar (NGP) villages.

The Total Sanitation Campaign (TSC) was launched in 1999 as a demand-driven, community-led programme. But the progress remains far from satisfactory. Open defecation by around 600 million people is our biggest national shame. Latest Census data reveals that the percentage of households having access to television and telephones in rural India in 2011 exceeds the percentage of households having access to toilet facilities and tap water. The APL-BPL distinction and the very low incentive under the TSC have played havoc with the programme.

Thus, the 12th Plan proposes a major shift in strategy. The APL-BPL distinction and the focus on individual toilets are to be replaced by a habitation saturation approach. The idea is not to sacrifice quality and sustainability of outcomes in the mad rush to attain targets, even if this means moving somewhat slower in reaching universal coverage. Through a convergence with MGNPREGA, the unit cost support for individual household latrines has been raised to Rs. 10,000. Toilet designs will be fine-tuned in accordance with local social and ecological considerations. In order to focus more centrally on sustainability of outcomes, the programme will be taken up in a phased manner wherein GPs shall be identified, based on defined criteria of conjoint approach to sanitation and water supply, for achievement of NGP status. This would progressively lead to Nirmal Blocks, Nirmal Districts and eventually Nirmal States.

6. Conjoint Water and Wastewater Management in Urban India

The challenges of safe drinking water and waste management are perhaps even greater in urban India. Nothing less than a paradigm shift is required in the 12th Plan if we are to move towards sustainable solutions to urban water and waste management:

(a) Investments in water supply must focus on demand management, reducing intra-city inequity and on quality of water supplied. This will require cities to plan to cut distribution losses through bulk water meters and efficiency drives. User charges should plan to cover increasing proportions of O&M costs, while building in equity by providing "lifeline" amount of water free of charge, with higher tariffs for increasing levels of use.

(b) Each city must consider, as first source of supply its local water bodies. Therefore, cities must only get funds for water projects, when they have accounted for the water apply from local water bodies and have protected local water bodies and their catchments. This pre condition will force protection and will build the infrastructure, which will supply locally and then take back sewage also locally. It will cut the length of the pipeline twice over - once to supply and the other to take back the waste.

(c) No water scheme will be sanctioned without a sewage component. Planning for 'full coverage and costs will lead cities to look for unconventional methods of treating waste. For instance, cities would then consider treatment of sewage in open drains and treatment using alternative biological methods of wastewater treatment. The principle has to be to cut the cost of building the sewage system, cut the length of the sewage network and then to treat the waste as a resource - turn sewage into water for irrigation or use in industry. Indian cities have the opportunity to leapfrog into new ways of dealing with excreta, which are affordable and sustainable, simply because they have not yet built the infrastructure.

Cities must plan for reuse and recycling of waste at the very beginning of their water and waste plan and not as an after thought. The diverse options for reuse must be factored in - use in agriculture, for recharge of water bodies, for gardening and for industrial and domestic use.

7. Industrial Water

As the economy industrialises, it is extremely important that the industry adopts the best international practices to improve water use efficiency. This can be broadly done in two ways:

- reducing the consumption of fresh water through alternative water-efficient technologies or processes in various manufacturing activities; and
- reusing and recycling the waste water from such water intensive activities and making the reclaimed water available for use in the secondary activities within or outside the industry.
- It is proposed to make it mandatory for companies to include every year in their annual report, the details of their water footprint for the year. This would include:
- the volume of fresh water (source-wise) used by them in their various production activities (activity-wise);
- the volume of water used by them that was reused or recycled (again activity-wise);
- a commitment with a time-line that the company would reduce its water footprint by a definite amount (to be specified) within a definite period of time (to be specified).

8. Non-Structural Mechanisms for Flood Management

In addressing the problem of floods, the central focus over the years has been on engineering/structural solutions. Apart from the massive investments in large dams, India has already constructed over 35,000 km of embankments. But these are rapidly reaching their limits. Recent studies show, for example, that "the existing storage infrastructure in Peninsular rivers is mostly designed to smooth out the southwest monsoon flows in, say, 9 out of 10 years. There may still be the 1 in 10 year flood, for which, however, there is no economic justification to invest in substantial additional infrastructure. Instead, better weather and flood forecasting is required, along with flood insurance and possibly the designation of flood diversion areas, whereby farmers are asked to temporarily (and against compensation) set aside embanked land to accommodate flood overflow.

Some State governments (such as Bihar) have decided to broaden their strategy of tackling floods by placing greater emphasis on rehabilitation of traditional, natural drainage systems, leveraging the funds available under MGNREGA (Samaj Pragati Sahayog and Megh-Pyne Abhiyan, 2012). Since this involves a process of complex social mobilisation and social engineering, civil society organisations will work in close partnership with the State government in this endeavour. The 12th Plan strongly endorses such a paradigm shift in flood management away from building more and more embankments and towards a "room for the river" kind of approach.

9. New Institutional Framework

State-level Regulators

The 12th Plan recognises the need to evolve an institutional framework backed by a legal regime that facilitates setting up of regulatory bodies that would enable resolution of water conflicts.

The water quality, environment and health standards set by the regulator have a bearing on tariffs. The final call on tariffs would, of course, be a political one but the regulators have a crucial role in advising governments on the objective basis for tariff determination (somewhat akin to what the CACP does for agriculture pricing). The basic requirements of drinking water and of the environment need to be determined and ensured in a transparent manner and kept as a "Reserve" (as it is called, for example, in South Africa). The determination of this level requires an independent regulator who can transparently, accountably, and in a participatory manner, conduct the processes and procedures required for this.

10. New Legal Framework

New Groundwater Law

Sustainable and equitable management of groundwater based on aquifer management requires a new legal framework to support efforts in this direction.

The 12th Plan has proposed a new *Model Bill for the Protection, Conservation, Management and Regulation of Groundwater*. It is based on the idea that while protection of groundwater is a key to the long-term sustainability of the resource, this must be considered in a framework in which livelihoods and basic drinking water needs are of central importance.

National Water Framework Law (NWFL)

The 12th Plan Sub-Group to draft a National Water Framework Law states that while under the Indian Constitution, water is primarily a State subject, it is an increasingly important national concern in the context of:

(a) the right to water being a part of the fundamental right to life;

(b) the emergence of a water crisis;

(c) the inter-use and inter-State conflicts that this leads to, and the need for a national consensus on water-sharing principles, and on the arrangements for minimising conflicts and settling them quickly;

(d) the threat posed by the massive generation of waste by various uses of water and the severe pollution and contamination caused by it;

(e) the long-term environmental, ecological and social implications of efforts to augment the availability of water for human use;

(f) the equity implications of the distribution, use and control of water between uses, users, areas, sectors, states, countries and generations;

(g) the international dimensions of some of India's rivers; and

(h) the emerging concerns about the impact of climate change on water and the need for appropriate responses at local, national, regional and global levels.

If a national law is considered necessary on subjects such as the environment, forests, wildlife, biological diversity, etc., a national law on water is even more necessary. Water is as basic as (if not more) than those subjects. Finally, the idea of a National Water Law is not something unusual or unprecedented. Many countries in the world have national water laws or codes, and some of them (for instance, the South African National Water Act of 1998) are widely regarded as very enlightened. There is also the European Water Framework Directive of 2000.

Having thus stated the case for drafting a National Water Framework Law, it is important to clarify the nature and scope of this law:

- What is proposed is not a Central water management law or a command-and-control law, but *a. framework* law, i.e., an umbrella statement of general principles governing the exercise of legislative and/or executive (or devolved) powers by the Centre, the States and the local governance institutions.
- But the law is intended to be justiciable in the sense that the laws passed and the actions taken by the Central and State Governments and the devolved functions exercised by PRIs will

have to conform to the general principles and priorities laid down in the framework law, and that deviations can be challenged in the court of law.

- The law incorporates all major legal pronouncements by the Supreme Court with reference to water such as the Public Trust Doctrine and the recognition of the fundamental right to water as also the principle of subsidiary, as explicated in the 73rd and 74th Constitutional amendments, the prevention and precautionary principles, most recently statutorily recognised in the National Green Tribunal Act, 2010 and the transparency principles of The Right to Information Act, 2005.

Given the present constitutional division of legislative powers between the Union and the States, the only way. a National Water Framework Law can be legislated is to follow the procedure laid out in Article 252 (1) of the Constitution. Thus, if two or more State assemblies pass resolutions in support of Parliament enacting such a law, Parliament can also accordingly enact it.

Conclusion

Putting in place this multi-faceted paradigm shift in the 12th Plan has been a massive challenge of initiating change that was in many respects long overdue.

What lies ahead is the even more difficult task of implementing this new approach. What gives hope is the fact that the process of hammering agreement on change has been deeply inclusive and has buy-in from key implementers, especially the State governments. They were a central part of fashioning this change and the most innovative ideas sketched out in this paper are based on examples of best practice from the States. There is also the fact that the emerging water crisis is forcing the pace of change from below, with a range of stakeholders no longer willing to countenance business-as-usual. Even so, the road ahead is a long and difficult one of confronting the recalcitrance of entrenched, attitudes and vested interests. The same preparedness of civil society, academia and government of closely working together that transformed the 12lh Plan agenda will now be required in its implementation, with close involvement of local communities, if success is to be achieved on this path.

References

Ackerman, R (2011): New Directions for Water Management in Indian Agriculture

Amarasinghe, UA et al (2007): India's Water Future to 2025-2050: Business-as-usual Scenario and Deviations, IWMI

Blackmore, D (2010): River Basin Management: Opportunities and Risks, Asian Development Bank

Briscoe, J and RPS Malik (2006): India's Water Economy: Bracing for a Turbulent Future, The World Bank

Central Ground Water Board (2009): Dynamic Ground Water Resources of India.

Climate Wire (2012): How the Dutch Make "Room for the River? By Redesigning Cities, http.www.scientificamerican.com/_article.cfm?id= how-the-dutch-make-room-for-the-river, January 20

Gol (2006): Report of the Technical Committee on Watershed Programmes in India, Ministry of Rural Development, Government of India

Government of Netherlands (2007): Spatial Planning Key Decision: Room for the River, www.ruimtevoorderivier.nl/ media/21963/pkb%204% 20approved %20 decision%20h01 -h086.pdf

Jarvis, T et al (2005): 'International Borders, Ground Water Flow and Hydro schizophrenia', Ground Water, Vol.43, No. 5

Llamas, R. and P. Martinez-Santos (2005): 'Intensive Groundwater Use: Silent Revolution and Potential Source of Water Conflicts", American Society of Civil Engineers Journal of Water Resources Planning and Management, 131, no.4

MoRD (2012a): Report of the Mihir Shah Committee on operational Guidelines for MGNREGA, Ministry of Rural Development, Government of India

MoRD (2012b): Report of the Mihir Shah Committee on Common Guidelines for Watershed Development Projects, Ministry of Rural Development, Government of India.

Ostrom, Elinor (1990): Governing the Commons. The.Evolution of Institutions for Collective Action, Cambridge University Press.

Rajamani, V., U.C. Mohanty, R. Ramesh, G.S. Bhat, P.N. Vinayachandran, D. Sengupta, Prasanna Kumar and R.K. Kolli (2006): 'Linking Indian Rivers vs Bay of Bengal monsoon activity', Current Science, vol.90, 12-13

Samaj Pragati Sahayog and Megh-Pyne Abhiyan (2012): Leveraging MGNREGA for Flood Control — A Case for Policy Reform in Bihar, National Consortium of Civil Society Organizations on MGNREGA

Shah, Mihir (ed) (2012): MGNREGA Sameeksha 2012 - An Anthology of Research Studies on the Mahatma Gandhi National Rural Employment Guarantee. Act (2006-2012), Ministry of Rural Development, Government of India

Shah, Mihir et al (1998): India's Drylands - Tribal Societies and Development through Environmental Regeneration, Oxford University Press

Tiwari, VM et al (2009):'Dwindling groundwater resources in northern Indian region, from satellite gravity observations', Geoph. Res. Lett., 36, L18401,doi.T0.1029/2009GL039401.

Wittgenstein, Ludwig (1953): Philosophical Investigations, Basil Blackwell, Oxford

Chapter –42

Rivers under threat-Strategies for Safeguard

Dhurjati Mukherjee

It is estimated that around 1.4 billion people live in 'closed' river basins where water use exceeds discharge levels, creating severe ecological damage. Symptoms of water stress include the collapse of river systems in western China, rapidly falling groundwater levels in South Asia (and India) and the Middle East and mounting conflicts over access to water. In fact, the resources of transnational rivers, aquifers and lakes have become the target of rival appropriation plans.

Examining the future of Asia's great rivers, it is understood that seven of these - Brahmaputra, Huang Hu, Indus, Mekong, Salween and the Yangtze - will be affected. These river systems provide water and sustain food supplies for over 2 billion people. According to the International Energy Agency (IEA) and the Organization for Economic Cooperation & Development (OECD), the following developments may occur:

(i) The flow of Indus, which receives nearly 90 per cent of the water from upper mountain catchments, would decline by as much as 70 per cent by 2080.

(ii) The Ganges could lose two-thirds of its July September flow, causing water shortages for over 5000 million people and one-third of India's irrigated land area.

(iii) Projections for Brahmaputra point to reduced flows of between 14 and 20 per cent by 2050.

(iv) In Central Asia, losses of glacial melt into the Amu Darya and Syr Darya rivers could restrict the flow of water for irrigation into Uzbekistan and Kazakhstan and compromise plans to develop hydro-electric power in Kyrgyzstan.

In such a situation, where rivers are underthreat and water scarcity looming large in most countries of the Third World, including India, there is need to have a national debate among scholars in the subject. But at this juncture Asian states need to invest more in transboundary basin resources but the harsh truth is that only 57 transnational river basins have a treaty covering water sharing. These are Mekong, Ganges, Indus and Jordan river basins.

Water Contamination

Another significant problem is water contamination affecting countries in South Asia, specially India and Bangladesh, further complicating the matter. One may mention here that the British journal ***Lancet*** reported way back in 2010 that 77 million Bangladeshis had been poisoned by arsenic - the largest mass poisoning in history. Thus for promoting environmental sustainability, water efficiency and stabilizing inter-riparian relations, it is necessary for the Asian nations to forge mutual understanding and develop partnerships.

It may be mentioned here that the International Water Management Institute predicted that by 2025 around 1.8 billion people will live in places suffering from severe water scarcity with land but no water. If this is taken to be the future scenario, one can easily imagine the effect it may have on human beings and on the whole environment.

As far as India is concerned, water crisis compounded with water contamination, including serious river pollution, have been major problems for the country. Rainfall is the main source of water in the country However, 85 per cent of the rainfall is confined to just 4-5 months of the year (or sometimes even less) and varies widely from just 310 mm in the Western belt to a massive 11,400 mm in Meghalaya.

Compared developed countries that capture and store 900 days rainfall in major river basins, India just captures 30 days. Consequently of the total precipitation, a mere 1123 BCM of water is available for utilization - 690 BCM in the shape of surface water and 433 BCM as groundwater resource.

Meanwhile cleaning of India's most important river, Ganga - the fourth largest river basin in the world - has now to be taken up in right earnest. Between 1985 and 2009, the government spent Rs 920 crores spent under the Ganga Action Plan (GAP)-I and GAP-II to clean up the river. Now the National Ganga River Basin Authority (NRGBA), constituted under Section 3(3) of the Environment Protection Act 1986 and headed by no other than the Prime Minister himself, has been trying to deal with pollution in a comprehensive manner. But problems still remain as discharges of domestic and industrial effluents have been increasing.

Delving into statistics, it may be pointed out that upstream of the confluence, where the Salori sewage canal meets the Ganga, biochemical oxygen demand (BOD) — a measure of organic pollutants - increased from an average of 3.5 milligrams per litre to 5 mg/l between 2006 and 2011, while the prescribed limit is 3 mg/l. moreover, coliform - an indicator of human and animal waste - reached an average of 15,000 mpn (most probable number) per 100 millilitres at Salori in September 2010, falling to 8875 mpn/100 ml by the time it reached the confluence a few miles away. Here also the limit in rivers is 500 mpn/100 millilitres. It is distressing that even today around 2900 million litres of sewage are being discharged into the river from various sources, including industries and municipal towns located along its banks.

The existing infrastructure has a capacity to treat only 1100 million litres per day, leaving a huge deficit. These alarm bells were sounded by Dr. Manmohan Singh at the recent (third) meeting of the NGRBA. It is understood that projects worth Rs 2600 crores have been sanctioned so far under the NGRBA in the states of Uttarakhand, Bihar and West Bengal for creating sewer networks, sewage treatment plants, sewage pumping stations, electric crematoria, community toilets and developing the river fronts, the Prime Minister stated but did not yield expected results.

Dr Singh asked the state governments to send proposals for new sewage treatment plants and assured that adequate funding would be available. An investment of around Rs 15,000 crores is estimated to be required in the next ten years to meet the 'Mission Clean Ganga'.

Meanwhile work entrusted to a consortium of IITs – Kanpur, Delhi, Bombay, Kharagpur, Guwahati and Roorkee – to prepare a comprehensive **'River Basin Management Plan for the Ganga'** has recommended comprehensive measures to restore and maintain the ecological health of the river giving due regard to the competing water uses. Discussions have also been held with the World Bank for long term support for NRGBA's work and its financial assistance of $ 1 billion is expected. In its affidavit before the Supreme Court last year, the National River Conservation Directorate under the MoEF stated: "An assistance of one billion dollars has been indicated by the World Bank. A project preparation facility advance of 2.96 million has been sanctioned by the Bank".

It is pertinent here to mention that according to a study titled **Changes in Continental Discharge,** conducted by the National Centre for Atmospheric Research in Colorado, the Ganga in 2004 had 20 per cent less water than it did 56 years ago. In the coming decades, it is likely to shrink even faster and could disappear in another 50 years. Ganga has obviously huge ecological and economic ramifications for India. It will reduce the country's supply of drinking water for irrigation. The region will lose a crucial vehicle for channeling sewage into the sea.

Meanwhile the Yamuna Action Plan have been has also been undertaken and around $ 500 million spent on trying to clean the river, yet pollution levels more than doubled from 1993 to 2005. The problem is that most of the sewage treatment plants are underutilized and a quarter runs at around 30 per cent capacity. According to the Centre for Science & Environment (CSE), 80 per cent of the river pollution is the result of raw sewage combined with industrial run-off that comes to more than 3 billion litres of waste per day, a quantity well beyond the river's assimilative capacity.

It is understood that resource constraints are a big hurdle. It needs to be mentioned here that the problem of the Yamuna river is that apart from sewage flowing into the river, the large-scale extraction of water in upstream Delhi for drinking and irrigation purposes has lead to negligible flow after Wazirabad and this has added to pollution, as per report of the Environment of the Environment Ministry.

The Ganga and the Yamuna aren't India's only polluted rivers, 70-75 per cent of the country's urban waste goes directly into rivers, many of which are so polluted that they exceed permissible limits of safe bathing. Discharge of untreated sewage and industrial effluents lead to a number of conspicuous effects on the river environment. The impact involve gross changes in water quality viz. reduction in dissolved oxygen and reduction in light penetration that tends to reduce self purification capability of river water.

Contamination by synthetic organic pollutants is a recent phenomenon which is even more difficult to demonstrate for lack of appropriate monitoring. The DDT content of the Yamuna river which flows through Delhi is one of the highest ever reported many other problems affect river water quality on a global scale. Very severe pollution by pathogenic microorganisms is still the prime cause of waterborne morbidity and mortality although it is difficult to establish reliable statistical correlation in each case. Many streams and rivers in South America, Africa and particularly on the Indian sub-continent show high coliform levels together with high BOD and nutrient levels. Eutrophication, which has spread widely to lakes and reservoirs of developing countries now also, affects slow flowing rivers.

It needs to be mentioned here that the Central Pollution Control Board (CPCB) has laid down new stringent environmental norms in the form of CREP (Corporate Responsibility for Environmental Protection) but only about 45 per cent of the grossly polluting industrial units have installed effluent treatment plants. Out of these, some have not been functioning properly and also did not meet the technical standards. The National River Conservation Directorate (NRCD) also have no mechanism to ensure that the installed these plants function properly. Therefore, punitive action should be taken against the violators of norms in this regard and defaulting industrial units should either be closed down or allowed to function only after they install effluent plants and ensure their proper functioning. Reliable sources pointed out that the contribution to the pollution load by various sources was estimated at 75 per cent and 25 per cent each for domestic effluent and industrial waste.

The requirement of water is increasing at a rapid pace due to the growth in population as also urbanization and industrialization. The swift economic rise of India has brought water resources under increasing pressure. In fact, the country's per capita water availability has declined from 2030 cubic metres in 1991 to below 1700 cubic metres (or a little above 1600 presently is 2011). The per capita water availability is declining and India is expected to fall in the list of water stressed countries by the year 2015. While the present situation is alarming, projections reveal that the per capita availability may decline to 1000 cubic metres by 2050 or even earlier.

Table 1. Environmental implication of Discharge of Sewage and Industrial Effluents

S.N.	Factor	Principal environmental effect	Potential ecological consequences	Remedial action
1	High biochemical oxygen demand (BOD) caused by bacterial breakdown of organic matter	Reduction in dissolved oxygen (DO) concentration	Elimination of sensitive species, increase in some tolerant species; change in the community structure	Pre-treatment of effluent, ensure adequate dilution
2	Partial biodegradation of proteins and other nitrogenous material	Elevated ammonia concentration; increased nitrite and nitrate levels	Elimination of intolerant species, reduction in sensitive species	Improved treatment to ensure complete nitrification; nutrient stripping possible but expensive
3	Release of suspended solid matter	Increased turbidity and reduction of light penetration	Reduced photosynthesis of submerge plants; abrasion of gills or interference with normal feeding behavior	Provide improved settlement, insure adequate dilution
4	Deposition of organic sludges in slower water	Release of methane and hydrogen as sulphide matter decomposes anoxically Modification of substratum by blanket of sludge	Elimination of normal benthic community loss of interstitial species; increase in the species able to exploit increased food source	Discharge where velocity adequate to prevent deposition

Other poisons				
1.	Presence of poisonous substances	Change in water quality	Water directly and acutely toxic to some organisms, causing change in community composition; consequential effect on pray-predator relation; sub-lethal effects on some species	Increase dilution
Inert solids				
1.	Particles in suspension	Increased turbidity. Possibly increased abrasion	Reduced photosynthesis of submerged plant. Impairing feeding ability through reduced vision or interference with collecting mechanism of filter feeders (e.g. reduction in nutritive value of collected material). Possible abrasion	Improve settlement
2.	Deposition of material	Blanketing of substratum, filing of interstices and/or substrate instability	Change in benthic community, reduction in diversity (increased number of a few species)	Discharge where velocity adequate to ensure dispersion

Moreover, with water pollution spreading rapidly in rivers, lakes and ponds, there is a grave threat to human life. It is a well known fact that most illness, specially in the Third World, are attributable to the use of untreated water, inadequate or faulty treatment or contamination. An increasing proportion of water-borne disease outbreaks are associated with non bacterial microorganisms such as enteric viruses and protozoan parasites because of their successful resistance to water treatment processes.

In such a critical situation, there is an imperative need to clean the major rivers so that it could be used for the welfare of the community. The use of groundwater in most cities has reached the optimum and there is now greater need to make judicious use of river water after purification.

Cleaning the rivers and ensuring proper safeguards to retain its health should be strictly imposed. More importantly, industrial sewage should not be allowed to enter and pollute the major rivers and the government should deal with an iron hand. If necessary, stricter enforcement norms - both of industry and of municipalities -need to be put in place. Apart from these, strategic planning and management of water resources would be absolutely critical for all the Asian states, including India.

Table 2. Major River Basins of the Country

S.N.	Name of the River	Origin	Length	Catchment Area (Sq. km.)
1.	Indus	Mansarovar (Tibet)	1114+	321289+
2.	a) Ganga	Gangotri (Uttarakhand)	2525+	861452+
	b) Brahmaputra	Kailash Range (Tibet)	916+	194413+
	c) Barak & other rivers flowing into Meghna, like Gomti, Muhari, Fenny etc.			41723+
3.	Sabarmati	Aravalli Hills (Rajasthan)	371	21674
4.	Mahi	Dhar (Madhya Pradesh)	583	34842
5.	Narmada	Amarkantak	1312	98796
6.	Tapi	Betul (Madhya Pradesh)	724	65145
7.	Brahmani	Ranchi (Bihar)	799	39033
8.	Mahanadi	Nazri Town	851	141589
9.	Godavari	Nasik	1465	312812
10.	Krishna	Mahabaleshwar (Maharashtra)	1401	258948
11.	Pennar	Kolar (Karnataka)	597	55213
12.	Cauvery	Coorg (Karnataka)	800	81155
Total				2528084

Source : Central Water Commission, W.M. Directorate (Reassessment of Water Resources Potential of India, 1933).

Water management, specially river water management, has emerged a key challenge before the nation. The institutional challenges in building the "new Indian water state" may be enumerated as under:

(i) the public sector must continue to have an important role to play in providing irrigation and water supply;

(ii) the vibrant non governmental sector, private sector and the cooperatives must also be given a role in providing formal irrigation and water supply services in a competitive manner with the state authorities;

(iii) as service provided by the mixed service sector improves, large number of people will move from the informal, self providing water economy into the formal service sector;

(iv) the public sector must play an expanded role in financing and provision of public services such as flood control, pollution control, sewage treatment etc.

(v) awareness generation regarding water use, water conservation; water recycling etc. needs to be geared up, specially by NGOs, in rural and semi urban areas;

(vi) the 'polluter pay' principle should be enforced and the river banks have to be kept environmentally clean; and

(vii) the government must deliver a set of laws, policies capacities and organizations for defining and delivering an enabling environment with special emphasis on the establishment and management of water entitlements and the regulation of services and resources.

Water is essential to life and living and plays a crucial role in socio-economic development of a nation. It is thus obvious that the government has to give proper focus to river water management and make more funds available during the 12th Plan period, as the Prime Minister himself indicated.